Avid Uncut

Workflows, Tips, and Techniques from Hollywood Pros

Avid Uncut

Workflows, Tips, and Techniques from Hollywood Pros

Authored by
Steve Hullfish

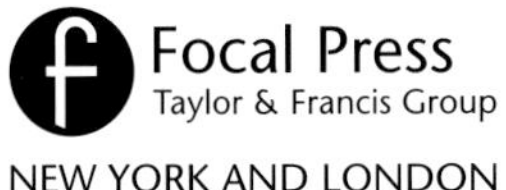
NEW YORK AND LONDON

First published 2014
by Focal Press
70 Blanchard Road, Suite 402, Burlington, MA 01803

and by Focal Press
2 Park Square, Milton Park, Abingdon, Oxon OX14 4RN

Focal Press is an imprint of the Taylor & Francis Group, an informa business

Notices
Knowledge and best practice in this field are constantly changing. As new research and experience broaden our understanding, changes in research methods, professional practices, or medical treatment may become necessary.

Practitioners and researchers must always rely on their own experience and knowledge in evaluating and using any information, methods, compounds, or experiments described herein. In using such information or methods they should be mindful of their own safety and the safety of others, including parties for whom they have a professional responsibility.

Product or corporate names may be trademarks or registered trademarks, and are used only for identification and explanation without intent to infringe.

Library of Congress Cataloging-in-Publication Data
Hullfish, Steve.
Avid uncut : workflows, tips, and techniques from Hollywood pros/Steve Hullfish.
pages cm
ISBN 978-0-415-82764-5 (pbk.)
1. Avid Xpress. 2. Digital video—Editing—Data processing. 3. Video tapes—Editing—Data processing. 4. Motion pictures—Editing—Data processing. I. Title.
TR899.H84 2014
777′.9028553—dc23

2013043502

ISBN: [978-0415-82764-5] (pbk)
ISBN: [978-0203-52223-3] (ebk)

Typeset in Meridien
Project Managed and Typeset by: diacriTech

Printed and bound in India by Replika Press Pvt. Ltd.

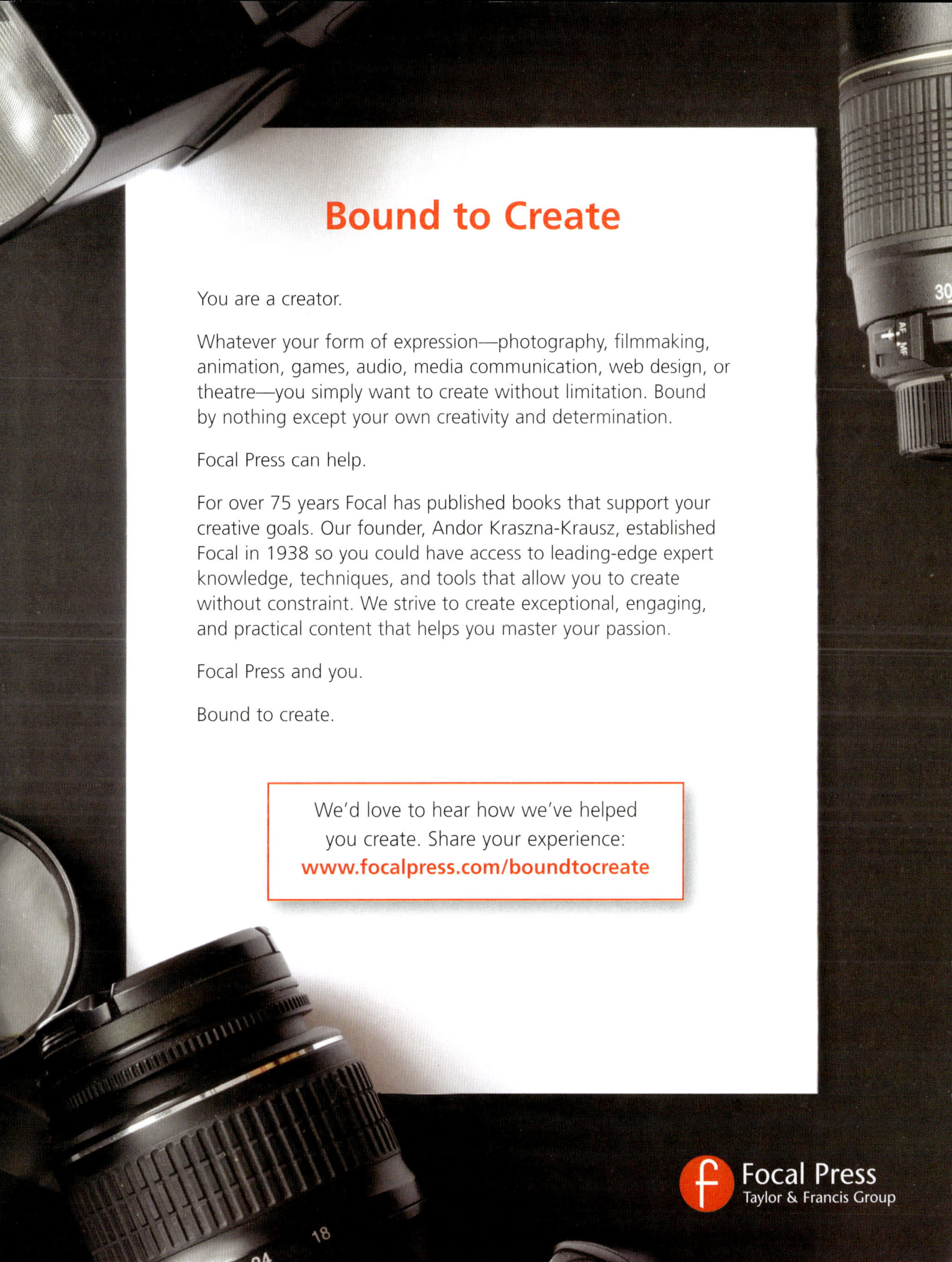
Bound to Create
You are a creator.
Whatever your form of expression—photography, filmmaking, animation, games, audio, media communication, web design, or theatre—you simply want to create without limitation. Bound by nothing except your own creativity and determination.
Focal Press can help.
For over 75 years Focal has published books that support your creative goals. Our founder, Andor Kraszna-Krausz, established Focal in 1938 so you could have access to leading-edge expert knowledge, techniques, and tools that allow you to create without constraint. We strive to create exceptional, engaging, and practical content that helps you master your passion.
Focal Press and you.
Bound to create.
We'd love to hear how we've helped you create. Share your experience:
www.focalpress.com/boundtocreate
Focal Press
Taylor & Francis Group

Praise for *Avid Uncut*

"A great book for experienced editors of all NLEs. Hullfish wrangled an impressive number of 'A list' editors who provided incredible insight into the specific ways that Media Composer empowers them. It makes a great read for those who wish to maximize their creativity and efficiency in editing."

—Ben Kozuch, Co-Founder, Future Media Concepts

"Like Hullfish's *Color Correction for Video, Avid Uncut* should sit on the shelf as a reference in every Avid suite. Condensing years of experience and real-world examples, *Avid Uncut* is easy to read yet packed full of tips, tricks and powerful time-saving Avid features. So whether you're a newbie needing to learn the ropes or a seasoned professional wanting to get up to speed on modern workflows and tools fast, this book is a must-have."

—Ra-ey Saleh, Senior On-line Editor/Colorist,
MythBusters; *Taboo*; *Hidden in America*

"*Avid Uncut* is one of the most comprehensive Avid Media Composer training manuals I've seen. Hullfish has years of Avid experience and is well-respected for covering some of the lesser known tips that make a Media Composer more productive. This book is no exception, covering not only software features, but numerous workflows. Plus, the reader gets to hear from top television and film editors throughout the chapters for glimpses at their creative solutions. Hullfish isn't afraid to include informative explanations of some of the lesser covered features, like ScriptSync, the Marquee titler and more, that other writers usually skip. *Avid Uncut* is a great handbook for editors moving over from other software, but even experienced Media Composer editors will be surprised at how much they'll pick up."

—Oliver Peters, Editor/Colorist, Oliver Peters Post
Production Services, digitalfilms.wordpress.com

"Over the years, Kathlyn and I have worked with many of the most respected Avid editors in the world, people who have edited some of the top-ranked movies and television shows in the market today. Among these prolific and respected editors, Steve Hullfish is often named as their go-to expert when they want to learn the Avid and master its many powers and nuances. You aren't going to master the Avid overnight but with Steve Hullfish showing you the way, you will learn its powers and maximize the workflow that makes the Avid the industry leader in high-end professional editing circles. If you are looking to excel in your understanding of Avid Media Composer, you have picked up the right book."

—Ronald & Kathlyn Lindeboom, Founders, CreativeCOW.net

"In today's adapt-or-die climate of non-linear editing, *Avid Uncut* is pre-requisite reading for every editor who needs to be relevant with complex file-based editing processes. The fact that this important book was organized with a creative mindset and written conversationally with commentary from the world's best editors also makes it a fun read."

—Mitch Jacobson, Author, *Mastering Multicamera Techniques* (Focal Press 2010)

Contents

Acknowledgments

To Marianna Montague, there is no adequate way for me to thank you for your contributions to this book. Marianna spun through her legendary Rolodex to line up all of the great experts you'll hear from throughout the book. Marianna, you were the "fixer" that made this book possible. Marianna's assistance is known far and wide throughout the Avid community. To us, Marianna, you are the heart and soul of Avid. Hopefully, you know how much we all love you. Thank you.

To the brethren of Curmudgeons: Bob Zelin, Oliver Peters, Terry Curren, Frank Capria, Jim Feeley, Greg Huson, Wilson Chao, and Jimmy Dutt. This is a "band of brothers" that shares advice and expertise with such unfailing generosity that I am humbled to be counted among them. I'm sure all of the Curmudgeons will see various gems of their wisdom spilled out on these pages with no credit given. So, I will say now to the readers of this book, "If the tip or advice seems really smart and wise and creative, it came from one of these guys, not me."

To the feature film and prime-time TV editors and their assistants who generously gave of their time and advice: J. C. Bond of innumerable blockbuster features; Wayne Wahrman of *I am Legend* among others; David Helfand of *The Mindy Project* and *Weeds*; Peter Chakos of *The Big Bang Theory*; Steve Audette and Tim Mangini of *Frontline*; Allyson Johnson and Beth Moran of *Smash*; Todd Morris of *The Big Bang Theory*; Richard Rossi of *The Mindy Project*; Michael Phillips of 24p.com; Robert Landua of CNN's *Crimes of the Century*; From the hit show, "SMASH," Director of Photography David Mullen, ASC; colorist Sam Daley and on-line editor Mike Nuget. Also thanks to Mary Cybulski, Focal Press author and script supervisor, for providing advice on script supervision. Also, thanks to fellow Avid book author, Steve Cohen for his advice and assistance.

Thanks to innumerable gurus who inhabit the various listserves and user groups that I frequently visit, lurk, lead, and contribute to. Thanks to David Ross for FCP help and to Scott Freeman. And to my friend, Rick "First Call" Mahal, for FCP advice.

For supplying RED footage and advice, thanks to Michael Cioni from Light Iron. For the footage from the ARRI Alexa, thanks to John-Michael Seng-Wheeler of Epsilon Films. Also, thanks to Chicago DP extraordinaire, Randy Riesen for helping capture the multi-cam project footage. And to Julie Hill and the good folks at ArtBeats for providing 4K and 3D footage for use as project footage and for images in the book.

For various images and contacts, thanks to Stephen and Alex Kendrick, Kris Raymer Fuhr, and others. Thanks to my business partner, Dave Metzger, for "starring" in my short film, *Avid Uncut*. Thanks to Stephen Marinaccio for his lined script pages. Thanks to Steve Shaw of Light Illusions for LUTS and the use of his LightSpace LUT app. And to Light Iron's Laura Pursley for images of Lily Pad and OutPost. Also, thanks to Mark Thompson and Terry Frechette of BlackMagic Design for DaVinci Resolve help. And to John Ladle and my other friends at AJA for hardware answers and the loan of an AJA IOXT. For numerous Avid PR requests, thanks to Sara Griggs. Thanks to her fellow Avid colleagues Mike Hale and Ian Krabacher.

To my Taylor & Francis colleagues: many thanks to Peter Linsley, Dennis McGonagle and Denise Power for shepherding this book along. I know you paid for my missed deadlines in your own blood and sweat. Also, again, to my friend and brother-in-arms, Terry Curren, for tech editing this book. It is because of him that you can be assured that what you are reading is truly sound advice and correct. And to Mary LaMacchia at diacriTech and her team for project managing the production of this book—thanks for making me look good.

To my parents, another huge thanks. You taught me that discipline and hard work are integral to any creative endeavor ... that creativity is not just born from some innate skill, but that it must be worked for. You also taught me that creative occupations are worthy goals. I am so appreciative of the love and support that I've always received from you, Sue and Bill.

Every time I write a book, I promise my family that I'll never write another one. Obviously, I'm a bad dad, because I have now broken my word four times since my first book. Thanks for giving me some space, Haley and Quinn. To my long-suffering wife, Jody, I know that the attention I lavish on book after book would be better spent on demonstrating how much I truly love you. Thank you for indulging me in my passion for sharing with these readers the hard-won knowledge I've gained through the years. Without you as the quarterback of the family, I couldn't do this work that I love so much. Thank you. And to my Heavenly Creator. Thank you for the gifts which allow me to create. I pray that I use those gifts to draw other to you, showing them the love of your Son, Jesus Christ.

Introduction

This book is not about learning how to edit. This is a book that takes a relatively experienced editor and helps him or her navigate one of two things:

1. Switching from another editing program to Avid, or
2. Helping an experienced Avid editor understand a bunch of fairly significant developments that have occurred in recent years *inside* the Avid software as well as developments in the broader production and postproduction environment that require new workflows and the exploitation of new tools.

This book is about breaking away from the old ways of doing things and embracing and leveraging the change that's happened in the industry at large and internally at Avid, as well. This book is about staying relevant and about using new tools and new workflows to overcome the hurdles of postproduction in a new age.

I've written a few other books on Avid editing. Those were based solely on *my* experience, which isn't a bad thing since I've been on Avid since 1992 and have worked at TV stations and post houses and as a freelancer, trainer, and consultant. I've edited sports, music, comedy, documentary, news, magazine shows, *The Oprah Winfrey Show*, trailers, behind-the-scenes footage, electronic press kits, marketing sizzle reels, national TV spots, local TV spots, and TV promos, and I even co-edited the Sony/TriStar feature film, *Courageous*. I've done linear, tape-based editing with U-Matic, 1″ tape, BetaSP, DigiBeta, and D2. I've edited on Avid, Quantel, Sony XPRI, Final Cut Pro, Premiere Pro, and even iMovie. I've done both off-line and on-line editing.

I have also written extensively about color correction. After I cowrote *Color Correction for Video*, I was feeling a little guilty about handing out advice, since my full-time job is not as a colorist. That's one of the reasons I wrote *The Art and Technique of Digital Color Correction*, which used more than a dozen high-level colorists from around the United States as the experts. I felt good that *Art and Technique* was not just about my opinion, but included the opinions of a wide gamut of the industry's leading experts. Since that method of "sourcing the experts" worked so well for my last book, I've decided to implement it as much as I can on this book as well.

Even with my considerable experience, I am simply not regularly exposed to ALL the workflows that so many people use in a myriad of editing situations. Editing a reality show is different from editing trailers, which is different from editing long form documentaries, which is different from editing scripted drama or multicam events. So, to deliver the best state-of-the-art advice across a broad spectrum, and to provide a serious reference for as many different types of editors and productions as possible, I'll be tapping into the best postproduction brainpower in the business.

The tone of the book is probably unlike the tone of many other application-specific books. Instead of writing tutorials with numbered, step-by-step instructions, I prefer to write with a sense that I'm sitting next to you, offering advice and counsel. Tutorials are written as parenthetical direction and are meant to be read and tried. If any of these ideas have immediate application to something you think you will use or need, then I urge you to walk through the process on your machine and try things out for yourself. The only way to get used to new things, processes, methodologies, and workflows is to get them "under your fingers." You need to try them and feel comfortable doing them. I eschewed the use of numerical, step-by-step instruction because with complex software that adapts to numerous workflows, there is not always a simple 1-2-3 process. Instead, there are decisions to be made and situations that need to be considered that preempt the ability to use a simple step-by-step process.

In trying to come up with a structure for this book that would be easiest to access and utilize, I broke the overall process of postproduction into separate, smaller processes. In doing this, I hoped I made a book that addressed the most common issues and that would be easy to reference at critical moments.

Chapter 1, Pre-Postproduction

The first part of the postproduction process is what I'll call "pre-postproduction." Obviously that's kind of an oxymoron. What I mean is that it's the part of the process before you actually get to edit. Prepping for the edit, getting the source footage into the system, getting the system itself ready and organized so you can sit down and do what we all really love: editing. Pre-postproduction has grown increasingly more complex with the advent of file-based workflows. When everything was on videotape, there were certainly challenges and there were problems to deal with, but it was a lot easier than it is now to deal with multiple file types, camera manufacturers, and ingest methods and maintaining the links from the original files all the way to the finished output. Nothing's worse than editing together a beautifully cut program only to find out that you can't get it off of your system and onto the big screen. Insuring that you have a useable end product starts all the way back at the pre-postproduction phase.

Chapter 2, Settings

Before we start to edit, it's important to understand that Media Composer is designed to be used by a wide variety of editors who have different styles and different project requirements. To address those issues, Media Composer is highly customizable. It's important to understand the many ways that Media Composer can be customized because properly setting up your Media Composer can massively leverage the effort of the editing process. Having a custom-tailored Media Composer is like having a custom-designed set of golf clubs or a hand-tailored suit or a custom-built home: it will work better, fit better, allow you to get more done, and make you comfortable while you're doing it. A word that I'm seeing more and more is *bespoke,* which is a high-end way of saying "custom tailored." That's the goal of this chapter. Having a custom environment for editing removes the impediments that can slow you down and sap your creative energy. The goal here is to "stay in the flow" while you're editing. That will help you think less about the technology and more about the editing.

Chapter 3, Off-Line

The editing section is really what it's all about. Experienced Media Composer editors will know a lot of this, and this book is not meant to be a user's manual. The main thing with this section is to help experienced editors from other systems, like Final Cut Pro, adapt to a new way of doing things and fill in some gaps for the experienced Avid editors who may have missed out on a few cool tricks. I'm a very, very experienced Avid editor, but I was watching an assistant editor do something on my system that I had never known was possible. I also was rereading my original Avid 101 courseware a while back and noticed a note that I had handwritten in the margins, and I SWEAR that I didn't know that the tip in that note was possible, even though

I had written it myself, more than ten years earlier. We all have gaps in our knowledge—even with software on which we consider ourselves experts. I'll also rely on experts from Hollywood to New York and beyond for their own experiences in unique workflows. Even if you may not find yourself using the specific workflow yourself, there will be ideas to internalize and assimilate into your own workflows. Hopefully these workflows will also open up some cool new worlds to you and explain exactly how some of your favorite film and TV shows get edited using Avid systems.

Chapter 4, Effects

Effects in Avid are, I believe, one of its weakest points. Media Composer was originally designed as a creative off-line editor back in 1988. By the time I started using it in 1992, even though it was capable of on-air work, it was still largely used as an off-line editor, generating EDLs to be finished in linear, tape-based editing suites. As Avid systems became capable of on-line quality work, effects became more and more important. Luckily many third-party effects plug-ins have stepped in to provide greater effects capabilities to Avid. There are also important internal effects assets, like Marquee, AvidFX and Animatte. For Final Cut Pro users, the way effects are handled inside Avid will be one of their largest sources of frustration, but as someone who started using NLE effects in Avid, I have found that there is a method to the madness, which will be explained in this chapter. Don't fight the way it works. Accept the way it works and make it work for you. I've seen some awesome effects work come out of the humble Media Composer effects engine.

Chapter 5, On-Line

The difference between off-line editing (creative editing) and on-line editing (finishing) has blurred to the point where talking about them as separate things is almost quaint: like Grampa explaining that he used to walk ten miles to school through the snow, uphill both ways. When storage solutions dropped in price to a point at which it was practical to store hundreds of hours of standard definition video at full resolution, the off-line/on-line division went away for many. With the advent of HD, most editors again had to go back to capturing low-res footage for the "creative edit" and then up-res when that was approved so that the finishing could be done. Storage solutions have continued to drop in price, so even with HD footage, editors are again capturing footage at on-line quality right from the beginning. However, now 4K, 5K, and 8K are potentially going to keep off-line and on-line separate for quite some time to come. The cost of storage for doing off-line at full 8K RAW are staggering, probably in the millions of dollars for features at the 2012 cost of gigs/dollar. In addition to this, there really is a basic difference between the off-line and on-line mentalities. In this section, we'll primarily discuss workflows and color correction. With the proliferation of software-based color correction options, like Baselight, Da Vinci Resolve, Adobe Speedgrade, and Apple Color, more color correction is being done outside of the built-in Avid tools. We'll be discussing doing color correction inside Media Composer (including the Symphony option) as well as how to exploit these other third-party grading solutions.

Conclusion

I am confident that you'll benefit from this book. We'll cover a lot of ground and you'll hear from a lot of experts that will provide their unique perspectives and solutions for specialized issues. This broad base of knowledge, across a spectrum of editors, should give you unique insights into how Avid Media Composer can be used in ways you may never have imagined. As you read this book, allow connections to form between the solutions that others have devised and solutions to your own editing issues. You'll rediscover an incredibly complex postproduction ecosystem that has allowed editors to create some of the finest media and entertainment the world has ever seen.

Companion Website, Project Files, and Tutorials

Make sure to visit the companion website (www.focalpress.com/cw/hullfish) for updates, corrections, project files, and video tutorials. Each chapter will have corresponding media and files that can be downloaded, so that you can follow along. Some media is provided in earlier chapters and is used throughout the rest of the book, so make sure to download all of the project files starting with Chapter 1. The companion website files are listed below:

Chapter 1
- ScriptSync video tutorial
- Avid Uncut short film footage and script
- Folder of Alexa footage called A001R2YS
- Folder of RED footage from LightIron
- ArtBeats 4K footage

Chapter 2
- Composer Settings video tutorial of options
- Command Palette video tutorial walk through of all tabs
- FCP mapped keyboard file
- Two Hullfish keyboard settings files—left and right

Chapter 3
- SmartTools video tutorial
- Short video tutorial on the Extend button
- Multicam video tutorial
- Multicam project files—four camera angles of interview
- Version 7 new audio functionality video tutorial
- Maintaining Sync During Trimming video tutorial including option-add edits
- Slipping and Sliding video tutorial
- ArtBeats 3D footage

Chapter 4
- Keyframing Effects Basic and Advanced video tutorial
- Avid FX Glow General and Glow Transition video tutorial
- Verascope Pictures logo file (.PNG)
- Animatte bin and footage—Steve MS clip from Avid UnCut and 4K ArtBeats footage of Capital dome

Chapter 5
- Pointer to the Alexa footage from Chapter 1
- Pointer to RED footage from Chapter 1
- Curves video tutorial using RGB Parade
- Secondary video tutorial

Stay in Touch

I urge you to contact me about issues in this book. If I was unclear on a process or something doesn't seem to work as described, let me know. I'm very interested in constructive criticism, and I always want to make my communications as clear and correct as possible. You can reach me at steve@hullfish.com. I hope your experience with this book is a positive and fruitful one.

CHAPTER 1

Pre-Postproduction

Preparing for an edit is not fun. It's not sexy. It doesn't impress clients. But if you do it incorrectly, it can lose you clients because it can affect every step of the process all the way to delivery. There are things that you need to know and understand in order to correctly ingest and organize the materials that you've received from production and that you will probably continue to receive throughout postproduction.

This is one of the steps of the process that separates the men from the boys and the women from the girls. Knowing the right questions to ask and understanding what the answers will mean for your end product has a lot to do with experience. But all that experience can be learned and distilled and taught.

Best practices for the entire postproduction chain start right here. In this chapter we'll discuss ingest, project and bin organization, and strategies to quickly locate and make sense of the vast number of assets that are involved in even the smallest of projects. But, we'll start the chapter "pre-Avid." This means the transition from camera department to post. One of the major elements of that transition for the editor is understanding and dealing with metadata. Not everyone needs to be deeply familiar with metadata, so if this section seems a little dense for you, move on to the more fun stuff. As a matter of fact, if you want, you can jump straight to chapter 3 for the actual editing process and then come back here when it suits you.

Metadata

Metadata, in its mainstream definition, is "data about data." In production and post, metadata is data about your media and data about your sequence. In the old days, metadata was scarce. You could argue that script notes and audio logs and timecode and reel names were, and are, metadata. As it applies to video and film, metadata can be almost anything from subtitling to timecode and much more. Many camera manufacturers are storing more and more metadata as part of their video files or in "sidecar" files that are linked to the video files. The camera manufacturer's metadata can include lens data (f-stop, focal length), shutter speed, GPS coordinates, time of day, user data (shot name, comments), timecode, camera ID (serial number), aspect ratio, and more. Most camera systems also include the ability to prepare metadata in advance of a shoot and enter it into the camera, so that the shots include that information. For instance, at a news station, a news editor can actually create metadata before the crew leaves for the day and import it into the camera. Then that metadata will be included in the files that will be shot that day.

This metadata is meant to ease the handoff from production to post and to help organize, archive, and access media.

To me, the balancing act with metadata is to have enough to make your job easier but not so much that it makes your job harder. The trick is to realize that a lot of metadata may not be necessary for *your* job, but it was created and needs to be tracked through your hands and into *someone else's* hands farther down the post or distribution pipeline.

While some metadata is automatically created by the camera, some of the most critical data can be added "live" with something like a linked laptop or tablet (like an iPad). Also, the audio department on larger shoots creates a lot of metadata in the sound reports. These reports are sometimes incorporated into metadata files. Another possible method for creating metadata is with camera-specific viewer software. This is software that allows you to view and log camera footage after the fact, adding metadata to the clips. An example of this is Sony's clip browser software, which allows you to browse proxies (see Figure 1.1). Supported metadata includes Clip Name, Title 1, Title 2, Comments/Description, Sony Status flags and Essence Marks. Titles can be used for scene descriptors and when these logged files are AMA linked into an Avid system, this metadata will be included in custom columns in the bin.

While some of this metadata can come from the MXF file (Materials Exchange Format) itself, for higher-end postproduction workflows, in the Avid universe, the metadata comes in a separate file, usually an ALE file (Avid Log Exchange), that is capable of linking its data to the imported media. On the other end of the postproduction process is another metadata-rich file: AAF (Advanced Authoring Format). AAF files are the new EDL files. They are replacing EDLs (edit decision lists) and even OMF files. AAF also competes in some applications with XML files. All of these files (MXF, ALE, AAF, OMF, and XML) are sources, carriers, and transmitters of important metadata.

For more on AAF: http://en.wikipedia.org/wiki/Advanced_Authoring_Format

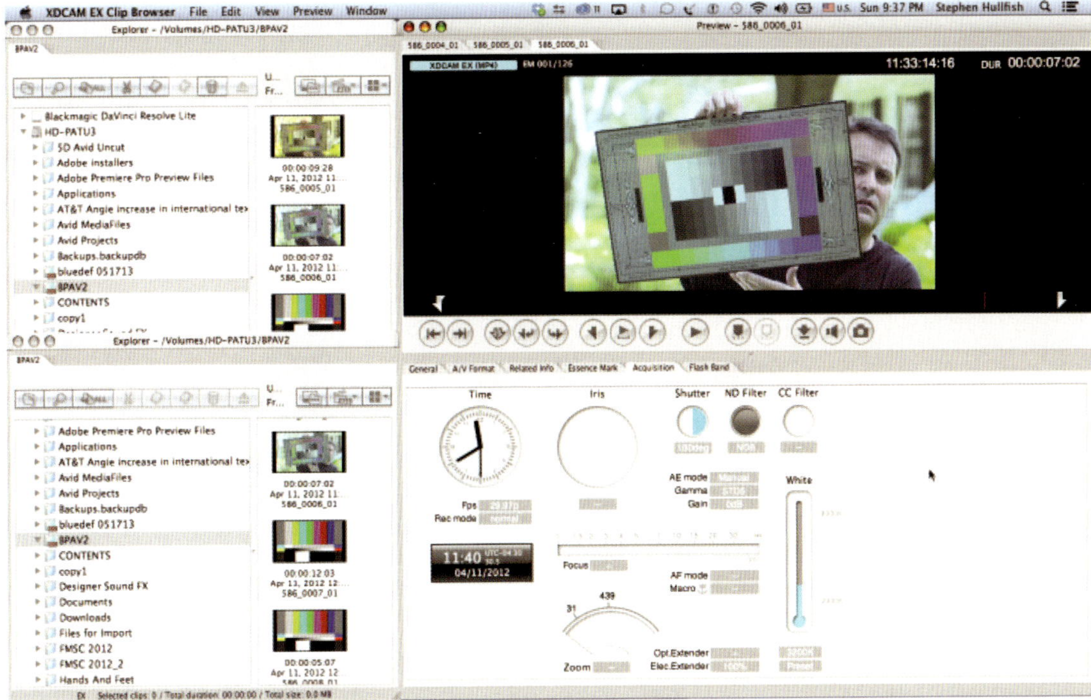

Fig. 1.1 Sony XDCAM EX Clip Browser. (Exact software names and numbering may change through the years.) This software allows you to view metadata and add it to files. This software needs to be pointed at a BPAV folder.

ALE

Avid Log Exchange helps to bring data from logs, such as shot logs, telecine logs (including FLX and ATN files), or Apple Cinema Tools. These are getting increasingly rare, but ALE can also be used with file-based digital cinema and HD files. In addition, ALE files can be used to draw logging data from other applications, like FCP bins. ALE can also ingest CMX EDL information.

ALE is becoming somewhat less used than it was in the heyday of telecine film transfers, but it is still valuable and useful.

ALE files are actually very simple database files that can be described, created, and edited as simple tab-delimited text files (see Figure 1.2). Consider each row in the ALE as a unique asset or media file, similar to a bin. The ALE matches a unique identifier in the media clip, like the original file name, and it matches that row in the ALE to the same unique identifier in one of the pieces of media. It can use file names. It can use clip names. It can use random bits of information. So when you bring the ALE in, Media Composer figures out: "In the ALE, you've got a column called 'Clip number' and in your bin you have Clip Number." If those two match up, that's the

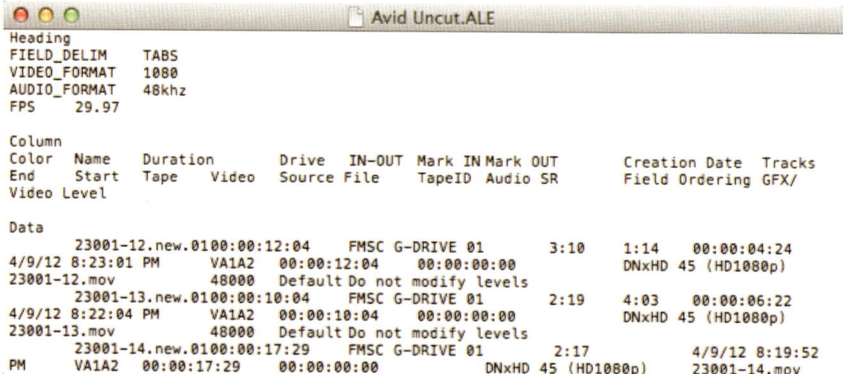

```
Avid Uncut.ALE
Heading
FIELD_DELIM	TABS
VIDEO_FORMAT	1080
AUDIO_FORMAT	48khz
FPS	29.97

Column
Color	Name	Duration	Drive	IN-OUT	Mark IN	Mark OUT	Creation Date	Tracks
End	Start	Tape	Video	Source File	TapeID	Audio SR	Field Ordering	GFX/
Video Level

Data
	23001-12.new.0100:00:12:04	FMSC G-DRIVE 01	3:10	1:14	00:00:04:24
4/9/12 8:23:01 PM	VA1A2	00:00:12:04	00:00:00:00	DNxHD 45 (HD1080p)
23001-12.mov	48000	Default	Do not modify levels
	23001-13.new.0100:00:10:04	FMSC G-DRIVE 01	2:19	4:03	00:00:06:22
4/9/12 8:22:04 PM	VA1A2	00:00:10:04	00:00:00:00	DNxHD 45 (HD1080p)
23001-13.mov	48000	Default	Do not modify levels
	23001-14.new.0100:00:17:29	FMSC G-DRIVE 01	2:17	4/9/12 8:19:52
PM	VA1A2	00:00:17:29	00:00:00:00	DNxHD 45 (HD1080p)	23001-14.mov
```

Fig. 1.2 This is a text representation of an Avid Log Exchange file (ALE). An ALE is basically just a tab delimited text file.

uniqueness that Avid needs to make the connection between the two. This system was developed originally by Michael Phillips, who was a long-time Avid employee and is now a consultant and has a blog called 24p.com. Check out Phillips's wealth of information at 24p.com/wordpress.

GLOSSARY

Sidecar file: A Sidecar file is a file containing additional data that cannot be stored in the format of the connected file. In film and video, a sidecar file is a metadata file connected to a video or audio file.

Another metadata source that Avid can use is the RMD .XML sidecar file to the RED .R3D media file. The AMA plug-in for RED media actually looks at the XML sidecar and also the metadata that's in the R3D file, so you don't need to do the ALE with RED anymore. It depends on what you're using to do the transfer. If you're using RED CINE X, which most people are, then RED CINE X puts that metadata into the R3D, and Avid can access it and reveal that information in the Source Settings of RED media (see Figure 1.3).

ALE is used to ingest ASC CDL information (American Society of Cinematographers Color Decision List). These values don't do you any good in a bin column, but if an incoming ALE file with CDL information is merged with the clips in the project, the values will alter the look of the associated clips, providing a primary color correction assigned on set by the director of photography and the DIT (Digital Imaging Technician) or perhaps by an on-set colorist or dailies colorist. The two types of CDL data that are transported in the ALE are ASC_SOP and ASC_SAT. These provide Slope, Offset, and Power and Saturation data. A sample CDL value would look like this: (1.0, 1.0, 1.0) (0.0 0.0 0.0) (1.0 1.0 1.0) 1.0. The first set of parenthetical numbers is from the ASC_SOP and indicates the Slope values of the red, green, and blue channels. The second set indicates the Offset's red, green, and blue channel values, and the third set is the Power values for red, green, and blue. The final number is the Saturation value from the ASC_SAT. If the ALE was created with CDL values, you can see them in a standard bin view within Avid by selecting a bin and, from the menu bar, selecting Bin > Choose Columns. Then, from the column list, select both ASC_SOP and ASC_SAT.

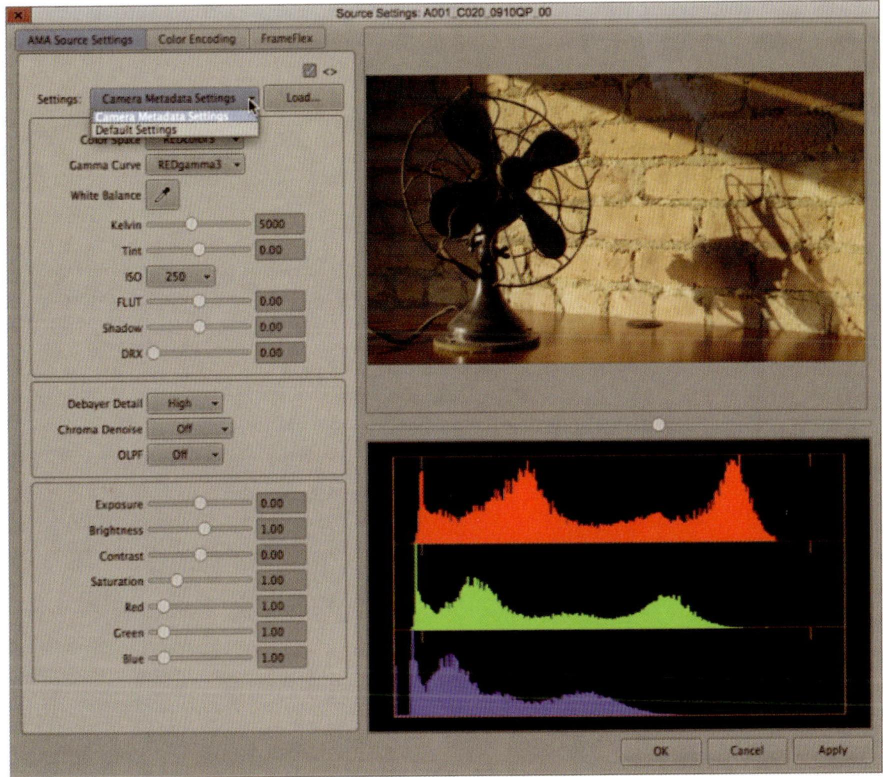

Fig. 1.3 With AMA-based RED footage, you can access some of the metadata available in RED footage through the Source Settings menu option (right-click on the Source window or clip and select Source Settings).

If you have a workflow or an exceptional situation to your regular workflow that causes you to first bring in the footage and then later import the ALE for that footage, you can select "Merge with known sources" under Shot Log options in the import dialog box. This allows the metadata stored in the ALE file to be merged with the MXF files imported or AMA-managed inside Media Composer. The big warning with this workflow is that if you have added any custom metadata to the media in the bins, that data will be lost when you merge the ALE file to the associated media, so if you intend at some later time to merge an ALE with the media in a bin, don't waste time adding information to that media in the bin columns (see Figure 1.4).

There are also methods of modifying bins by exporting them to ALE, modifying them outside of Avid, usually in a text editor or database program like Excel, and then reimporting them into Avid, where they're linked back to the original media.

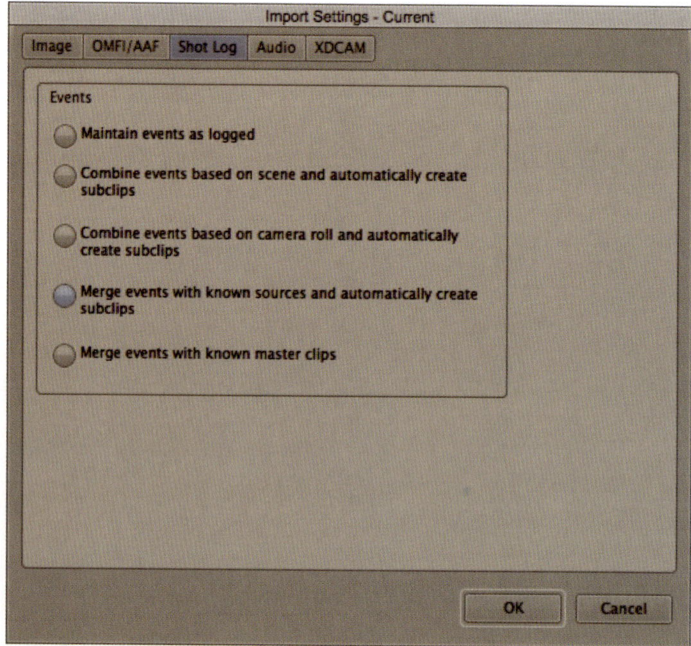

Fig. 1.4 You can merge the metadata from an ALE file—usually sent by a company doing dailies for a production—with the clips in a bin by selecting the bin and choosing File > Import. Choose the Shot Log option and, in the resulting window, choose Merge events with known sources and automatically create subclips.

Code Geeks Only!

Many of the high-end post-production workflows in Hollywood rely on some custom coding solutions. Often, custom applications are written to follow the trail of metadata through the entire post process. If you have coding skills or the resources to have something coded, you can see behind the curtain of AAF and MXF files. A white paper describing the creation of AAF files can be found here:

www.amwa.tv/downloads/whitepapers/AAFoverviewIBC2001A4.pdf

Software developers can find the SDK for AAF here:

http://sourceforge.net/projects/aaf/files/AAF-src/1.1.6/

To create MXF files, an application for the SDK Avid Media Toolkit can be found here: www.avid.com/us/partners/AvidMediaToolkitProgram

More on Metadata

I spoke to Ian Krabacher and Mike Hale, who are both Senior Principal Product Designers with the Interplay family of Avid products (see Figure 1.5), which are essentially media servers with deep Digital Asset Management capabilities. They explained that the key to Avid's legendary media management is the MOB ID, which is an identifier that creates the "uniqueness" of every media file and metadata file. As a matter of fact, there are multiple "uniqueness identifiers" for each file.

Hale explains, "We generate a MOB [media object] ID based on various things. Avid's got MOB IDs all over the place. Those are the unique IDs for things. If you took a single master clip and broke it down, you'd have the MOB ID of the master clip, you'd have the source ID, which is the MOB ID for the source, that's sort of the global source for all of the tracks. You'd have a MOB ID for all of the channels, and you'd have the file MOB, which is the uniqueness of the file. So all OPAtom media has a uniqueness in it, so it seems crazy that you'd have what seems to be redundant uniqueness. But they *ARE* unique. The source is different than the audio track. The master clips, which might have all the master tracks in it, is unique. If you stay within the Avid realm, we pride ourselves on the fact that we inject the uniqueness of the file. So the file MOB gets injected into the actual file. It's one of my biggest complaints about QuickTime: there's no predefined ID for the file.

GLOSSARY

MXF OPAtom: OpAtom is a media type designed to keep the audio and video elements discreet, yet tied together in a single clip. It is defined by SMPTE 377M and 390M and allows for easier editing of the file in a non-linear way. Multiple camera and NLE manufacturers are OPAtom compatible.

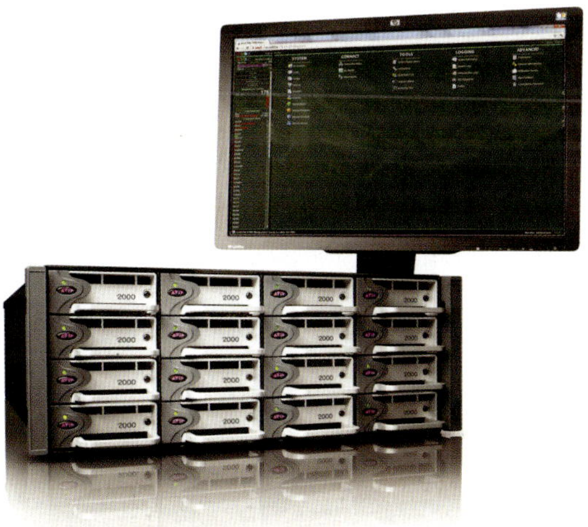

GLOSSARY

MXF OPAtom: OpAtom is a media type designed to keep the audio and video elements discreet, yet tied together in a single clip. It is defined by SMPTE 377M and 390M and allows for easier editing of the file in a non-linear way. Multiple camera and NLE manufacturers are OPAtom compatible.

Fig. 1.5 Avid's ISIS is shared storage that attaches to and can be controlled by the asset management capabilities of Avid Interplay.

GLOSSARY

Essence: Essence is defined as the actual visual or audio media, without its metadata.

In the Avid, every time you regenerate your media directories, you're basically creating a quick index of that stuff, but it's always there. The uniqueness allows you to have that conceptual break between metadata and essence. That break allows editors to switch between off-line and on-line resolutions linked to the same master clip."

Krabacher continues, "If at the camera we were generating uniqueness, like we used to with Keycode on film, which identified unique frames on unique reels and there was no other reel like it in the world—at least in theory, then we wouldn't need to do it [in Avid]. The MOB ID is our unique identifier in Avid, but we put other stuff in there for other systems to interpret, because if you read our MOB ID, you'd just say 'What is this?' It's just a global unique identifier. We put in things like the file name and the file path and the resolution and all sorts of other stuff so that the receiving system can say, 'I know this is a unique asset and this is where I thought it was and this is the resolution that it had and this is the various things I need to know about it,' so we're putting all those things into the AAF that we send over to the other system."

With Avid media, there's a certain amount of metadata that's exposed to the user, and there's a superset of metadata—MOB IDs, as a primary example or some camera manufacturer info—that Avid tracks behind the scenes (See Figure 1.6). The user never sees but it is what powers a lot of the multi-res workflows that Avid can deliver.

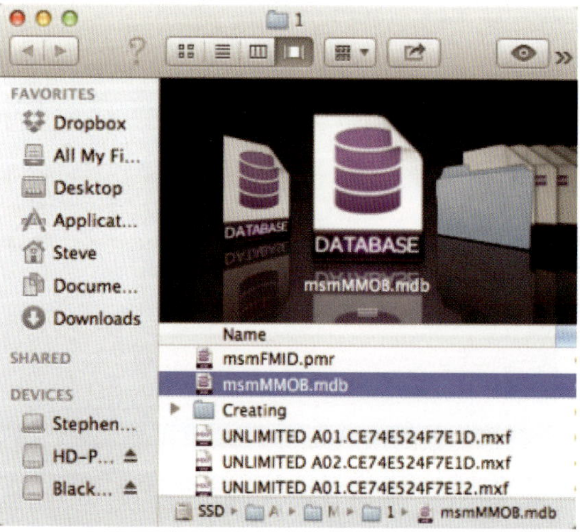

Fig. 1.6 Avid's MMOB file is a directory of all of the Avid Media Objects (MOBs) in the Avid Media Files folder.

Avid Organizational Concepts

Projects

The first line of organization in Avid Media Composer is the Project (see Figure 1.7). When you launch Media Composer, the Select Project dialog opens (see Figure 1.8). It shows you the previous projects that have been created on your system. In the old days, there was only one place that Avid Projects were saved, but now there are three radio buttons that allow you to create, save, and look for Projects in multiple locations. This

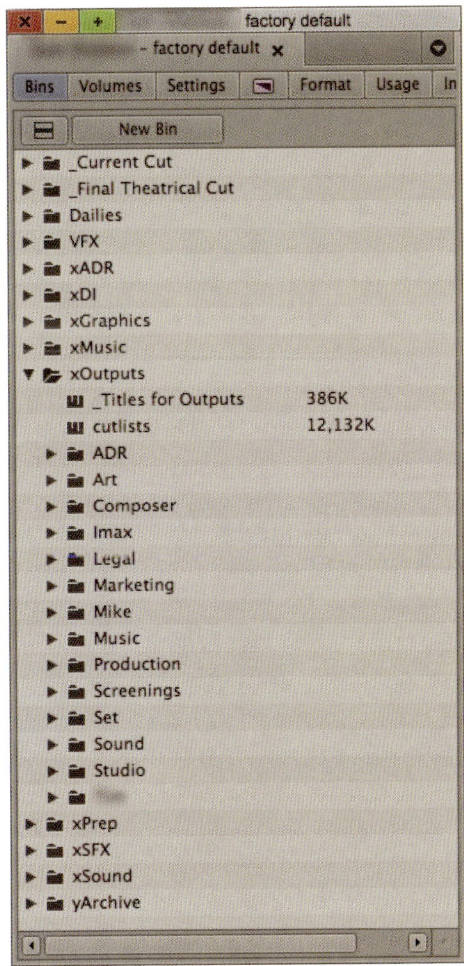

Fig. 1.7 This is the Bin tab of an Avid Project window from a major feature film. (The project title has been blurred out.)

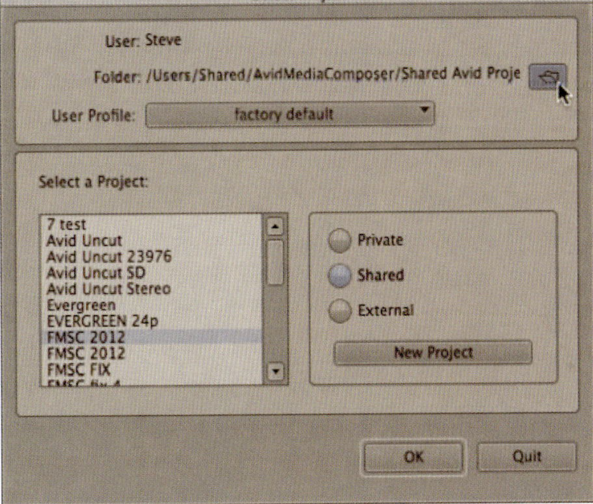

Fig. 1.8 The first thing you see after launching Media Composer is the Select Project dialog window.

is different from Final Cut Pro, which basically allows you to save your projects anywhere. Avid's method provides you with the same level of organizational freedom (using the External button), but also has an option that offers special privacy protections (Private) and another (Shared) that provides a common location that is easy to find and is common across Avid systems.

In the Select Project Window the first radio button allows you to create Private projects. On a Mac, the file path for these Private projects is:

/Users/(your username)/Documents/Avid Projects

On a PC, the file path for Private Projects is:

C:\Users\(your username)\Documents\Avid Projects

The benefit of this option is that the newly created project is accessible *only* to someone signed in to the computer with *your* username and password. There are many reasons why you might want to do this, but for anything other than personal projects, it's dangerous because no other editor can step in and even open a project that is saved as a Private Project. There is nothing actually special about the private project file itself, but because it is saved inside the folder for a specific computer username, access is restricted at the root level.

The second method, Shared, is the "old school" method of project saving for Avid. At one time, the path that this choice created was the only one available to Avid editors. This method saves projects to the folder that is shared between all computer users, so it is one of the best choices for projects that have any chance of being used by more than one editor. The other benefit to this method is that it creates new projects in a common, centralized location that any editor—even someone completely unfamiliar with the project, organizational structure, or computer system—should be able to find. This method would be my first choice for most of your new projects.

The file path for Shared Projects on a Mac is the following:

/Users/Shared/AvidSymphony/Shared Avid Projects

The file path for these Shared Projects on a PC is the following:

C:\Users\Public\Documents\Shared Avid Projects

The third method is called External and is most similar to the way FCP operates. When choosing this option, you are presented with a Choose option that allows you to browse through a path of folders and place the new Avid project anywhere on your system. This gives you a lot of freedom, but another editor stepping in to the project could find it a lot more difficult to locate and access the project. When you use the External option, you will see Project files from the last path (folder location) that was used. You can change this folder path to look for Avid projects in any folder on any drive on your system, including USB thumb drives or even locations out on the network. To access these paths, click on the little folder icon in the top right corner of the Select Project window and browse to the location you want in the Project Directory window that comes up.

Additionally, using the External Project choice, you can create folders of Avid projects to organize the Projects even further. For example you could create a folder for each client. Or if you are working on a TV series, there could be a separate folder for each season, and then each episode of the season would be its own project inside that season's folder. Or at a TV station, you could create a separate folder for news promos and one for entertainment promos. Or organize the files so that each season has its own folder, with just the promos for each season inside that folder, or each series could have its own folder with the promo for each episode as a separate project. The advantage of having smaller folders of projects is that it makes it easier to find and open bins from related projects. It also keeps the clutter down in the Project list. Give some thought to the file structure you want to use and make sure to document your thought process and structure so that other editors can easily track down old projects or bins if they have to.

Another important tip: The data in these projects is priceless. The work contained in the Project folder represents days and days of working on a project. I would strongly suggest that you create an automatic backup plan that saves the entire folder that contains your Shared Project folder (and any other External Avid Projects folder) to a separate hard drive AND to the cloud. Set a watch folder that automatically gets backed up to the cloud. There are many good solutions that do this; some of them are even free, like LiveMesh, DropBox, Box.net, or Apple's iCloud.

TIP

In a great tip I saw from fellow Avid book author, Benjamin Hersheder, you can even use this method to create and store your projects directly to the cloud using something like DropBox.

Avid Project Folder Structure

Avid projects are fundamentally different in root level structure than Final Cut Pro projects. For Final Cut Pro, a project is contained in an FCP project file that is a single file, not accessible by any other computer program. When you create an Avid project, the system creates a folder. (See Figures 1.9 and 1.10.) Inside this folder are the following items. The hash symbols (###) represent the name of your project.

SearchData folder, containing a Search database file
###.avp file
Statistics folder
###.avb
Settings.avs
Settings.xml file

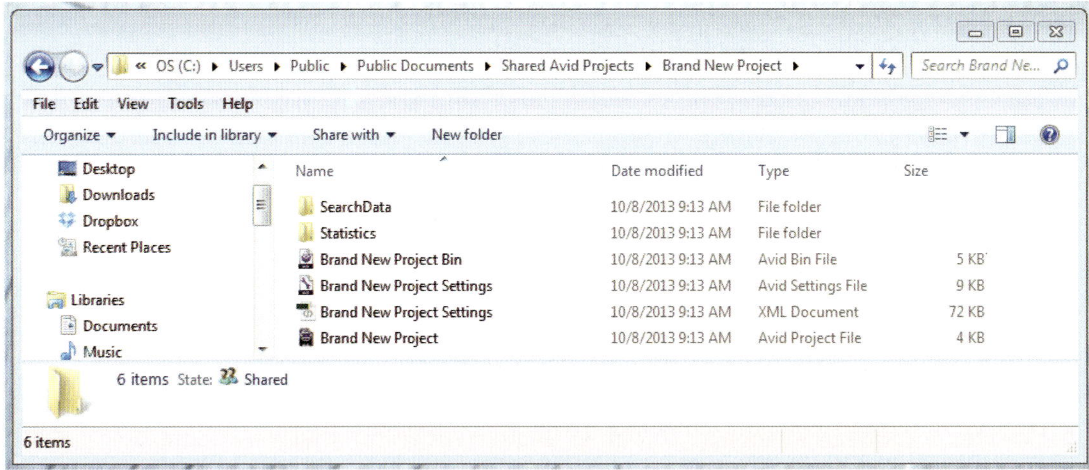

Fig. 1.9 This is a brand new Avid Project folder on a PC. The only bin created so far is the default bin that launches on start up of a new project.

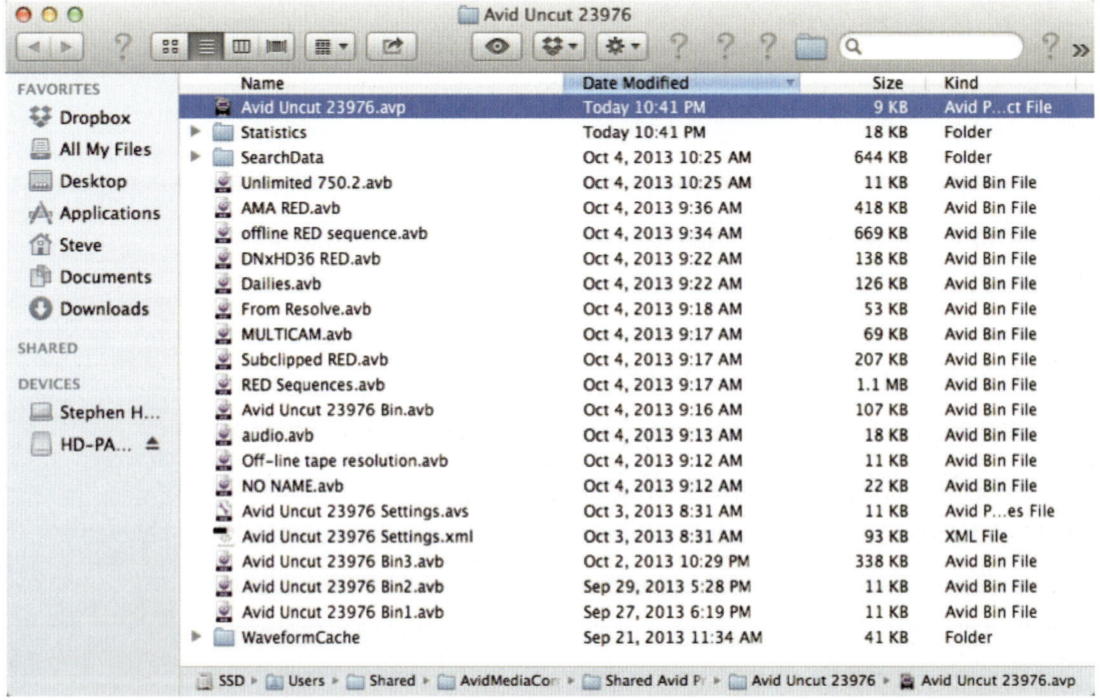

Fig. 1.10 This is an Avid Project folder for this book. Compare the files in this project folder with the image of the (Figure 1.11) same Project as seen in Media Composer's Project window.

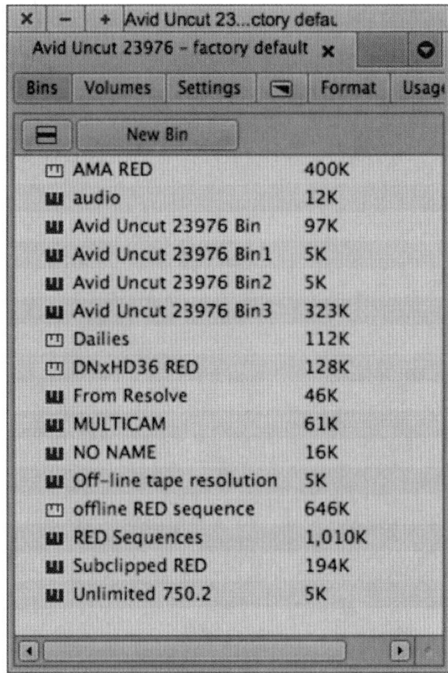

Fig. 1.11 This is the Avid Project window that corresponds to the finder-level Avid project folder seen in Figure 1.10.

You don't really have to worry too much about the specifics of these folders and files, but the basic breakdown is that the SearchData folder contains all of the text and metadata information for every clip, so that when you do a Find Command (searching for some piece of text associated with a clip or sequence in your project), the database that is searched resides in the SearchData folder. If you have the PhraseFind option, this also contains the phonetic search material for your clips.

The file with the .avp suffix is the Project file and contains the information about the project. Should this become corrupted—something that's never happened to me in twenty years—the bins will still work fine if you open them in another project. This is an advantage over FCP's project file system that puts all of the data into a single file.

A bin is created automatically when you create a project, and that bin is given the same name as the project. You can change this name at any time. Any other bins that you create by clicking on the New Bin button also are saved in the Project folder.

When the Project is first created, there is nothing in the Statistics folder, but each time the project is opened, a new Statistics file is created, which shows the date and time that the project was opened. The numbers shows the date (in European order—year, month, day), followed by a period and then the time (hours, minutes, seconds). If you ever decide to save or move the Project folder, this folder can be ignored or discarded to save space.

Two different Settings files are created and given the same name as the project. One has the extension .avs; the other has .xml. Basically, the .avs file is there only for backward compatibility with old versions of Media Composer. The .xml version is the "correct" version of the file. The cool thing is that if either version of the file is corrupted or deleted, it can be recreated from the other version.

The Avid Project folder is not special in any way. It operates like any other folder on your system. You can actually put anything you want inside the Avid Project folder. Some post houses actually have a standard operating procedure that dictates that all files that are going to be imported should be first placed inside the Avid Project folder. That way when the Avid Project folder is archived, all of the files that are needed are automatically inside the folder already. You can even place "readme" files inside Avid Project folders that other editors—or even you, years later—can use to rebuild or restore a project at a later time. The problem with this workflow is that your project folder can become unmanageably huge. A similar workflow would simply be to create an "import folder" for every project and put all the files—photos, graphics, motion graphic QuickTimes, etc.—into that folder so that there is some manageable way to move and archive the materials for an edit.

As for naming Avid projects, you can pretty much name projects any way you want. I would strongly suggest NOT naming a project, bin, or user with any characters other than simple alphanumerics, though the actual list of "bad" symbols consists of just / \ : * ? " < > | or leading spaces, trailing spaces, or trailing periods. Also, the old limit (from 6.0 and prior) is that names must be 27 characters or less. As of 7.0 (released the summer of 2013), the character limit for naming projects is 64 characters. To me, this is far more than you should ever need to use. A convention that I found helpful was used at one of the post houses that I worked at: each project was given a billing number before any editing was started. All of the Avid Projects used that billing number at the beginning of the name string. You could add any other part of a name you wanted, but by utilizing the billing number, it was a lot easier to find these projects years later and it was also easier for other editors to find and continue to work on these projects, especially when the list grew to be dozens and dozens of projects long. Occasionally, it's good to go in to the Shared path and archive "dead" projects to a CD, a DVD, or a shared location on a server. This data is valuable, but when the Project list gets too long, it's just a pain to find the project you want to work on. Manage the Shared

Projects list at the finder or OS level of the computer by simply locating the Shared Projects folder (do a search for it) and drag old or inactive project folders out of it.

Inside the Avid project window in the UI (user interface), everything has to be inside of a bin. This is a different idea for many Final Cut Pro editors. FCP has bins, too, but in FCP's browser, elements, like clips and sequences, can simply reside inside the browser with no other container. In Avid, the only thing that can be in the Project window, which is analogous to FCP's browser, are bins (and folders and scripts, but I'll get to that shortly). This really isn't a limiting thing. If you want to have just a single bin and put all of your elements into that single bin, you can certainly do that. But organizing footage and sequences and other elements into separate bins can definitely reduce the clutter during the editing process. It can also make it easier for collaboration with other editors.

Managing Bins Inside a Project

Managing Clips Inside a Bin

Bin Management. There are a number of reasons to have some discipline about how you organize your bins. One of the main ones is that open bins require RAM, and the more items in an open bin, the more RAM will be sucked up. So, by limiting the items in your bins and limiting the number of open bins in your project, your system will run more smoothly and you'll be able to edit faster. Also, with less visual clutter from multiple open bins, you'll keep your mind on the task at hand. As someone once advised, "Keep a clean machine."

> **TIP**
>
> In previous versions of Media Composer, there was a concept called the "Superbin," which let you open multiple bins inside the window of a single bin. Avid has swapped out that concept for the much more useable and easier-to-understand ability to grab the tab of a bin (which is the name of the bin with a little X next to it, located underneath the name of the bin at the very top of the bin) and drag it next to the tab of another bin so that two or more bins can share the same window (see Figure 1.12). This is very useful when you have limited screen real estate, as on a laptop. You can also use this for organizing groups of bins into the same space, for example, all of your audio bins. This should appeal to FCP editors.

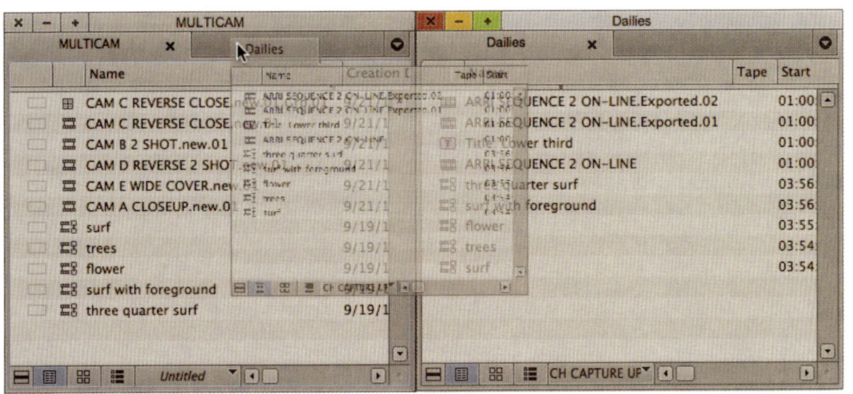

Fig. 1.12 This image shows the tab of one bin being dragged to another bin so that they occupy the same space. This helps maximize screen real estate, especially when editing on a laptop.

Bins Breaking Bad

The worst case of bin and project management I have ever seen on an Avid system was at a major market TV station that will remain nameless. They were having issues with full media drives and called me in to help them delete stuff and manage their process. Their problem was instantly apparent. They only had two projects, but they were MASSIVE. One was called "Programs" and the other "Promos." Inside each project were many bins. Each single bin had all of the media for a single ENTIRE TV show or promo. How is this bad? Let me count the ways. The main problem was that it was hard to find particular shots or resources inside the giant, overstuffed bins. Secondly, the systems launched and ran slower than molasses because of the bin and project sizes. Third—and I'm speculating here—I think the risk of disastrous bin and project file corruptions became an issue. Basically, you're putting all of your eggs in one basket. Even though I can count on a single hand the number of bins that have become corrupted on me over twenty years, it's not a happy thing when it happens. In chapter 2, I will discuss how the Avid attic can save you if this ever happens to you. Fourth, there are a lot of settings that you can—and should—set PER PROJECT, but since ALL of the TV shows and ALL of the promos had already been placed in a single project, this couldn't be really be done. Finally, many of the ways that you typically manage your hard drive space are project based. By having only two projects, none of these project-based storage management solutions could be utilized. The TV station's excuse for this mess was that they needed to be able to share stuff between the bins, so they couldn't use separate projects. This is wrong. You can easily share resources *between* projects. Bins from other projects can be opened in your current project, and you can copy and move resources between them as if they were in the same project.

Every editor is different and every project has different requirements. Therefore, discussing them all in a general way is very difficult, but, hopefully, even if some of the examples of how to manage your clips and bins don't apply directly to you, you can use these examples to spur your own thoughts and ideas of managing your next project.

Scripted Bin Management

When you have a script—or an unscripted project comprised of specific story elements—bin management is most easily managed by breaking each scene into separate bins with all of the material for that scene inside the bin. This method is easiest because when editing scripted projects, like feature films, editors tend to work on a single scene at a time. Even if, for some reason, they edit the scenes out of order, the general workflow is to consider only the shots and resources that were generated for that one scene. Exceptions make the rule, and with flashbacks and such you certainly might need to reference material from other bins, but that's easy to do.

The first Avid system I ever worked on at a post house was originally purchased and used to edit *The Fugitive*, directed by Andrew Davis and starring Harrison Ford. A copy of the script with the scene editing breakdown was tucked away for years in my desk. Editing it must have been a monumental task: six editors are credited on the movie.

Obviously, feature films use many more bins than simply one per scene (see Figure 1.7). Although you can't see the individual bins inside the numerous folders in the project in Figure 1.7, there are approximately 5,000 bins! Some examples of other bins for a feature film would be a bin with temp score music or sound effects. Also, most feature films will have double-system sound. It depends on how the bins were created, but the sound from the sound mixer might be in a separate bin or bins. They could also be organized so that the double-system sound for each scene would simply be included with the video and the synced clips for each individual scene.

For scripted TV dramas and comedies, the basics are similar, but the additional considerations include the use of "stock" exteriors shots that are used in multiple episodes. There would also be bins for music cues used in previous episodes.

Prep of *Smash*

Beth Moran was one of the assistant editors on the hit prime-time drama *Smash*. She also assisted on several feature films, including *Brokeback Mountain*, *The Manchurian Candidate* (2004 remake), and *Amelia*.

She described the preparation and organization of *Smash*: "The dailies files were brought in from Technicolor on a hard drive and were copied to the Unity. The dailies were all beautifully sunk to all of the audio tracks by Technicolor and they would deliver dailies bins [linked to media transcoded to DNxHD] and then we [the assistants] would just subclip out the mix." In other words, the dailies were delivered with eight or more isolated audio tracks, plus a mixed pair of audio tracks, and the assistants subclipped each take and sent only the mixed pair to the editors. "Then if the editors wanted they could match back to the original with all of the individual tracks," Moran explained, "but they normally edited [the offline] with the mix." With the subclips pulled from the dailies, Moran or the other assistants would write descriptions of the clips. Moran also said that the editor for whom she usually prepped footage didn't like having to watch everything before "action," so she'd put green locators on "action," then checked sync to make sure it was all right, and put the bin in frame view, which was the main editor's preference.

Smash employed elaborate and beautiful musical numbers. Prepping those was an additional step for the assistants. Moran explains her role: "We also grouped the multicams and multiple takes of the musical numbers. We could group multiple takes in multicams because they were all done to click tracks, so they matched. Tons of cameras, tons of takes, and tons of angles. And all of them were grouped to the song. Then, if there were pickups, we would just find a spot, and sync them up to wherever

GLOSSARY

Unity: This is an Avid project that works as a server and storage for large multi-seat Avid installations. It allows many editors, assistants and other post-production professionals to work on the same project simultaneously.

GLOSSARY

Click Track: Each musical number was recorded to a pre-recorded track, so the sync for each take was identical from take to take. Click tracks take their name from the countdown of clicks that precede the playback music on the track.

they started, so there would be one huge clip and the editors could just use the arrow keys to choose an angle and then, wherever they wanted to cut next, they just used the arrow key."

Moran was not part of the first season, and before she was a member of the team, she marveled at the ability to cut those musical numbers. "Before I started working on the show I watched *Smash*, and I wondered, 'How do they edit this so quickly on TV time? This seems like a movie!' But really, because it's organized this way they could edit the first cut of a musical number in an afternoon. They just worked really, really fast. So that was pretty amazing. I never thought about having different takes grouped, but since they're synced to the same song, it was easy."

Moran reveals another assistant editor trick that she'd used before on features but didn't have time to employ in the heat of a TV edit schedule. Moran explained the creation of something she referred to as "stacked dailies." Moran explains, "Supposedly Stanley Kubrick made up this system. I worked on a Noah Baumbach movie and the way Wes Anderson does it, I don't know if he still does it, he creates these sequences that have every line. The assistants break it down to the phrase or the line, then make a sequence with the close up, the single, the 2-shot, working out to the wide shot and then the line even on the off-cameras. So basically it helped you find the best reading for every line. I did this on *Margot at the Wedding*. Some of the assistants thought it was ridiculous, but you *really* get to know your footage. I put locators for all the resets, because this director, Noah Baumbach, he just goes 'Back to 1' without a reset, and he'd say 'Say it again. Say it again. Say it again,' like ten times within a take. So in a normal world, we'd just mark that with locators so we'd know there were ten resets in a single take. It is very helpful for them."

GLOSSARY

Off-cameras: A take or a section of a take where the actor who is speaking is not seen on camera.

For Smash, the bin organization was basically the dailies bins from Technicolor and then scene bins with the subclipped sections and the musical numbers, which would have their own bins.

We'll hear more about the off-line and on-line editing of *Smash* in chapters 3 and 5.

GLOSSARY

Script Supervisor: The script supervisor keeps track of what parts of the script have been shot and records the scene and take information, sometimes with corresponding timecode information. The script supervisor also tracks continuity issues among other responsibilities.

Prep of The Big Bang Theory

Todd Morris is the assistant editor for Peter Chakos on the CBS TV hit comedy, *The Big Bang Theory*. He describes how he receives materials from the set and prepares them for Chakos: "We shoot Tuesday night. I come in at about 5:30 or 6:00 a.m. on Wednesday. I receive the four cameras, each on an HDCam SR tape. Then what I do is I'll go through and, using the in timecodes from the sound recordists and the out timecodes from our script supervisor, I basically build my own clips. When I digitize, each scene might have three or four clips, one for each take. Typically they do three or four takes per scene. I take one camera and the script supervisor's script, I'll log every take for every scene using that one camera. Then I duplicate that information for each of the other cameras and build my own logs and bins. I digitize the one camera and batch-digitize the rest, all at DNxHD36. Then I import and sync up to the DVD-RAM audio files that come from the on-set audio mixer into a multigroup that goes into ScriptSync. I sync by source timecode. The timecode is run to all four cameras and to the DVD-RAM."

The rest of Morris' task in prepping the show is all done in ScriptSync, which we'll discuss in the Quick ScriptSync Overview section.

GLOSSARY

Multi-group: An Avid media item that allows multiple cameras and takes to be grouped together into a single multi-grouped clip.

Prep of The Mindy Project

There are many workflows for scripted comedy. Some depend on whether the shooting style is multicamera or single camera. The Emmys recognize these under two separate editing awards, actually. While *Big Bang* is shot with usually four cameras, another hit comedy, *The Mindy Project*, is shot in the "single camera" style with ARRI Alexas. However, though the style is single camera, there are almost always two cameras rolling. Richard Rossi is David Helfand's assistant editor on *The Mindy Project*.

Unlike *Big Bang*, the dailies are transcoded and bins are created at Universal Digital Services on the Universal Studios lot. Rossi explains, "We have a Dropbox, which is connected to UDS [Universal Digital Services]. They process five reels of dailies by 6:00 a.m., and I open it up and drop it onto a Unity drive designated for the episode."

Since the footage is already transcoded to DNxHD36 and organized into bins by UDS, Rossi's next task is to group the footage that was shot with the two cameras, allowing it to be edited using the multi-cam function. The rest of the process involves prepping for ScriptSync. "Every hour counts," Rossie explains. "They do a lot of improv on the show and the producers want to have access to that, and in ScriptSync; we have to know where it lives."

With Avid's optional ScriptSync workflow, you can actually use the script itself in lieu of a bin. I'll be discussing this in greater detail in a section devoted to ScriptSync, but because the pages of the script actually serve as a bin, I'll outline the basic concept of ScriptSync next and explain the rest of Morris's and Rossi's workflows. We'll also discuss the off-line editing process of these shows in chapter 3 and the on-line process in chapter 5, following the entire workflow from beginning to end.

Quick ScriptSync Overview

If you have a script, you can import it into Media Composer. With the script imported, it is then possible to link specific shots and takes to exact portions of the script. This is similar in appearance to a traditionally marked-up script from a script supervisor. Actually, you or an assistant editor can use those marked-up scripts from the script supervisor to match the shots in another bin to the script pages.

ScriptSync can be used in Avid without a license if you are willing to hand-sync each line of script with the matching audio. The licensed version allows you to automatically match the audio in the files with the phonetic sounds in the words of each line.

The basic process is to export a script as a .txt file from a program like FinalDraft. Then, in Avid, import the script. With the script open in Avid, listen to the various takes in a scene and then select the lines in the script covered by the take. With the lines selected, drag the clip onto the selected lines in the script. This links that take to the selected section of the script. You'll see a line, similar to the way a script supervisor marks up a script (see Figure 1.13), running down along those lines of script with a thumbnail of the take at the top. Continue to do this for each take. Eventually, you'll be able to see which takes cover which lines of the script (see Figure 1.14). When this is done, select the clips and choose ScriptSync. This uses phonetics in the audio file to match the exact words in the script to the exact moments in the clips that those words are spoken. With this done, you can click on any line of the script and audition the exact moment of that line in every take done of that line. A more complete description of how to do this is in chapter 3.

Check out the video tutorial I created, called ScriptSync Video Tutorial, in the on-line project files for this book. (www.focalpress.com/cw/hullfish)

When editing scripted content, it is very useful to be able to see the options for coverage on specific lines. Making informed and efficient choices using this method frees up your time and creative talents for the more demanding decisions of editing.

Todd Morris's workflow for *Big Bang Theory* moves from digitizing and logging to ScriptSync.

"Now that I have a multigroup for every scene, I use the Avid Script Integration program [ScriptSync]. I import a text file from the writers. I'll go through and I'll mark scenes and pages within it and put in a multigroup and mark each line on the fly, drag the multigroup to where in the script it relates. Put a mark in, hit record and mark each line on the fly. So I do every take in real time on the fly. You can do double time, which I can do, but I've found all of this to be faster than trying to get ScriptSync (the phonetic matching function) to work for me. I don't do ScriptSync's phonetic matching because when we do a rewrite, which we do a lot, ScriptSync can't find it. So what I do for Pete is color the take as a gray line. If they do a rewrite, I don't type in the rewrite, but what I do is mark the line blue. If they rewrite it again, I'll do it another color. If he sees a color, he knows that there's a rewrite. If we do wild lines I mark them

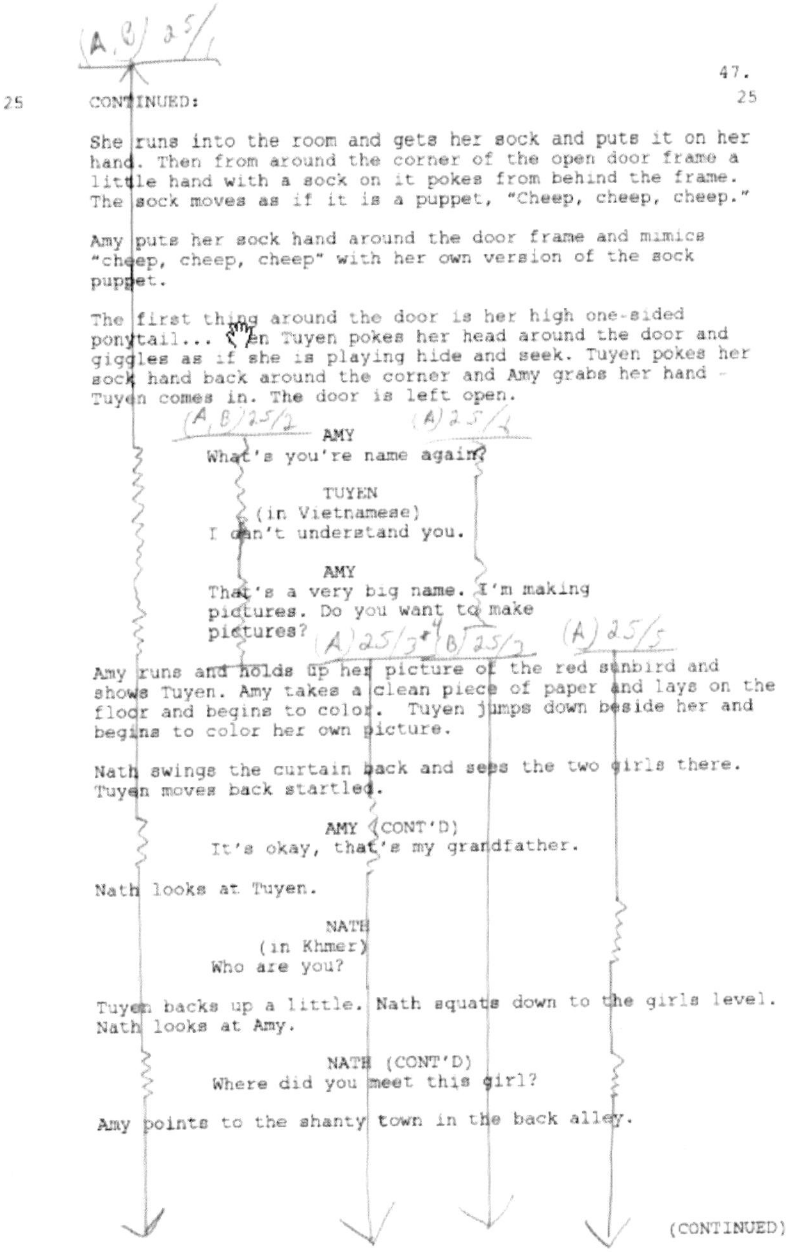

47.

25 CONTINUED: 25

She runs into the room and gets her sock and puts it on her hand. Then from around the corner of the open door frame a little hand with a sock on it pokes from behind the frame. The sock moves as if it is a puppet, "Cheep, cheep, cheep."

Amy puts her sock hand around the door frame and mimics "cheep, cheep, cheep" with her own version of the sock puppet.

The first thing around the door is her high one-sided ponytail... then Tuyen pokes her head around the door and giggles as if she is playing hide and seek. Tuyen pokes her sock hand back around the corner and Amy grabs her hand - Tuyen comes in. The door is left open.

AMY
What's you're name again?

TUYEN
(in Vietnamese)
I don't understand you.

AMY
That's a very big name. I'm making pictures. Do you want to make pictures?

Amy runs and holds up her picture of the red sunbird and shows Tuyen. Amy takes a clean piece of paper and lays on the floor and begins to color. Tuyen jumps down beside her and begins to color her own picture.

Nath swings the curtain back and sees the two girls there. Tuyen moves back startled.

AMY (CONT'D)
It's okay, that's my grandfather.

Nath looks at Tuyen.

NATH
(in Khmer)
Who are you?

Tuyen backs up a little. Nath squats down to the girls level. Nath looks at Amy.

NATH (CONT'D)
Where did you meet this girl?

Amy points to the shanty town in the back alley.

(CONTINUED)

Fig. 1.13 This is a lined script, prepped by a script supervisor. This image is used by permission of producer Stephen Marinaccio and is from a film called *Trade of Innocents*, written by director Christopher Bessette.

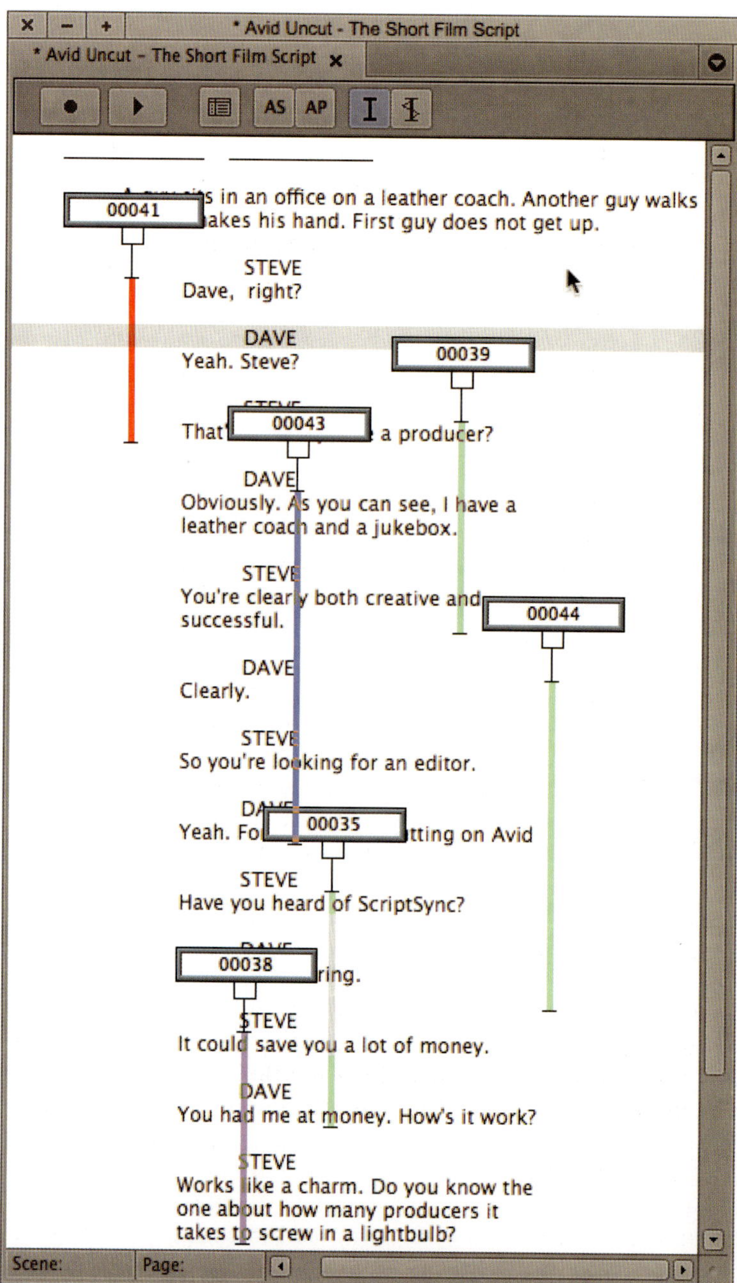

Fig. 1.14 This is a ScriptSync page from my thrilling blockbuster film, *Avid Uncut*. This project is available on the companion website (www.focalpress.com/cw/hullfish). By default, each of the lines on the page would be connected to a thumbnail of the clip, but the thumbnail can be turned off to better see the underlying text.

in red. So he can look at the script and it's all color coded and he knows where to find things."

All of this prep work takes a full day per episode, but to make the most efficient use of their time, Morris gets the first half of the show prepped between 6 a.m. and 10 a.m. so that Chakos can start editing while Morris finishes prepping the second half of the script.

The other big user of ScriptSync is David Helfand, who edits *The Mindy Project*. Helfand rarely worries about the organization of the bins because he recalls takes exclusively from ScriptSync, saying, "I find myself resistant to using bins."

Helfand's assistant, Richard Rossi, preps ScriptSync with a procedure that they've tailored to the unique needs of the show, mainly to accommodate the large quantity of ad libbing but also to deal with the change in production technique between film—when every take would stop before a reset to save costs—and digital, which tends to roll right through from one take to the next without resetting or stopping.

"In one take, there'll be three restarts," explains Rossi, "so I have to find the restarts. I give each restart its own line ("take" in ScriptSync). It also gets complicated when they ad lib. So for adlibs you have to retype the adlibs and cut and paste."

ScriptSync does not allow for retyping words or lines once the script has been imported into Avid, but Rossi reveals a special technique for dealing with these revisions. "David taught me that you can use TextEdit, which is a really simple text program, if you copy and paste it into TextEdit and you can retype the scene in TextEdit using the same margins and this way, when you copy and paste back into the script, it will cheat it." Rossi continues, "So you copy the section of the original script, which keeps the formatting, then paste it into TextEdit, where you do the revisions or add the ad libs, then copy that and paste it directly into the script that's already imported into Avid. As long as you've kept the same formatting it will recognize it."

The rest of their workflow involves Rossi adding a lot of markers for restarts or notes. But even with the tendency of the production to roll through from one take to another, Rossi and Helfand don't use subclipping to break up the takes. Instead they simply mark the specific takes with an in and an out and create a new line (take) in ScriptSync. Rossi also color codes these takes, indicating ad libs in blue and restarts in green, and if an actor skips lines for some reason, those missing lines are marked in red.

To allow Helfand to begin editing as soon as possible, the script is broken apart into as many as eight sections so that Rossi can deliver small chunks of the script to Helfand.

Helfand elaborates: "Actually ScriptSync is tough on this show because of the schedule. [On *The Mindy Project*] they use this technique of throwing stuff at the actors that's off-script and because of that, there's a lot of transcribing and modification of the script and it would be helpful if you could modify it or notate ScriptSync. We get all this footage and it's digitized by the lab and the lab will sync up the Alexa footage with sound and give us Avid dailies in DNxHD 36, which is what we edit in, then the assistants will sync up the script with the dailies, so there's hours of preparation of the script, and we break up the script into separate acts so I can start working before they're done prepping, and they'll type up the ad libs if they have time."

Helfand was an early proponent of ScriptSync and convinced many of his editorial colleagues and producers on his shows of the value of it.

EdiFlex and Script Mimic

Avid's ScriptSync feature is not found in other editing systems because it is patented. The patent was originally developed and marketed by Cinedco, which created the EdiFlex NLE (nonlinear editing) in the mid-1980s. EdiFlex had a feature called Script Mimic, which is very similar to Avid's ScriptSync. The patent for the feature was purchased by Avid when EdiFlex shuttered in the mid-1990s. Avid reintroduced the technology as ScriptSync in 1998. ScriptSync also includes technology for phonetic recognition and matching that belongs to Nexidia. Nexidia's technology allows for the automatic matching of the audio with the words in the script.

Unscripted Bin Management

When editing unscripted projects, bin and project management is much more of a free-for-all. Each project has unique needs that dictate the structure of the bins.

Some editors prefer to create bins when they have a need to fill them with something. I do this sometimes, but I also like to start with a series of bins that are the same on most projects. The whole point is to edit something, so I always create a bin that will contain nothing but my current sequence. This helps other editors if they have to jump in for some reason, because there's a Current Cut or Current Sequence or Current Edit bin. Also, because there will be versions and variations and revisions and even just progressively more complete edits, I also create an Old Sequences bin. When I get an edit to a certain point, I'll make a copy of it and name it with the date and time or some other explanation, and then put it into the Old Sequences bin, then continue to work on the original.

Other bins I like to create before I start editing are these:

Music
Voice Over

Titles
Graphics
Junk in My Trunk (I stole this from my editing colleague, Rick "First Call" Mahal.)

Documentary editors must be incredibly organized in order to quickly find the clips that they know are somewhere in the project. Robert Landau edits the CNN documentary series *Crimes of the Century*, executive produced by Ridley Scott. *Crimes of the Century* focuses on major news events that have had far reaching repercussions. It was also one of Tony Scott's last pet projects before his untimely death.

The source for each episode is about thirty to forty hours of stock footage, mostly from CNN's archives, plus another ten hours of new interviews. Because of the amount of material, it is all imported or captured at 20:1 standard definition into Avid.

Though the show has a script that includes the sound bites from the newly shot interviews, the archival sound bites and visuals are left to the editor. Landau organizes all of the clips using comments and clip names. He also labels the bins with descriptive names, separating sound bites into those bins.

Note Landau's Project organization of bins inside of folders (see Figure 1.15).

At the top of his project is a folder of cuts, or sequences (see Figure 1.15). This was forced to the top of the project window by starting the folder name with a number, which in this case needed to be preceded by a leading 0. Otherwise folders 10, 11, and 12 would end up in the wrong numerical order because the folders are actually being ordered alphabetically. This behavior differs between Macs and PCs. PCs can do this numeric ordering without leading zeroes. Note that there is a bin with nothing other than the locked final cut. There is a bin with just the current cut, and there are additional subfolders of various sequences. Inside the Current Cut bin are separate sequences for each act of the show (see Figure 1.16). In the locked cut, these acts are joined into a single sequence (see Figure 1.17).

Archival and interview footage are broken into two separate folders, each with multiple bins (see Figure 1.18). More important to the organizational scheme is the Selects folder that has many clearly defined selects (see Figure 1.19). Creating these Selects bins and naming the clips inside of them is the beginning of the process of culling the vast quantity of footage into something more manageable. Many of the bins were based on the description of the footage itself, but some bins were developed later as story elements became clearer. This is because Landau began looking at the footage before the script was delivered. After seeing the script, he could begin to create bins that would support the specific elements discussed in the script.

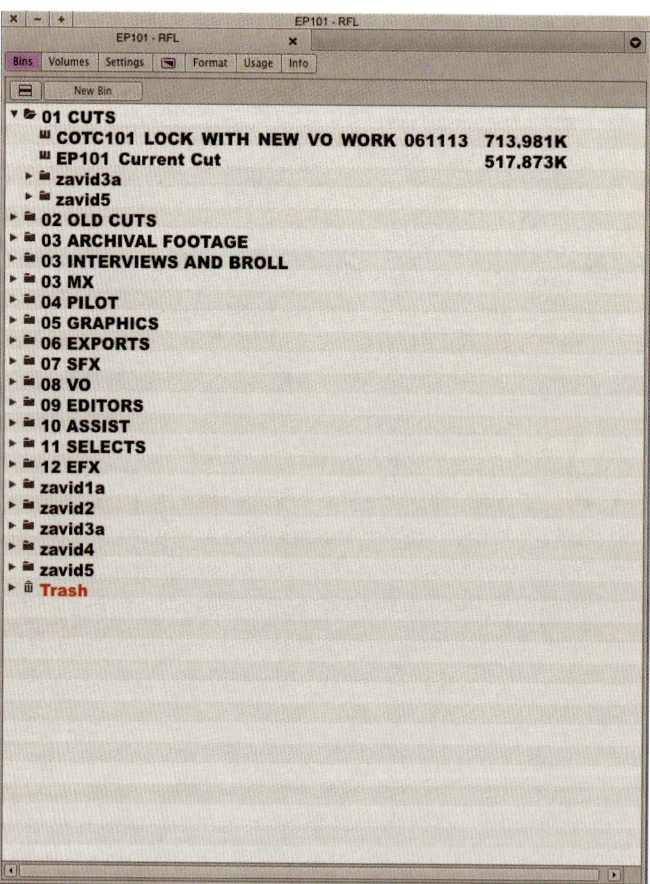

Fig. 1.15 This is the Project Window for an episode of "Crimes of the Century." Courtesy of CNN and editor, Robert Landau. Bins are not seen in the Project file because they've all been organized inside of numbered folders.

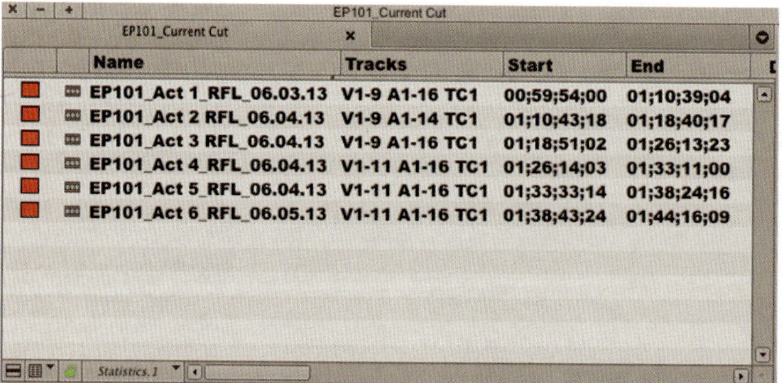

Fig. 1.16 This is the Current Cut bin from an episode of "Crimes of the Century." They are broken down into individual sequences for each act of the episode.

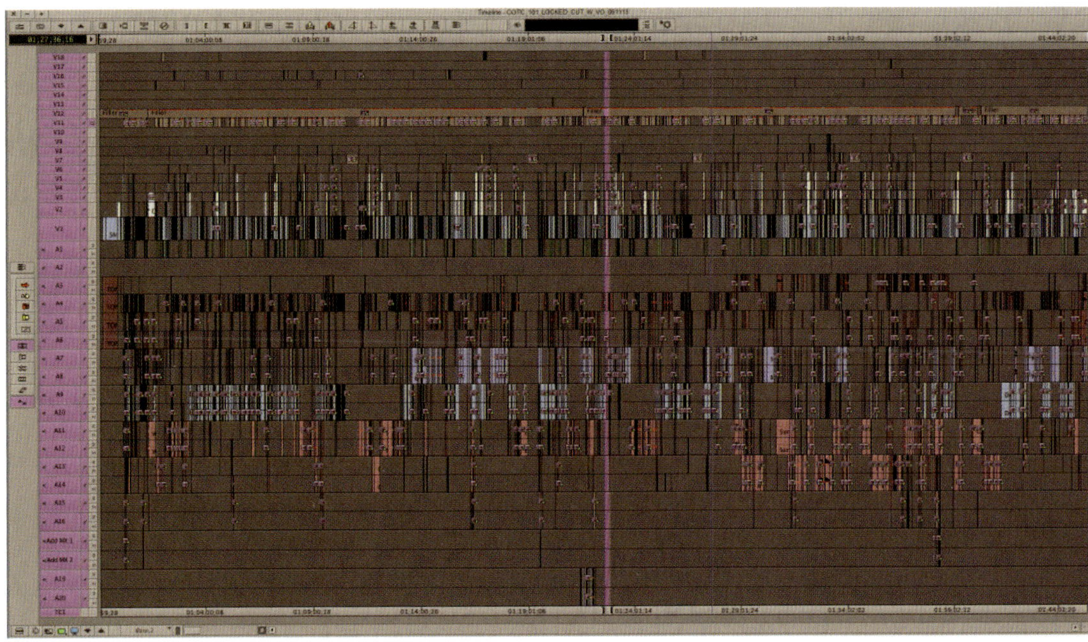

Fig. 1.17 This is the complete timeline for an episode of "Crimes of the Century" with the acts combined into a single, complete episode.

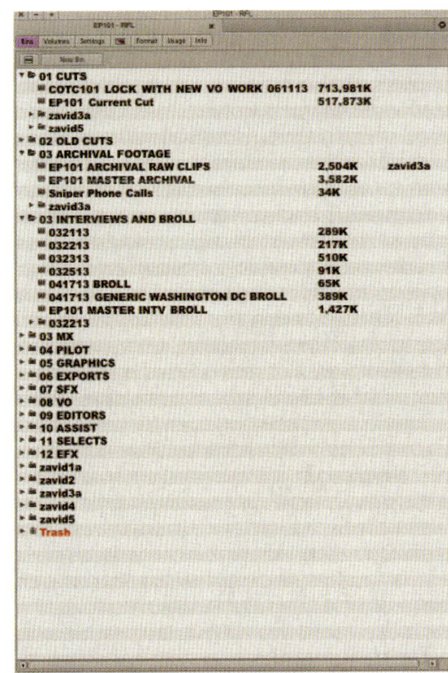

Fig. 1.18 This is a view of the Project window of an episode of "Crimes of the Century" with some of the folders "spun down" to reveal the bins inside.

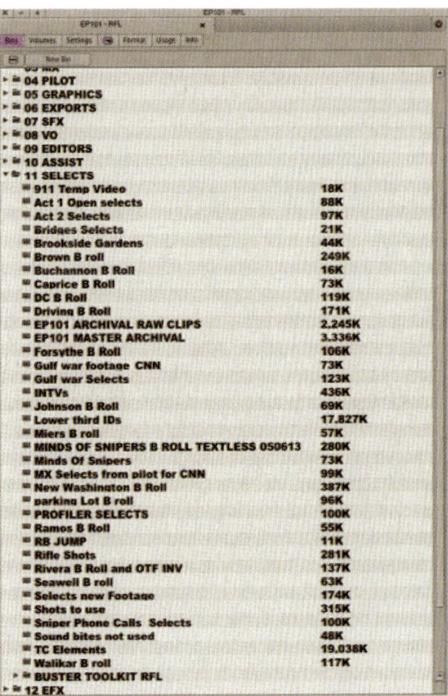

Fig. 1.19 This is the Project same Project window as in Figure 1.18, but with the "Selects" folder revealing all of the bins inside that folder. This gives a great sense of the thought behind the organization of the project as well as its scope.

There's also a folder for Graphics, which includes numerous bins devoted not just to story exposition but also to editing elements (film burns, for example) (see Figure 1.20).

Music is critical to the mood of documentaries, and Landau has a folder (MX; see Figure 1.21) that includes numerous bins of subcategories of music. Jingle Punks is a music library that delivered many of the elements for the show.

The bin of archival clips shows the descriptive logging of the content of the clips. Much of this notation came from CNN's original archive abstract (synopsis) of each clip (see Figure 1.22).

This careful text description is helpful in a number of ways. You can simply look through the bins to find what you remember typing. Or you can use the Find command.

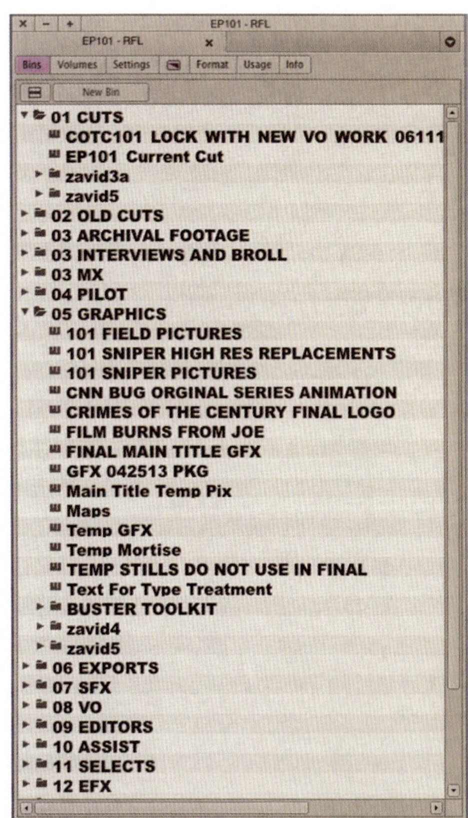

Fig. 1.20 This is the Project Window for the same episode of "Crimes of the Century" showing the folder for "Graphics" "spun down" to reveal the various bins of graphic elements.

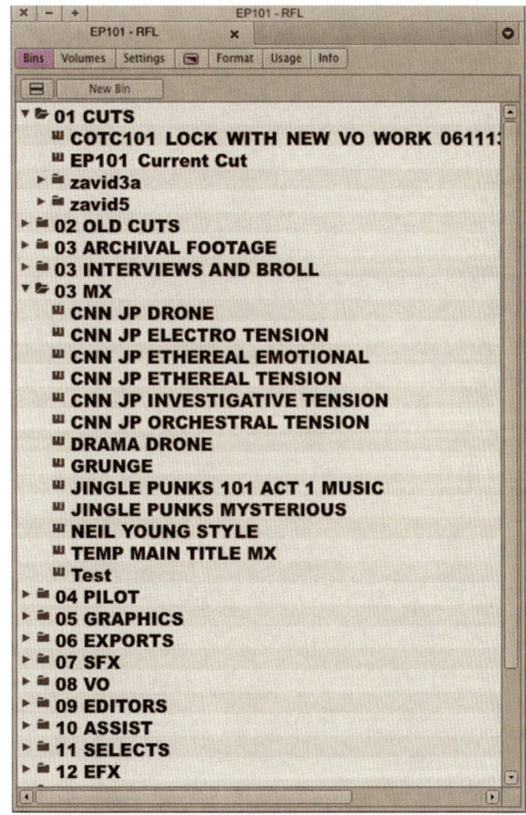

Fig. 1.21 This is the same Project window for the episode of "Crimes of the Century" showing the Music folder and the various bins of music and their organizational scheme.

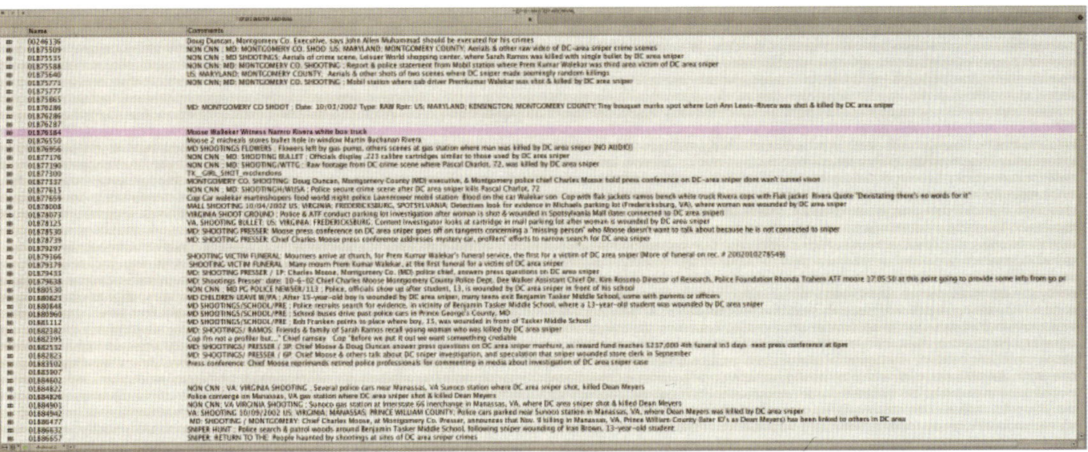

Fig. 1.22 This is one of the bins from "Crimes of the Century" showing the detailed description provided for each clip. All of this descriptive material can be used for searches.

Depending on what kinds of custom columns you've created and populated, you can also sort the bins by clicking at the top of the column to sort things alphabetically. For example, using a column of interviewees, you could easily sort it based on which interviewee you were looking for. Landau also uses Avid's Custom Sift functionality.

If I'm trying to narrow down my options for trailer sound effects (over 500 effects), I can go to the menu bar to Bin > Custom Sift or in the bin's Fast menu. (see Figure 1.23). If this is something you think you'll be doing a lot, you can also map the Custom Sift functionality to a key. (See chapter 2, page 109, for more information.)

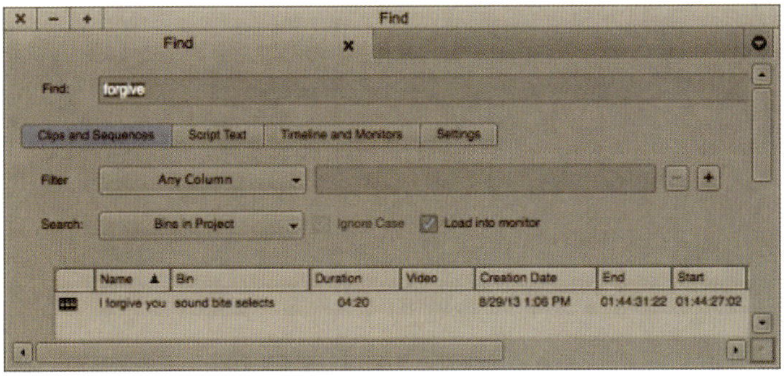

Fig. 1.23 Command-F on a Mac and control-F on a PC call up the Find function. You can quickly search for any text in any column or any bin.

The Custom Sift command calls up a window that allows you to do very complex searches (see Figure 1.24). These are searches that include not just a simple word—though searches can be that simple—but also more complex combinations of words. My search for "soft whoosh" didn't bring up any hits, but searching for "soft" AND "whoosh" did the trick (see Figure 1.25). These searches can also be done on exact matches or just that the clip "contains" the word in the sift parameter. They can also be limited to specific columns. Note that after I did the Custom Sift, the name at the top of the bin indicates that the bin has been sifted. This is very important because the bin now looks as though it has only about fifteen clips, when in reality the bin has hundreds of clips that are now

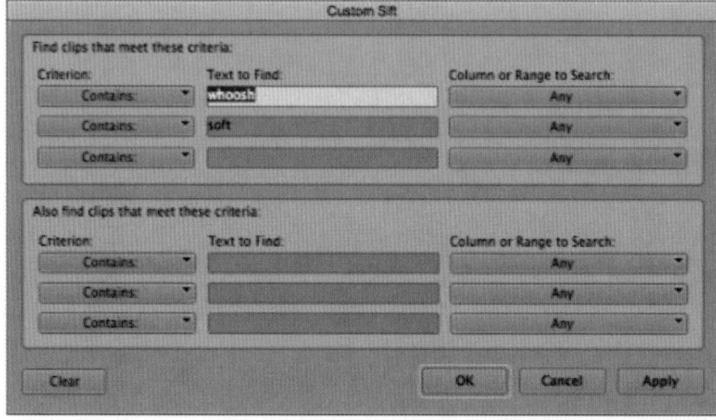

Fig. 1.24 The Custom Sift command with a simple operator. Here we're sifting results based on the word "whoosh" in any column.

Trailer SFX (sifted)

Name	comments	Creation Date	Duration	TapeID
WHOOSH-SOFT_Velvet.wav		8/29/13 10:13:15 AM	5:01	WHOOSH-SOFT_Velvet.wav
WHOOSH-SOFT_Treater.wav		8/29/13 10:13:15 AM	8:06	WHOOSH-SOFT_Treater.wav
WHOOSH-SOFT_The_Comedown.wav		8/29/13 10:13:14 AM	13:05	WHOOSH-SOFT_The_Comedown.wav
WHOOSH-SOFT_Subtle_Wave.wav		8/29/13 10:13:14 AM	6:07	WHOOSH-SOFT_Subtle_Wave.wav
WHOOSH-SOFT_Spirit_Of_Freedom.wav		8/29/13 10:13:13 AM	7:18	WHOOSH-SOFT_Spirit_Of_Freedom.wav
WHOOSH-SOFT_SpaceBy.wav		8/29/13 10:13:13 AM	6:18	WHOOSH-SOFT_SpaceBy.wav
WHOOSH-SOFT_Slightly.wav		8/29/13 10:13:12 AM	8:06	WHOOSH-SOFT_Slightly.wav
WHOOSH-SOFT_Rumble_Spell.wav		8/29/13 10:13:12 AM	6:20	WHOOSH-SOFT_Rumble_Spell.wav
WHOOSH-SOFT_Passenger.wav		8/29/13 10:13:11 AM	5:04	WHOOSH-SOFT_Passenger.wav
WHOOSH-SOFT_Custer.wav		8/29/13 10:13:11 AM	4:23	WHOOSH-SOFT_Custer.wav
WHOOSH-SOFT_Flee_You_Zany.wav		8/29/13 10:13:11 AM	8:05	WHOOSH-SOFT_Flee_You_Zany.wav
WHOOSH-SOFT_Deathbreath.wav		8/29/13 10:13:10 AM	7:17	WHOOSH-SOFT_Deathbreath.wav
WHOOSH-SOFT_Darth_Breather.wav		8/29/13 10:13:10 AM	5:23	WHOOSH-SOFT_Darth_Breather.wav
WHOOSH-SOFT_Chimera_Breeze.wav		8/29/13 10:13:09 AM	6:18	WHOOSH-SOFT_Chimera_Breeze.wav
WHOOSH-SOFT_Bottom_Rocker.wav		8/29/13 10:13:09 AM	7:13	WHOOSH-SOFT_Bottom_Rocker.wav

Untitled

Fig. 1.25 With the Custom Sift set for "whoosh," the bin only shows clips with the word "whoosh" in any of the columns of metadata. This makes it much easier to find material in very large bins.

hidden. To reveal the hidden clips from the menu bar, which you should do before you forget, select Bin > Show Unsifted.

Using Bins from Other Projects

You may notice that I don't have an effects or sound effects bin in the list of "starter" bins on the top of page 25. That's because there are a number of bins that I like to use across ALL projects. Instead of creating a separate effects bin for each project, I like to have an effects bin that I use over and over, collecting effects as I go. Obviously some projects that are effects-heavy might require their own effects bin as well as the "universal" effects bin used across multiple projects. I also have a sound effects bin that I like to use across multiple projects. Similarly, if you like to use a lot of color correction, you may want to create a color correction bin in which you can save and share color corrections and sampled colors from multiple projects. Finally, I also have a bin that contains the basic opening elements of a sequence, like bars, tone, slates, and countdown.

I am not alone in this organizational method. Robert Landau, discussed in the section above, also uses this method for organizing bins for use across multiple projects. You can see his Effects bin (Figure 1.26), Looks bin (Figure 1.27), and Audio Effects bin (Figure 1.28). Note that all of the effects in these bins are labeled with clear, easy-to-find, descriptive names.

It may be a good idea to create an Evergreen project that has the bins that are commonly shared. That way you always know where to go to find them.

Opening bins from other projects is simple. It is done from the File menu or from a keyboard shortcut, Command-O (Mac) or Control-O (PC). The trick is that the File menu and the keyboard shortcuts are dependent on the state of the Project window. To open bins, the Project window has to be in the Bin tab, not the Settings, Effect, Info, or Status tab.

In order to open a bin from a specific project, you have to know where that project was saved. This is a strong argument for using the Shared project choice when creating new projects because it will organize all (or at least most) of your projects in a single, reliable location that should be able to be found even by other editors unfamiliar with your system. Use the options in the Open Bin dialog to navigate to the project that you think the bin is in and open the bin you want.

Be aware that if you *move* a clip from a bin that comes from another project, it will not be in the old project anymore and could be very difficult to track to this new project, so it's important that any files that are going to be used in a new project be *copied* or *duplicated* from the old bin into a bin in your current project. It's possible to use clips directly from the old bin from the old project, but this can make tracking these assets

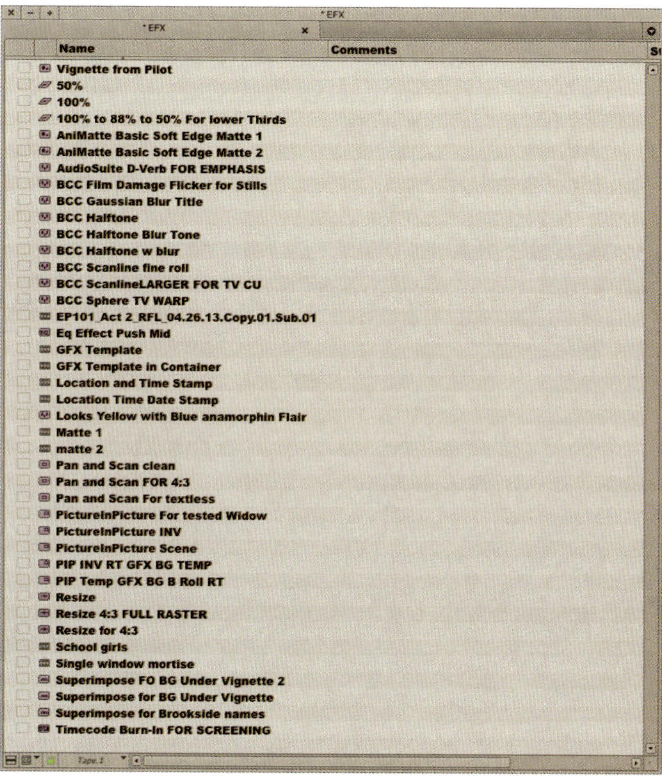

Fig. 1.26 Landau's EFX bin with various pre-built effects saved in it. Each effect is thoroughly described to make it easy to find and search for.

more difficult farther down the road. It's also smart to consolidate any media pulled from old projects because this consolidated media will be identified as part of the current project. Otherwise, if someone deletes all of the media from the old project, it will also be deleted from the new project. Consolidating will prevent this. Consolidation is a process that copies media selectively. A consolidated clip *can* be the entire clip, but more commonly it is only a portion of the clip, such as the section that was used in the sequence, plus some handles. Avid's excellent media management allows this portioned copy to maintain its relationship with the original full-length clip. It will also protect the "old media" by identifying it as part of the new project when the new project media is archived or saved in some way.

If you decide not to consolidate, it may be wise to name a new bin something like "Clips from Project X" and copy the clips from Project X into this new bin. That way, you will have some reference as to where they came from.

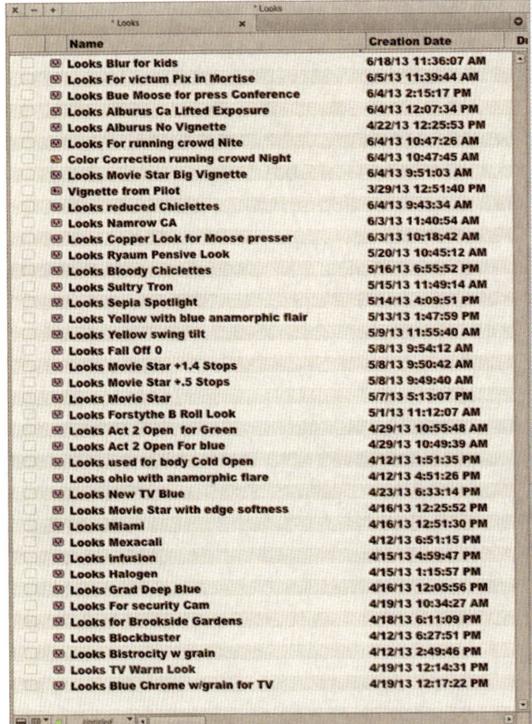

Fig. 1.27 Landua's Looks bin containing various Magic Bullet Looks presets, all named descriptively. It's possible to have a bin like this with straight Avid color correction looks.

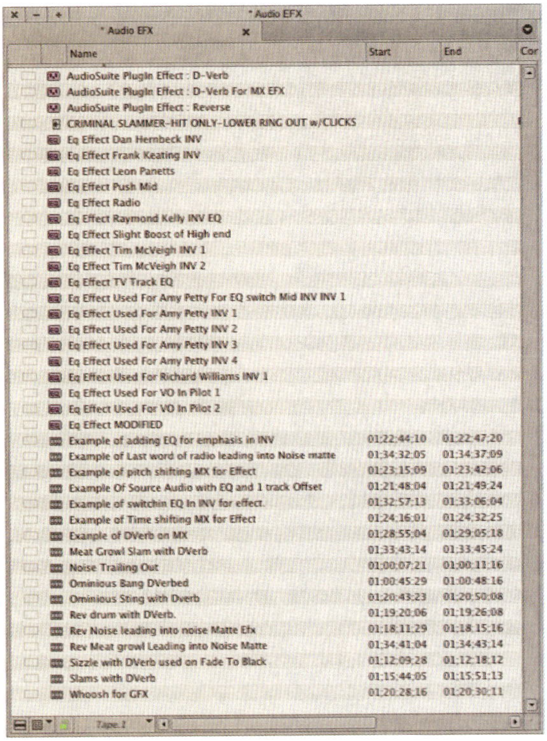

Fig. 1.28 Landau's Audio EFX bin includes various AudioSuite and EQ presets and sound effects.

Understanding Avid MediaFiles Folders

The first time you create media on a specific drive, an Avid MediaFiles folder is created and a few critical files and folders are created and placed inside (see Figure 1.29). For Final Cut Pro users, the Avid MediaFiles folder is similar to FCP's "scratch disk." This is where Avid stores all of the media that is captured, imported, or rendered. The only real difference is that when FCP imports files into the system, they are not placed on the scratch disk, it merely points to the original location of the file. This is why FCP imports are so fast. When Avid imports files, it actually converts them to "Avid Media" (MXF files) and places that converted file in the Avid MediaFiles folder. This is slower, but safer and improves media management. The MediaFiles folder also contains several folders and database files that are used to track all of the media within the folder. This database program makes accessing and locating the media much quicker and more stable. The Avid Mediafiles folder can be seen by Media Composer if it is at the root level of the drive. That's the only place that Avid will create the folder. And if you move the folder to the root level of a different drive, Media Composer will see it and link to all of the files

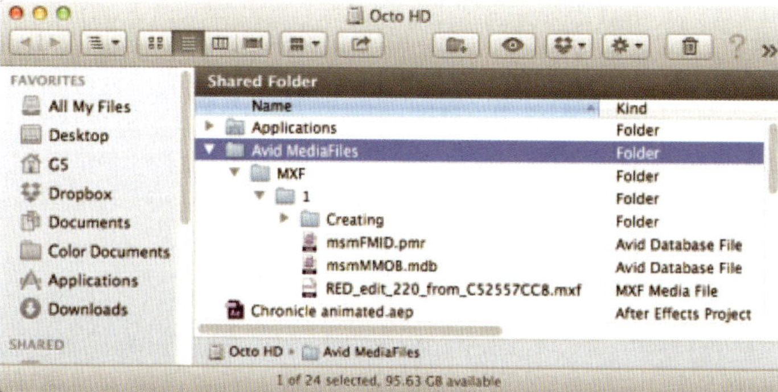

Fig. 1.29 This shows the Avid MediaFiles folder (on a Mac) at the root of the drive (called Octo HD), with a hierarchy of folders, starting with the MXF folder. Inside the MXF folder are a series of numbered folders. In this case, there is only a single numbered folder, "1." Each folder holds 5,000 media objects. Then another folder is automatically created and populated with more media. Inside the numbered folder are: a folder called "Creating," an msmFMID.pmr file, an msmMMOB.mdb file, and the various MXF files you ingest or create. In this case, there is only a single rendered dissolve from a sequence called "RED_Edit."

in it automatically. But if you move it to a spot that is not on the root of a drive, then Media Composer will not be able to see any of the media inside. Nor will Avid see the media inside if the folder is renamed.

Avid MediaFiles are meant to stay out of sight and out of mind. With any luck, you shouldn't have to bother with them much. But understanding how they work and what is contained inside the Avid Mediafiles folders will help you comprehend the way media can be recovered, relinked, deleted, archived, and moved.

The Avid MediaFiles folder should reside on a high-speed RAID connected to the computer. Barring that, it should be on a high-speed drive *other* than the drive that runs the computer's OS. It is possible to have the MediaFiles folder on the main OS drive. Though this is not recommended, it happens often on laptop systems. The danger of placing the MediaFiles folder on USB or FireWire drives is that drive failure is a real possibility. These drives also have significantly slower throughput than is necessary for most HD video, especially when dealing with multiple streams of HD.

There should be only a single Avid MediaFiles folder on any given drive or RAID. If you want to have another Avid MediaFiles folder on a drive, it is possible to rename the current MediaFiles folder by simply appending a letter or a number to the folder name. Unless Avid sees the EXACT name of the Avid MediaFiles folder, it does not recognize it as a MediaFiles folder. By appending a letter to the name of the folder, Avid does not see the folder anymore and will create another folder with the correct structure and files. Any media from the original MediaFiles folder will appear as off-line. To get the Avid system to recognize the original folder again, simply append a letter or number to the second folder and remove it from the first. At Alpha Dogs, run by my good friend, Terry

Curren, in Burbank, California, they use a similar trick, copying the Avid Mediafiles folder to another folder, to archive projects. They copy the Avid project folder to a separate media drive. Then they consolidate to that drive and drag the media folder into the Avid project folder. To restore a project, they drag that project's media folder back out to the root level. That way there is one folder for archiving everything in a specific project.

The Structure and Files of the Avid MediaFiles Folder

Inside the Avid MediaFiles folder, there should be a single folder called "MXF" (see Figure 1.29). On some older systems, there may also be a folder called "OMF," which contains OMF media, but this is increasingly rare.

Depending on how much media and how many projects you've created, inside the MXF folder are a series of numbered folders, or at least a single folder named "1."

Inside each of the numbered folders is a folder called "Creating," two database files called msmFMID.pmr and msmMMOB.mdb, and then a long list of all of the media that has been captured, imported, or rendered into your system. The video files will all have the suffix .mxf.

The msmFMID.pmr file lists all the online media files. PMR stands for Persistent Media Record. When a media file is created, it gets added to the PMR file, and when media is deleted, it's removed from the PMR file. The msmMMOB.mdb file is the media database file. The .mdb file stores all of the information that is critical to the Media Tool and is crucial to the relinking process. These are not files that you can do anything with. But if you are having relinking, Media Tool, or off-line media troubles, you can delete both of these files and they will be rebuilt the next time the Avid is activated or launched. If you are on a shared storage solution, like Isis, deleting the .pmr or .mdb files should be done with caution and probably not without the advice and consent of the system administrator. Again, these files and folders should all really remain behind-the-scenes and should not be anything you need to worry about under most circumstances.

TIP

Dragging the msmMMOB file into an empty bin in Media Composer will create all the master clips in that media folder.

Managing Storage

The media (.mxf files) inside the Avid MediaFiles folder will continue to build as you capture, import, and render. Avid creates these media files and tracks them. In Final Cut Pro 7 and many other systems, the NLE simply points to the original file, wherever it is on the drive. For example, if you have a file from a camera or After Effects or maybe a piece of imported stock footage, other NLEs simply read the original file as it sits on your hard drive. Some of them claim to "import" the file, but this just means that they create a pointer to it. If you rename, move, or delete that file, the NLE will not be able to find the file again without operator intervention. With Avid, when those types of files are imported, they are actually transcoded or "rewrapped" into an MXF wrapper and stored separately in the Avid MediaFiles folder. If the original file is renamed, moved, or deleted, Avid still has the link to the new file that has been created within the MediaFiles folder.

Every piece of media in the Avid MediaFiles folder is assigned a unique name by the Avid system and that file is also tracked using the database files (.mdb and .pmr), which are continuously and automatically updated as media is added and deleted from the system.

At some point, the drive containing the Avid MediaFiles folder will become full. No one ever manages this ahead of time, but it can be done. There are several methods of managing drive space on an Avid system:

- Root level purging of files—The Tactical Nuke
- The surgical strike of the media tool—Make Room, Not War
- Consolidate and store—The Thinking Man's Archival Solution

Root level purging is the fastest, easiest, and most dangerous. Basically, you can go into the Avid MediaFiles folder and simply delete all of the files in the folder or all of the files within a certain date range. There is no way to tell which files in the MediaFiles folder belong to which project at the Finder level. Depending on the type of work you do, it may be possible to know that ALL of the media in the folder—or at least all of the media between today's date and some previous date—belong to a single project that you never intend to get back again. Personally, I have never been sure I was at the end of a project. Even projects with "hard deadlines" often come back in one form or another, or the assets from that project are repurposed for another project. Or, if you have a lot of time and external drive space, it could be possible to archive all of the media in the Avid MediaFiles folder to an external source, possibly an LTO drive, and once that archive is made, you could simply want to delete all of the media. This is a pretty broad solution to the problem of archiving and managing drive space. If you're going to go this route, you can simply delete the entire Avid MediaFiles folder, and Avid will rebuild a brand new empty one as soon as the app becomes active again.

If you are considering this method, know the shortcuts to selecting large numbers of files. On a PC, Control-A will select all files in a folder, or even all folders in a folder. For Mac, it's Command-A. The other important shortcut to know is to sort the files on an attribute, like creation date, then click on a single file at one end of the list (like a start date) and Shift-click on another file at the other end of the list (like an end date). This will select all of the files in between.

Note that when these files are deleted, Avid has no way to recover or relink to these files. If you archived them before deletion, then restoring them is simply a matter or dragging or copying the files back into one of the numbered folders inside an Avid MediaFiles folder. The Avid system will automatically relink the returned files to the original sequence and clip information in the bins. This is one of the awesome features of Avid's robust media management. I cut a lot on Final Cut, and there was simply no way to easily do this with FCP. The relinking process always took some intervention and a whole lot of fervent prayer!

For many years, the savior for many archiving workflows was Tony Black's MediaMover—actually, Random Video's MediaMover. It's still working great with MC7! Check out www.randomvideo.com. The program is very basic. It analyzes all of the media on your system and displays a list of Avid Projects for which media is online. You can then select a project and move, archive, or delete all of the media for just that project (see Figure 1.30). I've been using this product for more than a decade, and I highly recommend it. I would hate to have to do media management without it.

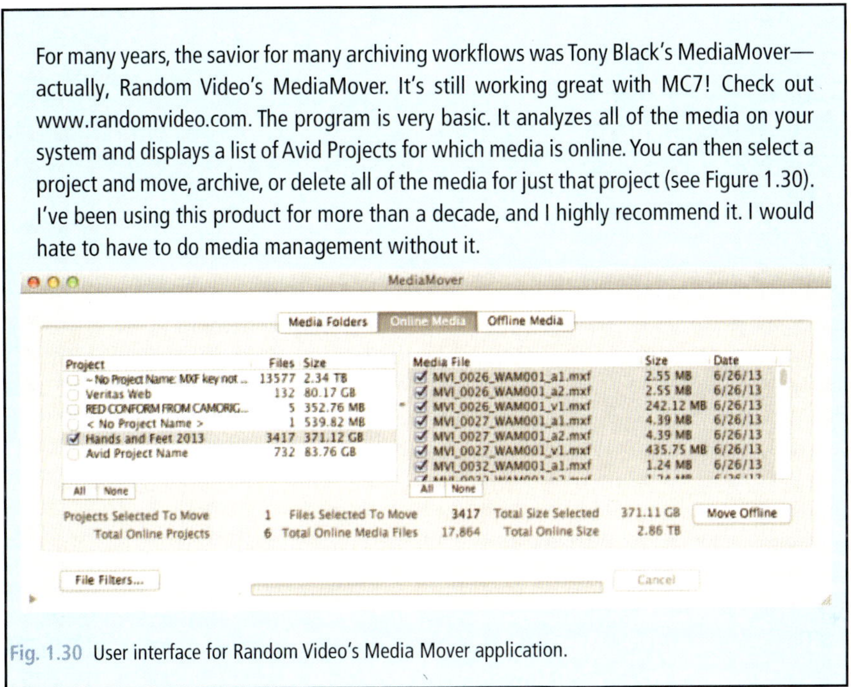

Fig. 1.30 User interface for Random Video's Media Mover application.

Using the Media Tool to Delete/Find Unused Media

The second method utilizes the Media Tool and gives you a lot more control than root— or Finder-level purging. It takes a bit longer to do, but your chances of losing critical media are greatly reduced, though nothing is truly foolproof. According to an old quote from Douglas Adams, "A common mistake that people make when trying to design something completely foolproof is to underestimate the ingenuity of complete fools."

Media Tool (see Figure 1.31) is one of the tools that is selectable from the Tools pulldown menu in Media Composer. It allows you to search for actual media files based on a number of attributes. The things that you can search on are Project name, hard drive, Master Clips, Media Files, and precomputes. Once the search has been completed, Media Tool allows you to delete these files selectively or en masse.

When you call up the Media Tool, there is a small delay as Media Composer actually queries the databases (one for each drive that has an Avid MediaFiles folder) to display a list of all of the Avid Projects that still have

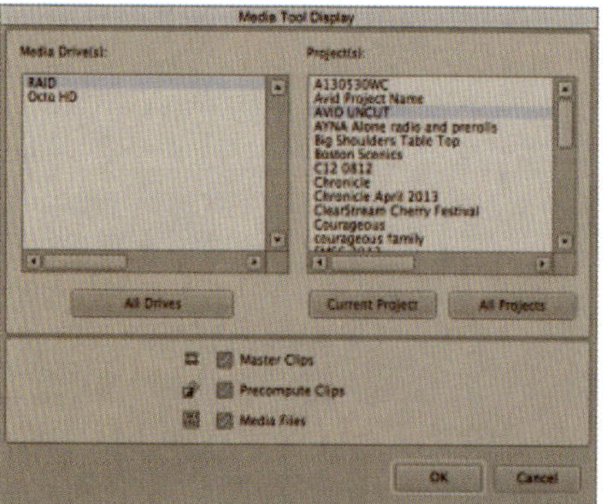

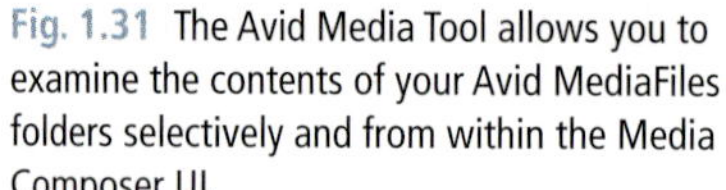

Fig. 1.31 The Avid Media Tool allows you to examine the contents of your Avid MediaFiles folders selectively and from within the Media Composer UI.

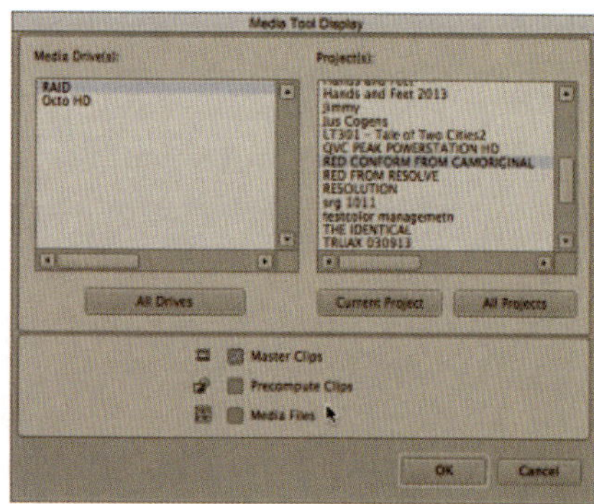

Fig. 1.32 The Media Tool with only Master Clips selected.

TIP

Media Tool's ability to recall all of the Master Clips for a given project or projects is very helpful because it allows you to basically create a single bin with all of a project's media in it. When Media Tool completes a search, the results are actually displayed in the form of a temporary bin. If you select only the current project and all drives, then select Master Clips and deselect Precompute Clips and Media Files (see Figure 1.32), you will end up with a bin of all the Master Clips for the current project. On some larger projects, where all of the clips are scattered throughout multiple bins, it may be helpful to create a bin that has all of your Master Clips so that you can do things like global searches or sifts across all of your media. Once the Media Tool has returned the results, simply create a new bin and drag all of the clips from the Media Tool into the bin. It's also possible to recover "lost" clips that you can't find, but you can see the media in the Media Tool.

media in any of the Avid MediaFiles folders on the drives. Once this is complete, the list of projects is shown to the right of the dialog window and the list of drives with media on them is listed to the left (see Figure 1.31 or 1.32). With many modern Avid systems, this Media Drive list can be as small as a single RAID, or it could be the internal drive of a laptop. In the old days, Avid media was typically spread out over numerous drives.

Below these two lists are a series of buttons that allows you to make additional selections: Master Clips, Precompute Clips and Media Files. As explained in the Media Tool tip nearby, selecting Master Clips can be helpful in gathering all of the Master Clips, including the metadata from them. However, when selecting material for deletion to open up drive space, Master Clips are only very small text files, so they are not worth deleting and, in fact, may contain important information that may be needed later. So unless you are looking to use the Master Clips for a specific purpose, like creating a Master Bin for a project, you should deselect Master Clips.

Precomputes (see Figure 1.33) is basically another name for a rendered file, though Precomputes also include imports with alpha channels. Any time you render anything—whether it's a simple audio dissolve, a large stack of video effects, color corrections, or even Title Media—the Avid system creates a Precompute that it stores with the source media for the effect. For many projects, this rendered media can take up a lot of space because it is not usually deleted as the effects are modified. Or, if you duplicate a sequence and continue revising a sequence, all of the precomputes from that old sequence are still on the drive, just in case you want to play it back. On effects-heavy sequences with a lot of revisions or longer project deadlines, this can amount to a lot of space. One way to very safely gain back that space is to do a search on old projects or your

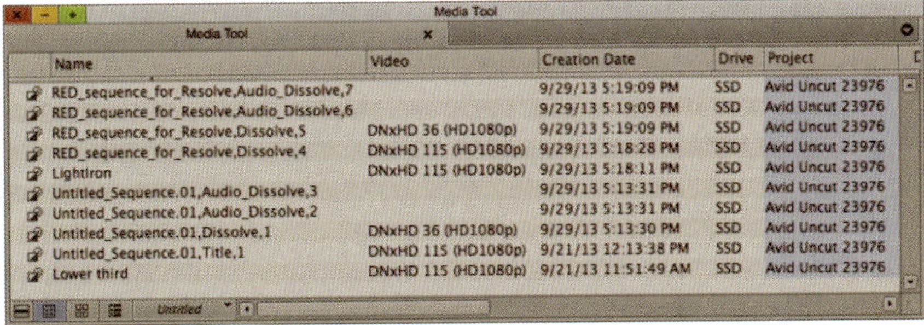

Fig. 1.33 Precomputes are rendered effects or imported graphics with alpha channels. There are often many precomputes on your system that you don't need because they belong to effects that have been modified or deleted, but the media of the render still remains on the drive. The little icon in the column on the far left indicates that these files are all Precomputes.

current project and choose only Precompute Clips from the checkboxes at the bottom. This will return a list of all rendered media for a given project. You can choose either to delete all of this media (since, it can be rerendered if you end up needing it) or you can selectively delete older precomputes. The best solution, however, is to do a search for Media Relatives on the precomputes. The concept here is to pick the current version of your sequence and delete all of the precomputes that are not associated with that current sequence. The only warning with this is that any old sequences will have their renders deleted, so if you want to go back to them, you will have to rerender.

Here's the recipe for this great "Media Relatives" space saver.

Launch Media Tool from the Tools Menu

Select all drives and the current project. Making sure you only have the *current* project selected is critical because otherwise you may delete needed rendered files from other projects (see Figure 1.34).

Check ONLY the precomputes check box. Uncheck Media Files and Master Clips. This is important because otherwise you might delete important clip information and imported or captured media.

Select OK to start the search.

This will open a window that displays the search results. The window looks just like a bin. As a matter of fact, it basically IS a bin (see Figure 1.33). You can even utilize any saved Bin views to sort or create specialized displays of the search.

Open the bin with the current sequence in your project and select the current sequence.

Click on the Fast Menu (sometimes called the Hamburger Menu) in the bottom left corner of the sequence's bin. Choose Select Media Relatives from the resulting pulldown menu (see Figure 1.35).

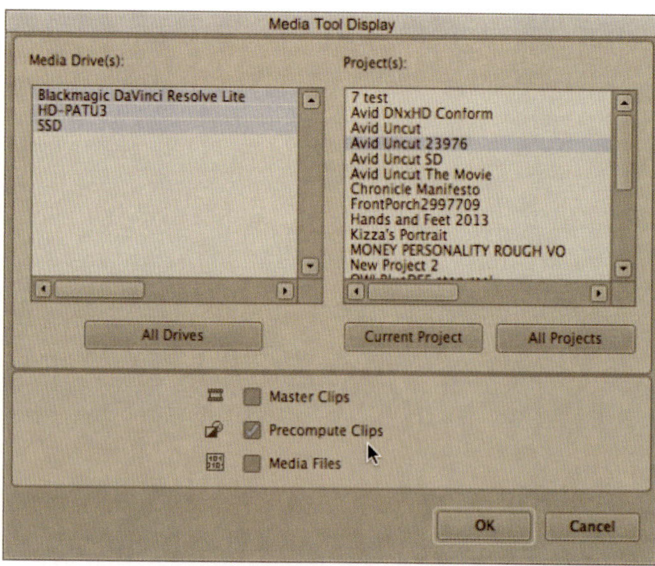

Fig. 1.34 The Media Tool prepped to delete just the Precomputes (rendered files) from a single project (Avid Uncut 23976).

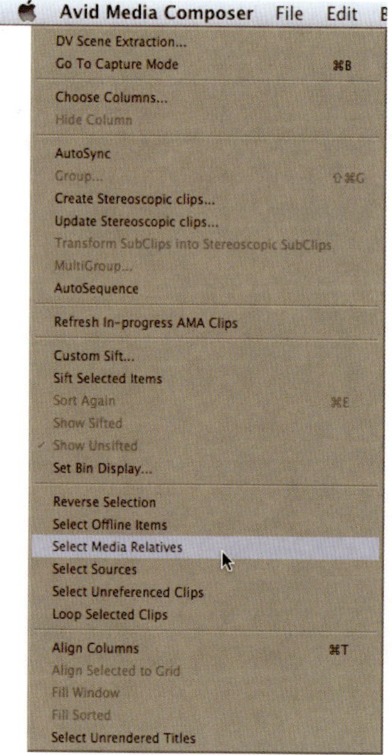

Fig. 1.35 From the Fast Menu of the Media Tool, choose the Select Media Relatives option—shown selected in blue here.

In the Media Tool bin—and also in the sequence bin itself—any file that is related to that sequence will be selected (see Figure 1.36). In the Media Tool, the selected items are the precomputes or rendered files that are part of that sequence. These are the files you want to keep.

In the Media Tool bin, click on the Fast Menu and select Reverse Selection (see Figure 1.37). This will mean that all of the selected files in the Media Tool are now the files that are NOT part of the sequence. All of

Name	Video	Creation Date	Drive	Project
RED_sequence_for_Resolve,Audio_Dissolve,7		9/29/13 5:19:09 PM	SSD	Avid Uncut 23976
RED_sequence_for_Resolve,Audio_Dissolve,6		9/29/13 5:19:09 PM	SSD	Avid Uncut 23976
RED_sequence_for_Resolve,Dissolve,5	DNxHD 36 (HD1080p)	9/29/13 5:19:09 PM	SSD	Avid Uncut 23976
RED_sequence_for_Resolve,Dissolve,4	DNxHD 115 (HD1080p)	9/29/13 5:18:28 PM	SSD	Avid Uncut 23976
LightIron	DNxHD 115 (HD1080p)	9/29/13 5:18:11 PM	SSD	Avid Uncut 23976
Untitled_Sequence.01,Audio_Dissolve,3		9/29/13 5:13:31 PM	SSD	Avid Uncut 23976
Untitled_Sequence.01,Audio_Dissolve,2		9/29/13 5:13:31 PM	SSD	Avid Uncut 23976
Untitled_Sequence.01,Dissolve,1	DNxHD 36 (HD1080p)	9/29/13 5:13:30 PM	SSD	Avid Uncut 23976
Untitled_Sequence.01,Title,1	DNxHD 115 (HD1080p)	9/21/13 12:13:38 PM	SSD	Avid Uncut 23976
Lower third	DNxHD 115 (HD1080p)	9/21/13 11:51:49 AM	SSD	Avid Uncut 23976

Fig. 1.36 Any media files that are related to the original sequence will be highlighted in blue, showing that they are selected.

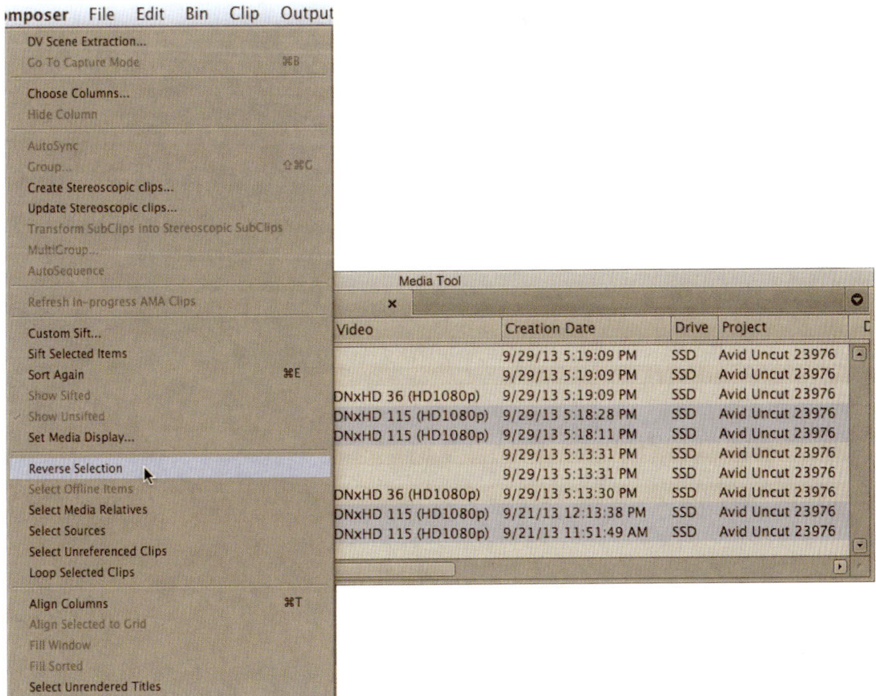

Fig. 1.37 In the previous step, all of the "related" media to the sequence was selected, but since we want to delete unneeded renders, we want to reverse the previous selection so that the un-related renders are selected, so they can be deleted.

these files can be deleted safely, though this can also cause the deletion of imported matte or alpha keys. If these are deleted by accident, they can be Batch Reimported. Any Title media that is deleted can also be reconstructed.

Press the delete key on your keyboard. This calls up the Delete Media dialog box. None of the boxes at the top of this dialog should have checks in them because the top is for imported or captured media. Depending on how many unneeded precomputes you have, at the bottom you'll see a specific number of audio or video precomputed files with a checkmark next to them (Figure 1.38). To commit to deleting these files, press the OK button, which will bring up a confirmation dialog box. Choose delete, and the files will be deleted and the space from those files will be instantly available.

This method of using the Media Tool can also be used with regular media files. When you launch Media Tool, go through the same steps as above, but check the Media Files box in addition to the Precomputes box.

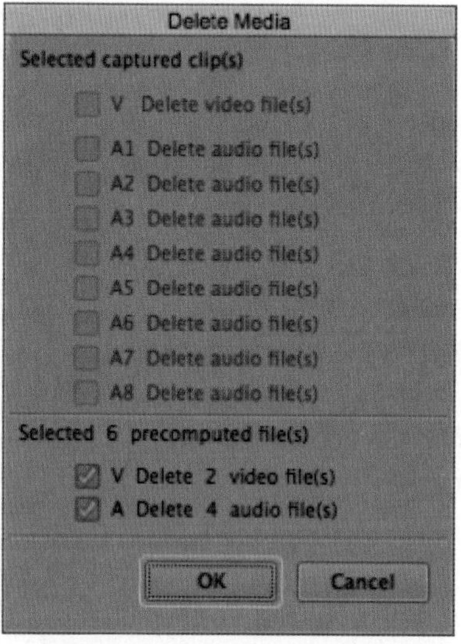

Fig. 1.38 The Delete Media confirmation window allows you to see how many audio and video files and precomputes you are about to delete, giving you a last chance to consider what you're about to delete.

Consolidating

Consolidate and Store requires the most time and the most effort but is the safest option while still opening up a lot of space. Consolidate is an awesome tool. In general terms, Consolidate allows you to save specific parts of a larger file. It's possible to use Consolidate to duplicate entire clips for archival purposes, but in this specific instance, it allows the Avid system to look at a finished sequence and then save only the parts of the media clips that are actually used in the sequence, plus a little "heads and tails" for safety and future editing possibilities. So let's say that you have a thirty-second spot that you cut from maybe twenty clips totaling over an hour of media. Let's say that some of those clips weren't used at all and some of the clips had several sections used in the final spot. When you consolidate a sequence Avid looks at the sequence, locates all of the media files that are related to or used in the sequence, and then makes copies of JUST the few seconds from a much longer file that were actually used in the sequence. Once this consolidated media is created, the original files can be deleted. To give an example, a 1-hour documentary might have used 130 hours of source material. After consolidating, the amount of material will be cut down to maybe only a couple of hours, leaving you with something that's still editable and revisable but without the bulk of unused material. The reason that a consolidated 1-hour documentary would still be several hours of material is due to the fact that each clip is saved with a couple seconds of heads and tails to allow for trimming and possible additional time for transitions or multiple-layered video. The amount of heads and tails is selectable in the Consolidation window (see Figure 1.39).

TIP

Consolidating sequences is a great way to save space, but Consolidating works for two other good purposes: (1) selectively copying media for archival purposes, and (2) permanently reassociating certain media to a new project. The value of the second purpose is that media is always associated with the project in which it was *created*. If you decide to use that media in another project, which can happen frequently, then if you delete that media from the *original* project, you'll also lose it in the subsequent project to which it was copied. To prevent this, Consolidate the other project media inside your new project and that consolidated copy will *not* be deleted if the original media from the first project is deleted. As for the archival purposes, the value is that a long clip with perhaps multiple takes can be consolidated by choosing only the best take and saving that portion alone.

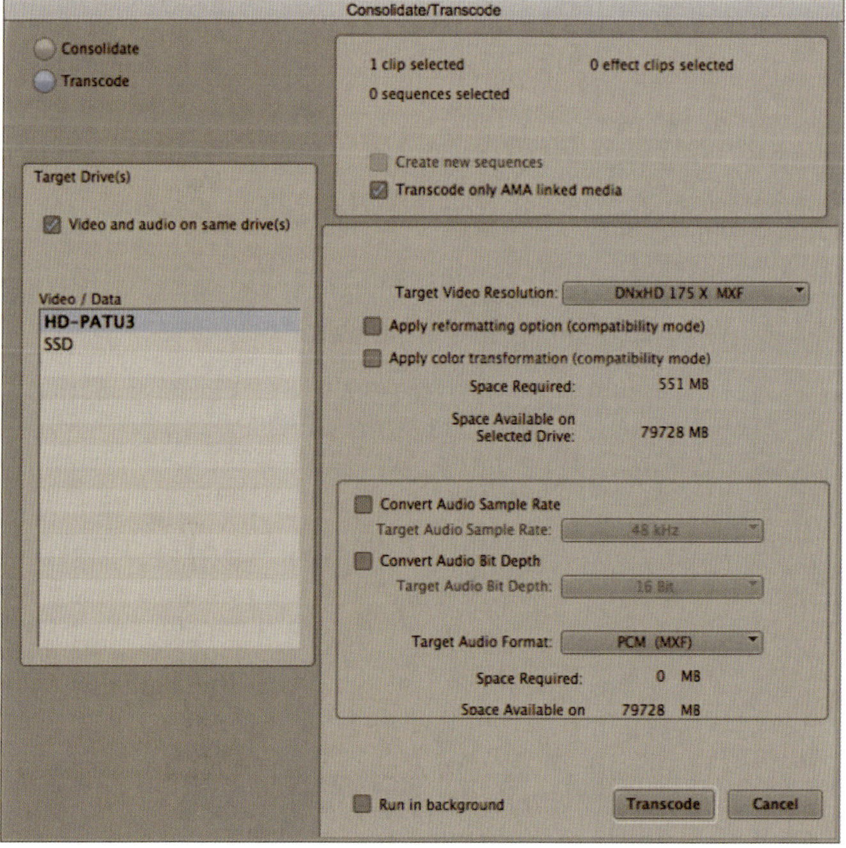

Fig. 1.39 The Consolidate/Transcode window.

It's much easier to store 2 hours than 130. Obviously, there is still valuable footage in the 130 hours, so doing something like creating a selects reel of unused footage and consolidating and archiving that as well could be an important thing to do—especially in today's file-based world.

Decomposing

Decomposing sounds disgusting, but in Avid terms, it's very useful for tape-based media. (More on file-based media in a moment.) In some ways, decomposing is similar to Consolidating. Decomposing is something you do to a sequence, usually before you up-rez it. The concept is that if you are editing footage that is low res and you want to up-rez to a higher quality, you don't want to have to bring all of the project's footage back into the Avid system at the higher resolution when you reaccess the original footage. You want to only recapture or reimport the specific portions of those files that are actually used in the sequence. That's what decomposing helps you do. When you decompose a sequence, it breaks the sequence down into new, shorter clips that reference only the portions of the original master clips that are part of the sequence. These new clips still maintain all of the original metadata that links them back to the original media without the need to access the entire length of the original media. If you created a sequence from a single captured file, for example a trailer for a movie, the trailer might have 100 edits totaling two minutes. When you decompose the file, you will end up with 100 new clips totaling two minutes, plus some heads and tails on each new clip that allow you to make future revisions.

On an AMA file-based job, instead of decomposing, relink to the AMA files (if you've already transcoded) and consolidate.

Decomposing is best done to a copy of the sequence and is best done in a new bin, empty of everything except the copied sequence. So, duplicate your final sequence by selecting it in its current bin and pressing Command-D (Mac) or Control-D (PC). This will create a new sequence, probably at the top of the bin, with the same name as the original and the word "copy" as a suffix. You can rename this file if you want and then drag it into your new bin. I usually call this bin "DECOMPOSED FINAL" or something like that. Alternatively, instead of duplicating the sequence, you can Option-drag it (Mac) or Alt-drag it (PC) to the new bin, and the sequence will be duplicated directly into the new bin in one step.

With the duplicated sequence in your new bin, select Decompose from the Clip menu. This process may take a minute or two as Avid processes the sequence, breaking it down into its component pieces. At the end of the process, the bin will be filled with possibly hundreds of new clips, all appended with the suffix ".new01." When this is done, this entire sequence will be off-line. This is one of the reasons for duplicating the sequence. The original sequence will still be associated with the original media and will remain on-line at its old resolution.

The Decomposed bin can now be sorted and sifted. When I used to do a lot of on-line work for outside production companies, we'd decompose the sequence and then sort it based on tape name. (Yeah, I'm old!) Then we'd print out the list, and my assistant would verify the list with the production company so that we made sure all of the tapes we needed would be sent over to the session. Those tapes would actually be loaded onto the cart in the exact order of the Decomposed list. That way there was no searching for the next tape to load into the tape deck.

Ingest from File-Based Workflow—AMA

At this point, it is safe to say that file-based workflows have completely taken over the known universe. Tape, especially for capture and ingest, is gone except for archival footage and some high-end backup/on-line, like the *Big Bang Theory* workflow described earlier in this chapter on page 18. If you are someone who regularly needs to capture from tape, then chances are you don't need me to tell you how to do it.

> The first 24p digital HD camera was introduced in 1999. The adoption of digital cinema techniques was very slow at first. Skeptical cinematographers doubted that digital could ever produce the artful images of film. Some still feel that way. But, the rapid switch from film to file-based media was hastened by the threatened 2008 SAG strike. SAG actors needed to be used when film cameras were used, but AFTRA actors could be used if the cameras were not film based. That year, many of the TV pilots chose to shoot file-based so that they could work with AFTRA actors. There were many other factors involved, including several big feature productions shooting with HD or file-based and the growing quality strides of camera manufacturers, such as ARRI, Panasonic, SONY, and RED. The other contributing factor was that more and more digital theaters came online around this time. The fact that the theaters were digital doesn't have a direct impact on needing digital acquisition, but the bulk of motion picture film manufactured is actually for film prints, and when the theaters no longer needed film prints, the economics of the entire motion picture film manufacturing was thrown out of balance. The collapse of Kodak's and Fuji's motion picture film business was not because file-based acquisition took over, but because the film *print* business disappeared.

File-based workflows mean importing. The unfortunate part of this is that it can take a good deal of time to import nonnative formats. That's why Avid came up with Avid Media Access (AMA). AMA allows you to "import" files from RED, P2, XDCAM, Alexa, Canon, and others directly from the capture media (P2 card, for example) instantly. You can view, log, and edit with this footage straight off the cards or, more likely, from a copy of the cards on an attached RAID.

There are a few warnings with this seemingly magical capability. The primary warning is simply that with AMA, the media sits, unchanged, on whatever drive or media it started on, so if that media is removed, you lose your footage. Also, if you decide to use AMA directly from a camera card instead of copying the card over to a high-bandwidth RAID, the card or connection to the card may not have enough throughput to smoothly play back your footage. Finally, many cameras are capturing video or storing video in formats that are very difficult for any and all NLEs to deal with. A good example of this is MPEG-4 footage from Canon 5D Mark IIs or IIIs. These files are not designed for editing. They are meant for easy storage and playback. Because of their Long-GOP (group of pictures) structure, these video files are notoriously difficult to edit with because not every frame of video contains all of the information necessary to re-create it. Because of all three of these variables—loss of files, media drive bandwidth, and difficult codecs—AMA files should not be used for the entire lifespan of the edit. The files should be transcoded to Avid DNxHD, which is Avid's native HD codec, similar to ProRes in FCP. The advantage of starting the process as AMA is that you can see and log your footage instantly before the lengthy Transcode step. It allows you to start your edit and find clips that are, perhaps, completely useless and not bother to transcode those at all, saving transcoding time and drive space.

Once the edit has begun with the AMA clips, the transcoding can happen overnight if the edit is going to continue the next day, or perhaps it can be done over lunch or while working on another aspect of the edit. Also, with the introduction of MC7, Avid now media-manages (keeps track of) AMA footage and it allows for background media transcoding of footage while you continue to work in Avid. MC7 also introduced the concept of a Dynamic Media Folder. We'll get to the specifics of the Dynamic Media Folder before the end of this chapter, but the basics of what it does is to serve as a "watch folder" that executes a series of custom functions on incoming camera files whenever they are dropped into the Dynamic Media folder. These functions can include copying the media to multiple locations and running unsupervised, automated transcodes on the media, delivering the completed transcodes to specific folders. The Dynamic Media folders work independently of the Avid Media Composer application.

Using AMA does require some technical preparation. Each camera manufacturer has its own codecs and third-party AMA plug-ins that you need to download from their website. I have always tended to download these as the need arises, but you could download and install all of them so that you're ready for anything.

To prepare for AMA, start with Avid's AMA site to find out which AMA codecs and plug-ins are available and where to download them.

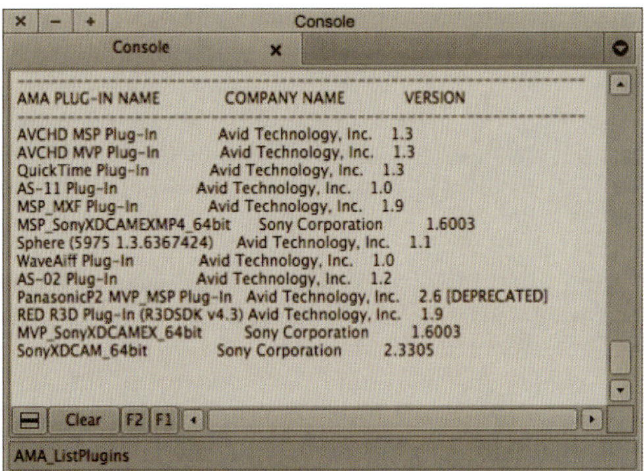

Fig. 1.40 This is the Avid Console, displaying the installed AMA plugins. This is described in the Tip to the right.

TIP

There's a little-known console command that lets you check which AMA camera plug-ins are loaded on your system. Open the console (Command-6 on a Mac, Control-6 on a PC), and type "AMA_List plugins" This will return a list of the plug-ins that are correctly installed on your Media Composer (see Figure 1.40). This is a great troubleshooting step if an AMA ingest doesn't seem to be working.

At the time of the publication of this book, Avid's AMA information was available at www.avid.com/US/products/Avid-Media-Access/Plug-ins.

Avid doesn't keep the actual camera plug-ins on this website because these are owned and maintained by the camera manufacturers, but the information about tracking down these plug-ins is on the website above.

Installation and specifics for each AMA plug-in is slightly different. Each plug-in has its own PDF guide, and all of them are available at the link above. Since these instructions and the plug-ins themselves are updated occasionally, and because a complete list of the instructions for all of the plug-ins would be too lengthy for this book, I suggest you download and read the guides for the cameras you plan to use. However, like most editors, I just downloaded them and ran the installer without reading anything.

Each camera manufacturer has a specific folder and file structure for its recorded media. Being able to tap into these camera files using AMA means that, by and large, you can ignore the structure altogether *except* to know which level of the structure AMA needs to be pointed at. Each type of camera has a specific level that you need to navigate to: for P2, it's one level above the Contents folder. For XDCAM disks, it's a level above the Clip folder. For XDCAM EX, it's one level above the BPAV folder. For RED, select the root directory of the RED card. For QuickTime, go to the QuickTime folder that holds the QuickTime files or to the file itself if you want to bring them in one at a time.

Ignoring Dynamic Media folders for a moment, how do you actually bring file-based media into the Avid environment through AMA?

With Media Composer launched, select File > AMA Link from the menu bar, (see Figure 1.41). MC7 does away with the need to choose AMA File or AMA Volume that was required in pre-MC7 versions. A dialog window opens and provides the ability to navigate to the proper folder level of your AMA media. The example in Figure 1.42 shows a folder of RED footage. You can also type in the name of a specific file. This is useful for QuickTime files. You can use the Files of Type pulldown menu to automatically select the correct AMA plug-in type, or to select a specific camera plug-in from the list. You can also use Profiles created in the Avid DMF support services (Dynamic Media folders) to choose specific AMA profiles for ingest. Or you can use the Profile editor to create profiles from scratch. We'll get into that a bit more later in the chapter when we discuss DMF. So with your file or folder of media chosen and your settings selected, click the Open button.

If the proper plug-ins are loaded and you opened the right level of the folder, your bin will immediately populate with clips. (Each media type or camera manufacturer is different and the file structures are all listed below.)

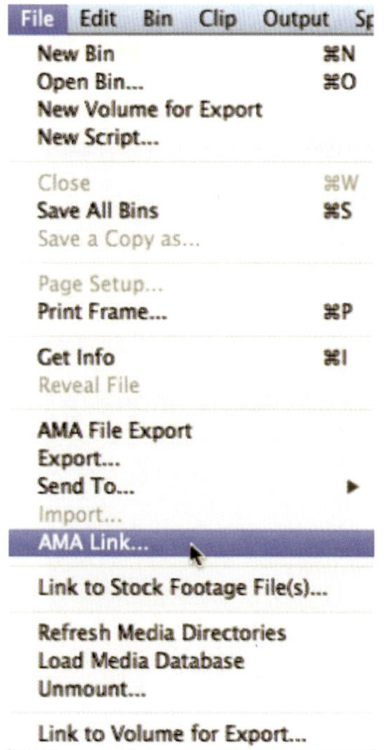

Fig. 1.41 The AMA Link option is available from the File menu.

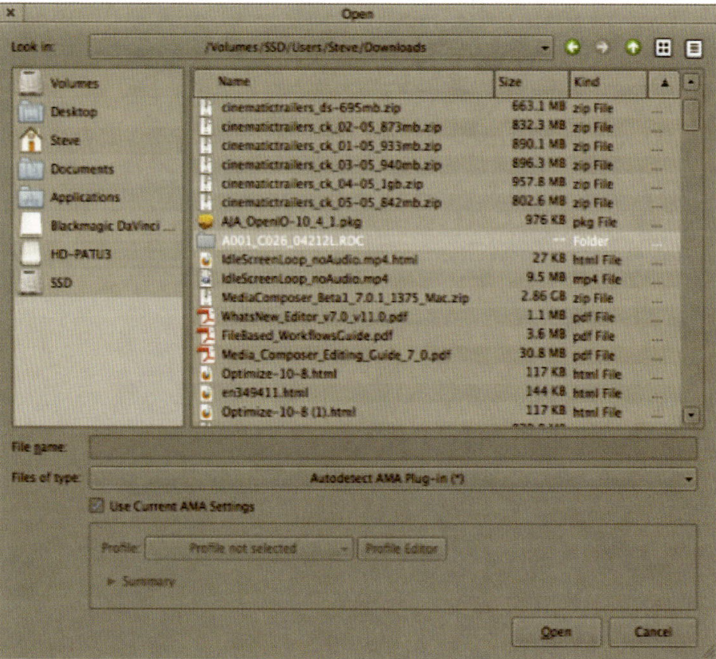

Fig. 1.42 The AMA Link window allows you to link to camera footage and QT files at their native resolution and codec.

If you didn't have a bin selected or you didn't assign a bin in the Profile, then the default is to create a new bin based on the Volume name. This is actually not that critical. You can always copy or move the AMA clips from this bin to another bin if you want.

The AMA clips will appear with a Master-Clip-like icon, except that there will be a small "link," or "chain," image next to it. This icon can be seen in the bin in Figure 1.52, later in the chapter.

RED

For RED cameras, the folder structure (see Figure 1.43) is the following:

.RDM, is the top level and stands for RED Digital Magazine. All of the recorded files go into this folder. If you're going to copy RED files to a drive or RAID, make sure you grab this level in order to maintain all of the folder structures.

.RDC is the Clip-level file. This is created every time the record button is pressed.

.R3D contains all of the visual data and metadata for each clip. This is the RAW data.

.MOV are standard QuickTime files. They're proxies, but they require the .R3D files to play back. These may seem like the easiest files to use for editing because they're a "common" file type, but because they're proxies, they shouldn't be used for editing.

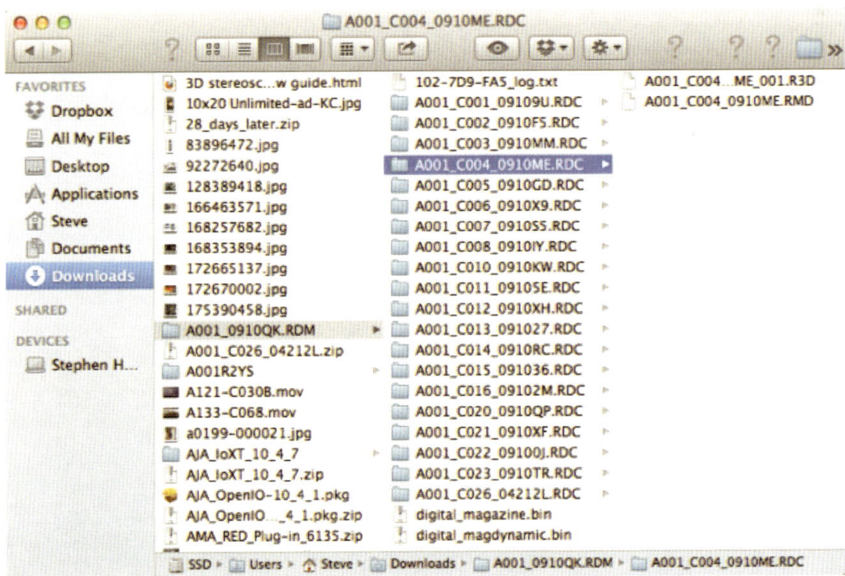

Fig. 1.43 This is the folder structure for RED folders.

When .R3D files are processed by RED CINE X, it generates an XML-based sidecar file called an .RMD file (different from the above mentioned RDM). The .RMD sidecar file is filled with metadata that is linked to the .R3D media (essence). The link that binds the two files is nothing more than the fact that they share the same file name, but with two different extensions.

GLOSSARY

LUT: Look up table. This is a method of converting color spaces and/or compensating for the "color personalities" or defects of certain cameras. Instead of using a mathematically challenging algorithm to alter the colors, the camera or a third-party application or hardware creates a 1D, 2D, or 3D table that converts colors from their recorded values to the look with which they are displayed (see Figure 1.44).

ARRI Alexa

Alexa can be recorded in-camera in a number of ways. If you, as the editor, have any advance communication with the production team, you can see if the DP is amenable to recording in OP1a MXF format. It may be preferable for some projects to record RAW, but for easiest ingest into the Avid at a high quality, OP1a MXF is going to be the editor's best option. The other choice is OP-ATOM MXF, which is similar to OP1a MXF but has the audio and video files actually broken out as separate files. There's also OP1a, which is for streaming. For the Alexa STUDIO camera, this is a standard recording format. For other Alexa cameras, this is a paid option, purchased and installed per camera. The other preproduction choice that needs to be determined is whether to use the (30fps) DNxHD145 codec, which is 8-bit, or the DNxHD220x codec, which is 10-bit. All DNxHD

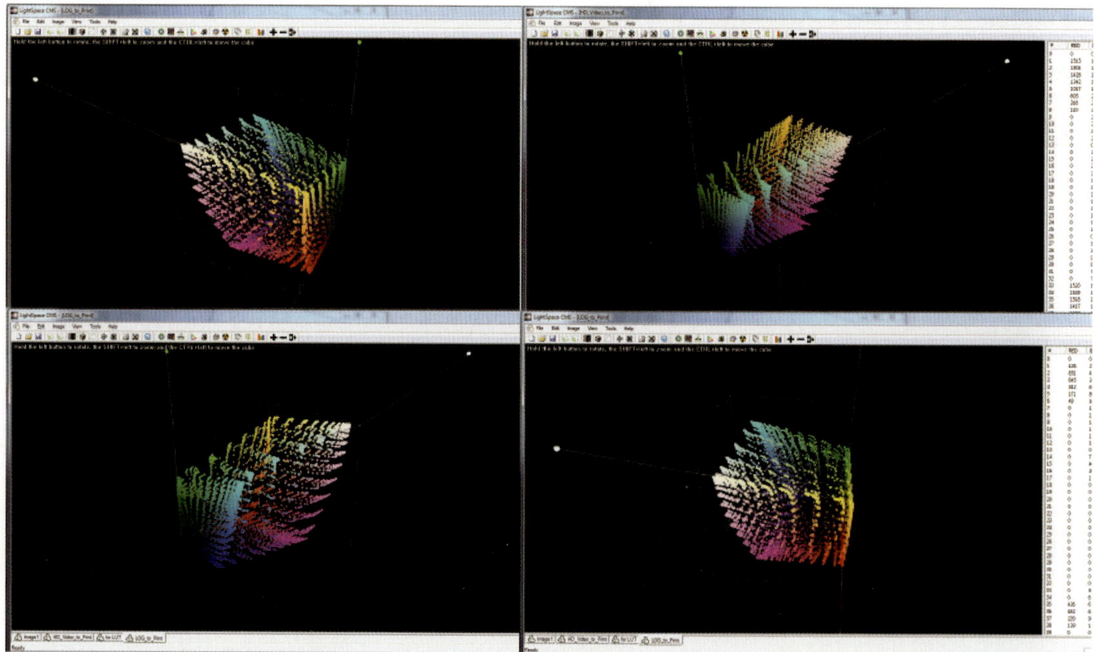

Fig. 1.44 LUTs are easier to see in three dimensions. These are two different LUTS seen inside LightIllusion's LightSpace CMS application, which does LUT management and display calibration. Each LUT is seen from two angles. Check the companion website to see the LUT in motion in 3D space (www.focalpress.com/cw/hullfish).

resolutions that are followed by *x* are 10-bit. The others are 8-bit. Each frame rate has different DNxHD options. At 24fps, the options would include 115 and 175. At 25fps, the options would include 120 and 185. Finally, the choice of recording in the REC709 or Log-C color space must be determined. For "broadcast ready" images, which will look "correct" on import with no LUT, the best choice is REC 709. The DP, however, will probably want to shoot with Log-C, which will look flat and desaturated without a LUT or grading applied, but will yield the best footage for the maximum latitude for use by a colorist. But editing with these images will require use of an ARRI LUT (or some other LUT) to look the way the footage was envisioned by the DP. The great thing is that with the latest version of Media Composer, you can apply these LUTs inside Avid if they haven't already been "baked in" in the dailies.

There are other options for capturing with the Alexa, but the options above are the most common and best for editing. Discussions, compromise, and budgetary issues for these options and others need to be worked out between production and post.

The file and folder formatting from an Alexa starts with a folder that is named for the current camera reel name (see Figure 1.45). Inside that are the individual .MXF files (or QT .MOVs). There will also be an .ALE file (Avid Log Exchange) and an .XML file (for Final Cut Pro) containing clip metadata.

The file naming convention for the .MXF files is [CameraIndex] [Reel-Counter][ClipIndex][ClipCounter]_Date_CameraID.MXF (or .MOV).

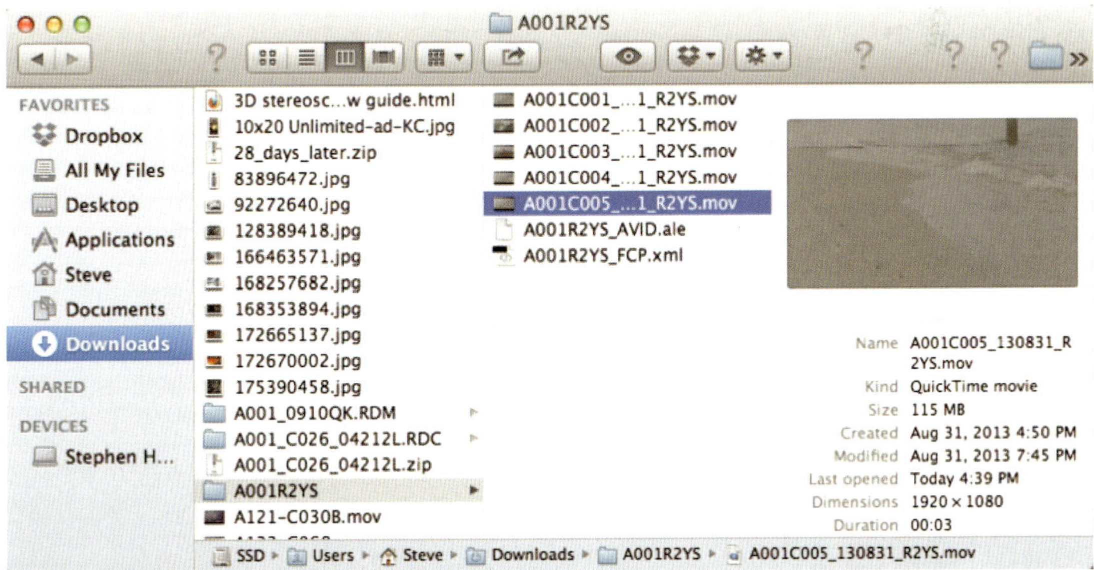

Fig. 1.45 Alexa's folder structure is much more basic than RED. But note that the folder includes an Avid ALE file and an FCP .XML file containing metadata.

The camera index is a one-letter identifier of which camera the clip came from.

The reel counter is three digits that auto-increment with each CARD (001–999).

The clip index is almost always C, unless there are more than 999 clips, in which case, the C auto-increments to D.

The clip counter is the next three digits, which auto-increment with each new CLIP (001–999).

The date is after the first underscore and is provided in the more logical European method of year/month/date.

Finally the camera ID includes a stereoscopic designation of L or R for left or right (defaulting to R for nonstereo clips), followed by the camera serial number.

The XML or ALE file includes a bunch of metadata about the clips in the folder. This information may include Clip Name: IN/OUT Time code, Duration, FPS, UUID, Sup_Version, Exposure_Index, Sxs_Gamma, Whitebalance, Cc_Shift, Look_Name, Look_Burned_In, Sensor_Fps, Shutter_Angle, Camera_Model, and so on.

Despite the fact that the clips may appear to be Avid-native .MXF files, they are not directly readable by Media Composer. AMA is the way to bring these files in.

TIP

Alexa .MXF files can generally be read only from within Media Composer. But if you're looking for a free viewer that will allow you to view these files, check out Blackmagic Design DaVinci Resolve Lite.

You'll first need to download the ARRI AMA plug-in from www.arri .com/alexa/downloads.

This site may not be active by the time you try to use this, doing a search on the arri.com website should yield the same results. Otherwise, you'll need to do some searching on the ARRI site. It is possible to AMA link directly from the SxS cards, but the typical workflow is to copy and verify the complete contents of the cards to a RAID first.

XDCAM

Sony's professional camera systems record using some flavor of XDCAM. Basically, this means either XDCAM HD, from Sony optical disks, or XDCAM EX, from SxS cards. Using the disks allows for some interesting hybrid methods of getting the media into an Avid system, including using a Sony disk reader being controlled like a VTR (video tape recorder). The EX cards have to be used with the basic file-based media methods, similar to those already described for RED and ARRI.

The XDCAM files are MPEG-2 Long-GOP, with low-res MPEG-4 proxies. On the disks, the MPEG-2 files use an .MXF wrapper.

The folder structure of a camera like the Sony F3 has a BPAV folder on the card (see Figure 1.46). Inside the BPAV folder are three other folders and two .XML files. The folders are the CLPR folder, the General folder, and the TAKR folder. The .XML files are the CUEUP file and the MEDIAPRO file. The

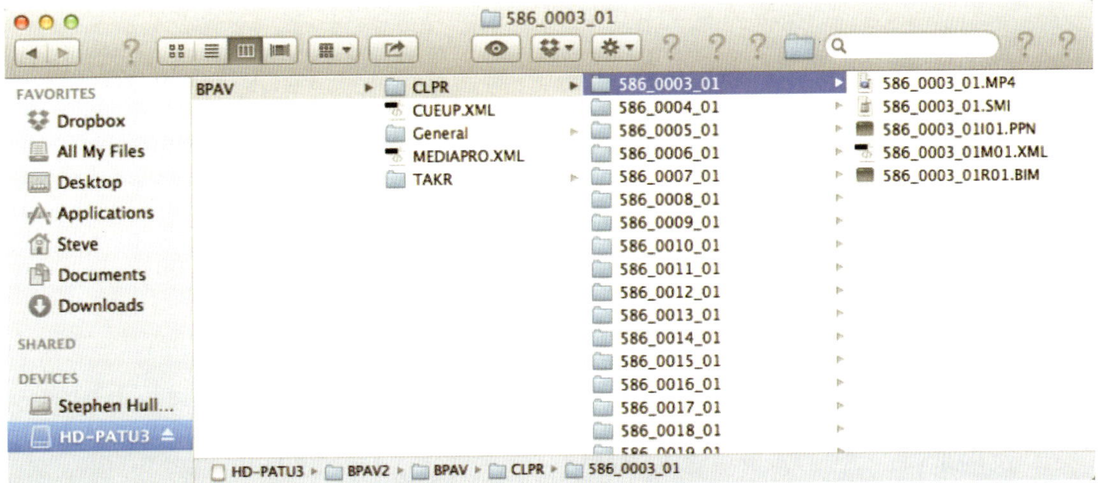

Fig. 1.46 XDCAM folder structure. BPAV is the top-level folder inside the card.

most important of the folders is the CLPR folder, which contains folders for each clip. Inside each of these are files, including the .MP4 file, which is the main file that is the actual media, but this .MP4 file is not a standard .MP4 file that can be read by QuickTime player. The .XML file has metadata for the clips; the other files are basically unimportant but should be maintained.

This section is about using AMA linking, but if AMA is disabled, you can actually import from XDCAM using File > XDCAM Proxy directly.

As with the previous examples, you can link via AMA directly from a connected SxS card or optical disk, but copying these cards to a RAID or to multiple external drives for protection of the media is recommended.

For EX footage (from an SxS card), select the folder or drive letter above the BPAV folder copied from the card. For example if you simply copied the BPAV folder from the card directly to a RAID, you'll need to point at the RAID. If you copy BPAV folders from multiple cards, you'll want to put each BPAV folder into another folder, possibly with some identifier about what was on the card or identifying the specific camera that the card was in, because the BPAV folder needs to maintain the name. If you follow this workflow, then you'll want to point the AMA Link dialog to that custom-created folder that contains the BPAV folder that you want to link to. For example, I shot using SxS cards in a single Sony F3 on a recent shoot. We ended up shooting two cards worth of material during the day. I used the Sony card reader to copy the cards to my laptop, placing one of the BPAV folders into a folder I called "F3_093012AM" and the BPAV folder from the second card to "F3_093012PM." Then I pointed the AMA Link at each of those custom-created folders. When I attempted to point directly at the BPAV folders themselves, the Media Composer couldn't see them.

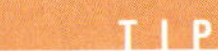

TIP

If clips are linked directly to external camera cards or disks, the files will be highlighted in yellow as a warning that the media can come off-line easily if the cards or disks are unplugged.

With the files linked into an Avid bin, I was able to review the clips and even edit with them immediately. Instead, I reviewed them, and labeled any of the clips that had only bad takes or the clips where the record button was accidentally pressed. That way I could save the time and drive space of consolidating those clips. When I stopped working for the day or took a break for lunch, I started the consolidation process, keeping the Avid system working while I was taking a break. These consolidated clips are much easier to edit with. Editing with AMA linked clips (prior to consolidating) is processor intensive and also increases the possibility of losing the link.

Another method of consolidation is to consolidate only partial clips. To do this, go through the linked clips and edit only the potentially useful sections that you want to edit with into a sequence. Then right-click (or option-click on the Mac), and choose Consolidate/Transcode from the contextual menu. Select OK, and each clip in the timeline will be consolidated into a new clip that is only the length of the clip cut into the sequence (plus some handles). For some projects, this extra effort will pay off with a big savings in consolidation time and used disk space. If you discover that you need to edit with some of the footage that you didn't consolidate, you can pull it from the AMA linked media and choose to consolidate those clips as you go.

P2

P2 media comes from Panasonic cameras and are usually recorded on P2 cards, which are similar to SxS cards but are proprietary to Panasonic. The media on these P2 cards can be DVCPro20, DVCPro50, DVCPro HD, AVC-Intra 50, or AVC-Intra 100. These codecs can have varying bit depths (8- or 10-bit), progressive or interlaced scanning, and differing frame rates, including VFR (Variable Frame Rates) for slow motion or fast motion.

AMA linking is similar to the above methods for XDCAM. As with those other methods, it is possible, and expeditious, to link directly to the media on the P2 cards themselves using a USB or FireWire card reader or a PCMCIA slot or adapter. However, it is usually wise to copy the entire P2 card structure over to a RAID or to multiple drives in order to secure the safety of the media.

The P2 card contains a Contents folder and a LASTCLIP.txt file. Both of these need to be copied from the card to the RAID.

As with previous camera formats, it is necessary first to download and install the AMA plug-in from the camera manufacturer.

If the media files from the camera card are copied to a RAID or other external drive, use File > AMA Link, selecting the folder level just above the Video and Audio folders from the P2 card, which is the level of the

Contents of the P2 Card (see Figure 1.47)

At the top level, there are a Contents folder and a LASTCLIP.txt file, which is simply a text file containing the name of the last clip recorded onto the card.

Inside the Contents folder are six folders: Audio, Clip, Icon, Proxy, Video, and Voice.

The Clip folder has .XML files for each clip containing the metadata for each clip.

The Icon folder has .BMP files, which are thumbnails for each clip (80x60 pixels). If there are more .BMP files than clips, it's because one or more clips are "spanned" because they were too long to record in a single file. Each of these "spanned" clips will have multiple "icons."

The Video folder contains the high resolution .MXF files.

The Audio folder contains the high-resolution files for every track of every clip, so you can have 4x, 8x, or more audio clips than video clips.

The Proxy folder contains low-resolution (standard def) MPEG-4 files. This is an optional capability, based on the camera, so you may not have proxies.

The Voice folder can contain voice annotations that a media logger can record while using the P2 Viewer software. Avid systems do not recognize or support these files.

TIP

If AMA is disconnected in the Settings menu, it is possible to import directly from P2 using the File > Import P2 command. Point to the folder level above the Contents folder for each card to import all of the files inside it. This native method works quite well with P2, so you may wish to bypass the AMA method in this case.

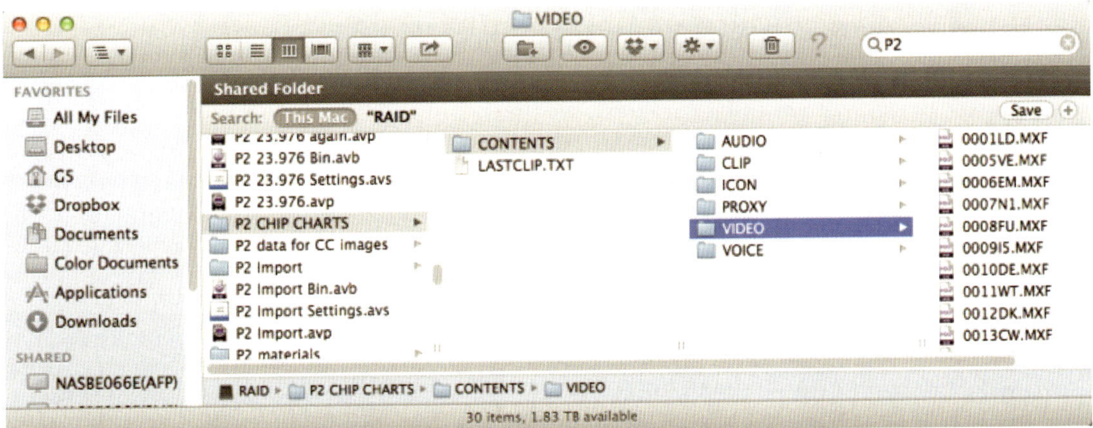

Fig. 1.47 The folder for Panasonic's P2 format is relatively complex.

Contents folder. The Video folder does have .MXF files inside it, but these are not directly readable. The Contents folder and the LASTCLIP.txt file should be placed inside another folder that you can create and name to identify the card or camera from which the media came. Obviously if you have multiple P2 cards, the Contents folders can't be copied to the same folders or drives without separating them into their own folders, or copying them would cause the newer Contents folders to overwrite the older Contents folders. This is something to be very mindful of.

As with all of these camera formats, trying to AMA link to the wrong folder level causes failure of AMA to find any files to link to. If you know the media is there, try a level above or below the one you're on, or place the folder into another folder and attempt to link to the new folder.

If you are going to create an off-line/on-line workflow, create a separate bin for each P2 card. The bin names must match the names that contain the P2 card contents. In each bin, use AMA Link in the File menu (or contextual menu) to link to the media. Then in each bin, transcode to an off-line resolution, such as, DNxHD36. If you want to edit in SD, go to the Format tab in the Project window and change the Project type to 30i NTSC. Move these off-line-quality transcoded clips to new bins. These are the clips you want to edit with for the duration of the off-line process.

When it's time to up-rez, change your Format tab to the HD flavor you want to use for the on-line. Make sure the high-resolution AMA volumes (the original P2 card media) are mounted. Relink your sequence to this high-resolution media using the Clip > Relink tool. In the resulting dialog box, there is a Relink Method option. Choose Highest Quality from the pulldown menu. In an Avid ISIS/Interplay environment, use the Multi-res feature.

Canon 5DMII, 5DMIII, 7D, T21, and Other Canon DSLRs

Canon DSLR cameras have pretty simple folder structures (see Figure 1.48). Most of these systems use CF (compact flash) cards. On the CF cards are two folders labeled DCIM and MISC. The MISC folder usually appears to be empty, but it can contain an invisible file that catalogs the card for photo printers that print directly from the cards. Inside the DCIM folder is another MISC folder, which is empty, and another folder named something similar to "100EOS5D," depending on the make of the camera.

Inside the 100EOS5D (or similarly named) folder the DSLRs usually record .JPG files and RAW .CR2 files for each still image, depending on

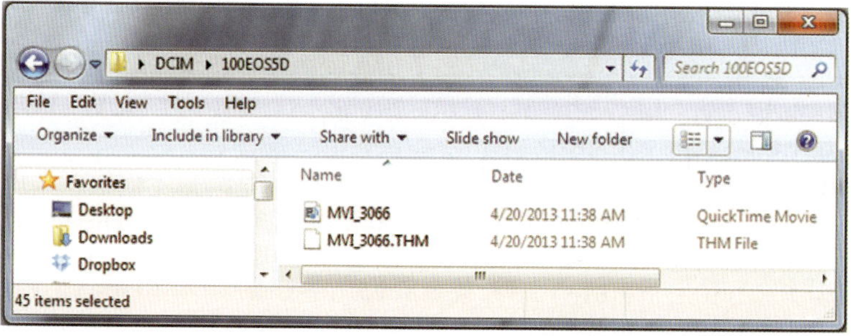

Fig. 1.48 The relatively simple structure of the Canon DSLR folders. This is from a PC.

the settings of the camera. For motion clips, it records a .MOV file and a .THM file for each recorded clip. For AMA linking, the only files in the folder that are really needed are the .MOV files. The .THM files are thumbnail images of the .MOV files and contain EXIF (Exchangeable Image File Format) metadata associated with the .MOV. If you rename the .THM files to .JPG, any JPEG reader can view the thumbnail. The EXIF metadata can be read by the Avid system to provide information, like timecode. What links the .THM file and the "essence" file of the .MOV is simply that the file name before the extension matches.

To AMA link to the files on the card—or with the card copied to a RAID—choose File > AMA Link. Navigate to the data one level down from the DCIM folder, which should be named something similar to 100EOS5D, then click Choose. This will create AMA links to the correct files. As with the other AMA files described earlier in the chapter (like RED, XDCAM, and P2), it is possible to view and edit these files immediately, even straight from the cards, but it is better to consolidate the media to generate native DNxHD files before editing.

TIP

To see the EXIF metadata, copy the .THM files and change the .THM file suffix to .JPG. On a Mac, double-click to open the .JPG in the Preview application. Use Command-i to open the Info tab. In the Info tab, open the EXIF tab and the metadata for the clip will be revealed (see Figure 1.49).

It is also possible to link directly to the .MOV files on the card using AMA Link, but this will not utilize the metadata available from the .THM files.

If you want to utilize the data on the .THM files, there are a few applications that will convert them for you. I've used VideoToolShed's MXF Creator app or MetaCheater. I have also heard of, but not tried, workflows including Squeeze, Compressor, or Adobe Media Encoder to marry the .THM files with the QT files. The main useful data from the THM file is the time-of-day timecode from the shoot.

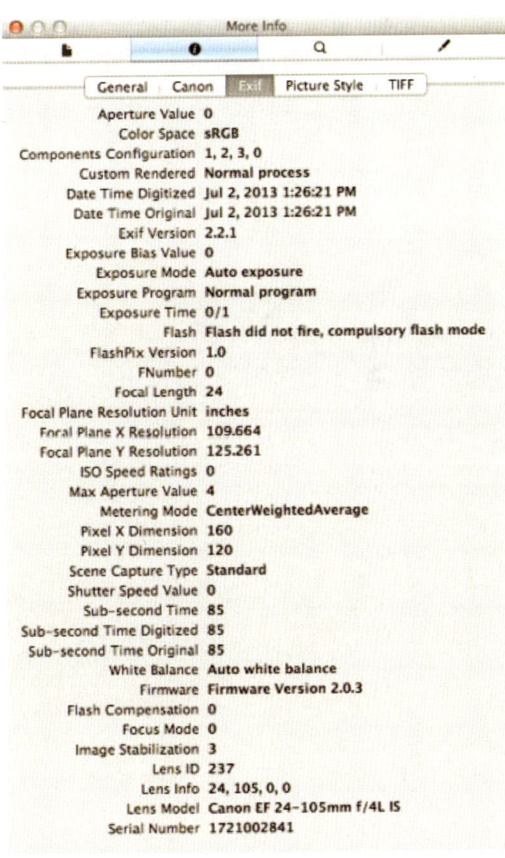

Fig. 1.49 This is the information revealed from the EXIF data of the .THM files described in the TIP on this page.

QuickTime (.MOV)

QuickTime files, including ProRes files, can also be linked via AMA. This allows for easy transfer of projects from Final Cut Pro.

With a bin selected, choose File > AMA Link and navigate to the QT file or files that you want to link to. As with the previous camera files, you can begin screening or editing with these files immediately, but it is wise to consolidate these files prior to editing.

Ingest from File-Based Workflow—Non-AMA

Ingesting from file-based without using AMA is certainly possible. Some of the same plug-ins will be required to be able to read the camera codecs and allow Avid to understand the folder structures. This is easy to do with simple files, like QT files. But many of the choices for bringing in raw files from RED, ALEXA, C300, and others are possible only if the files are AMA-ed or run through some kind of transcode before reaching Avid. This is the way many of the dailies programs work.

Some post houses have fairly lengthy pre-Avid workflows that prep the files in specific ways before they are even imported into Avid. These can involve third-party software like MPEG Streamclip to convert to DNxHD outside of Avid, Cinema Tools to change or conform the frame rate, QTChanger to add metadata (like TC and reel name), MXF Creator to transcode, and Better Renamer to change the names of multiple clips in a batch. There are also industrial strength dailies systems, like those by Assimilate Scratch, BlackMagic DaVinci Resolve, DFT Bones, Gamma and Density 3cP, Mistika, ColorFront, Baselight, Flexxity, YoYo, Pomfort, Convey, Light Iron Digital (lightirondigital.com), MTIFilm's Coretex Dailies (mtifilm.com; see Figure 1.50), Technicolor's Global Dailies (technicolor .com), and i-Dailies (i-dailies.co.uk).

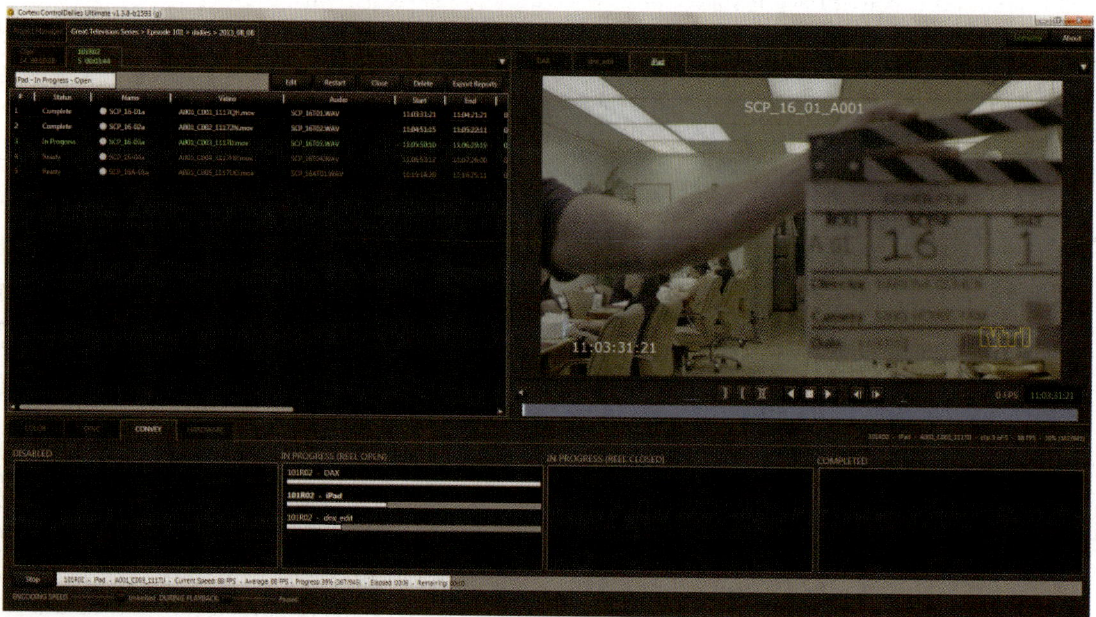

Fig. 1.50 This is the CoreTex UI for dailies. A piece of software like this can be used to generate Avid-ready .MXF files and create useful metadata for post.

Moving from AMA to Transcoded Media

It's not good to edit with AMA footage for very long unless your sequence is very short and very simple. Beyond that, in order to have a smooth and responsive editing experience, it is suggested that you transcode the AMA'd media as soon as possible. If you do not have to edit immediately, I would suggest that transcoding is the first step.

To save time and disk space, you can view, log, and make notes on your clips before you transcode them. This can save time and disk space because any clips that have no useful footage in them don't have to be transcoded. You can mark these bad clips with a color code or drag them to a separate bin.

With your good AMA clips selected in a bin, from the menu bar choose Clip > Consolidate/Transcode. This opens a Consolidate/Transcode window (see Figure 1.51). The main two choices are shown on the buttons, Consolidate and Transcode, in the top left corner. Many AMA files can't be consolidated because of their native compression format,

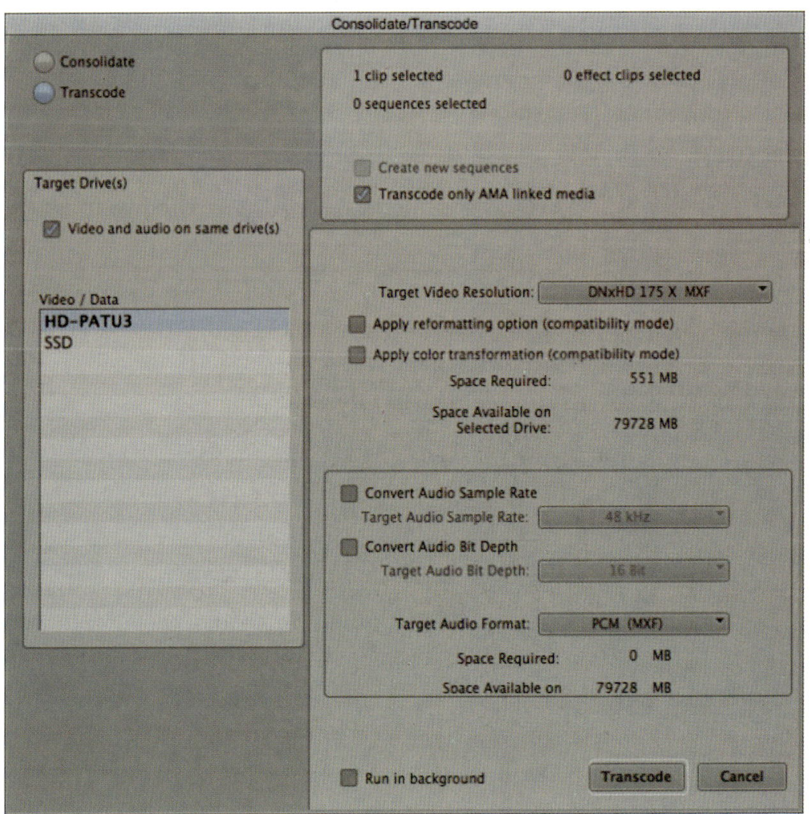

Fig. 1.51 This is the Consolidate/Transcode window.

so your choice here is to Transcode. Below that choice you can decide which drive you want to Transcode the media to. Transcode converts from the native format to another format, making a converted copy of the media.

On the right side of the window, you have a pulldown menu for selecting the format of the transcoded footage. I would generally suggest a flavor or DNxHD. If you're doing an off-line edit, go with DNxHD 36 or something small. For on-lining and high quality, choose 175 or 220. As mentioned earlier, DNxHD resolutions that end with *x* are 10-bit. If you're doing color correction, you'll definitely want 10-bit. There are even ProRes options and SD options. These choices depend on computer horsepower and RAID size and speed as well as quality considerations. For my typical projects, I stick with 115 and 145. Below that, you can choose to add color transformations—from LUTs—and reformatting. You can choose to convert your audio sample rate, bit depth, and the format the audio is stored in. I like to stick to my project rate, which is usually 48k/16-bit. On larger projects, you may want to discuss this option with your audio house. If they are relinking to original media, they won't care. But if they're using the Avid audio media, they will have specific instructions and desires. Finally, you can choose the check box for Run in background. If you do this, you'll be able to continue to edit during the transcoding process, but background transcoding runs much more slowly than foreground (dedicated) transcoding. You may want to test this based on the power of your own machine. Also, if your footage and project do not have matching frame rates, an additional dialog box will appear asking you if you want to convert the footage to the project frame rate or leave it as is.

The transcoded versions of the AMA files appear identical to the originals, except that the clip icon will be the standard master clip without the "link" next to it (see Figure 1.52), and if you look at the video

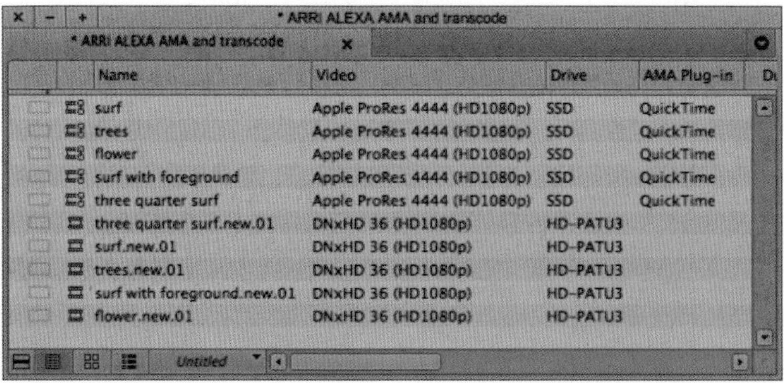

Fig. 1.52 A bin with AMA clips (top five clips) and transcoded clips (bottom five clips).

column in the bin, the new clips will have the transcoded media type next to it. Also, if you have an AMA column viewable, you'll see that the AMA version has the native camera file format next to it and the transcoded files will have nothing in that column. I usually tend to have the transcoded files sent to the bin where the original AMAs were saved. This will be confusing to edit from, so you should move the AMA files to a bin that is clear that it contains AMA material, and then move the transcoded files to a bin with a descriptive name, identifying the footage.

Dynamic Media Folders

Dynamic Media Folders (DMFs) were introduced with MC7 in 2013. They are basically just watch folders that wait for media to be dropped into them and execute a series of commands once that media arrives. Dynamic Media Folders help facilitate bringing material into the Avid system in the background, automatically. You can have several different Dynamic Media Folders, each of which executes different commands, known as Profiles. Once a Profile has been created for a specific workflow, the media is automatically processed and the editor is informed by an icon that lights up in the timeline that new footage is ready.

To create a folder within Dynamic Media Folders, make a folder and name it something like "Dynamic Media Folder_test1" on the desktop or anywhere. The big tip for getting this to work is that you have to make sure that you enable Avid Editor Services before you launch Media Composer. On a Mac, Avid Editor Services are in the series of icons on the right side of the menu bar, near the icons for enabling WiFi or launching Creative Cloud or seeing battery life or even Dropbox (see Figure 1.53). It's a little icon that looks like dotted arrows both facing to the right. If the icon is green, then it's active. If not, click on it for a pulldown menu and choose Start Avid Editor Services; then launch Media Composer. On a PC it's on the right side of the task bar (see Figure 1.54).

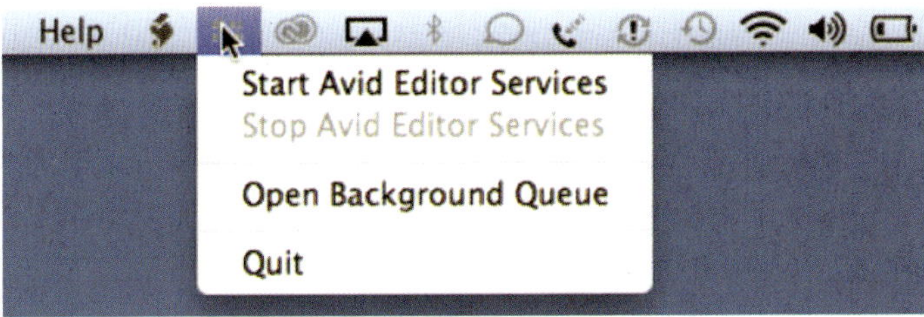

Fig. 1.53 You must start Avid Editor Services for DMA folders to work.

Fig. 1.54 Avid Background Services icon in the toolbar of a PC (circled in red). On a Mac, this is at the top of the screen. If the icon is red, the services are turned off. If green, they are enabled.

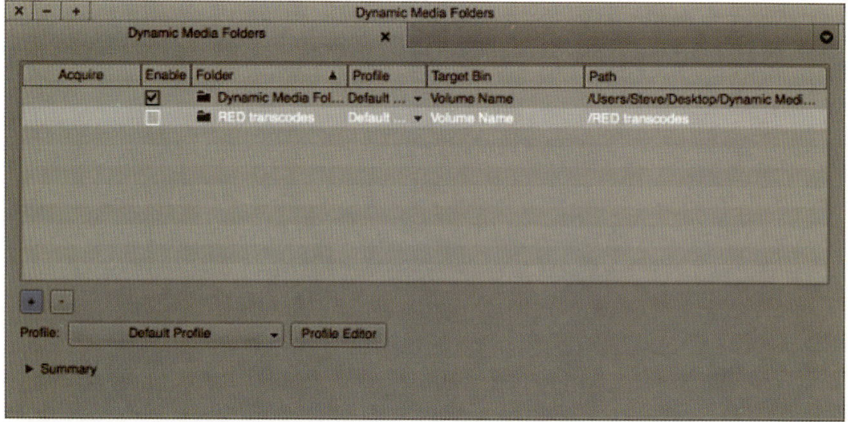

Fig. 1.55 This is the UI for Dynamic Media Folders.

In the menu bar of Media Composer, choose Tools > Dynamic Media Folders. The Dynamic Media Folders window appears. Add a folder using the + icon on the left side, under the empty folder list (see Figure 1.55). Select the folder you just created (or add a new folder using the New Folder button at the bottom left corner) and click the Choose button from the bottom right corner.

The folder will now populate into the Dynamic Media Folder list at the top of the window. Click on the Profile Editor button under the list to launch the Profile Editor window. There is a Summary of the Profile at the top of the window, followed by an AMA Settings and Actions spin-down menu. Spin down the right-facing arrow next to AMA Settings and Actions and then spin down the options for AMA plug-ins, Bins, and Link Options (see Figure 1.56).

Under AMA plug-ins, I would stick with AutoDetect AMA plug-in, but you can select a specific plug-in from the spin-down menu (see Figure 1.56). Under Bins, the default is Volume name. You could also choose Default Bin Naming Convention. I would caution against Use Active Bin or Specify Bin name because Dynamic Media Folders are supposed to be designed to automatically work in the background, and the Profiles should be able to be used under multiple circumstances, so having options that Media Composer can execute on its own based on those circumstances seems like a good idea.

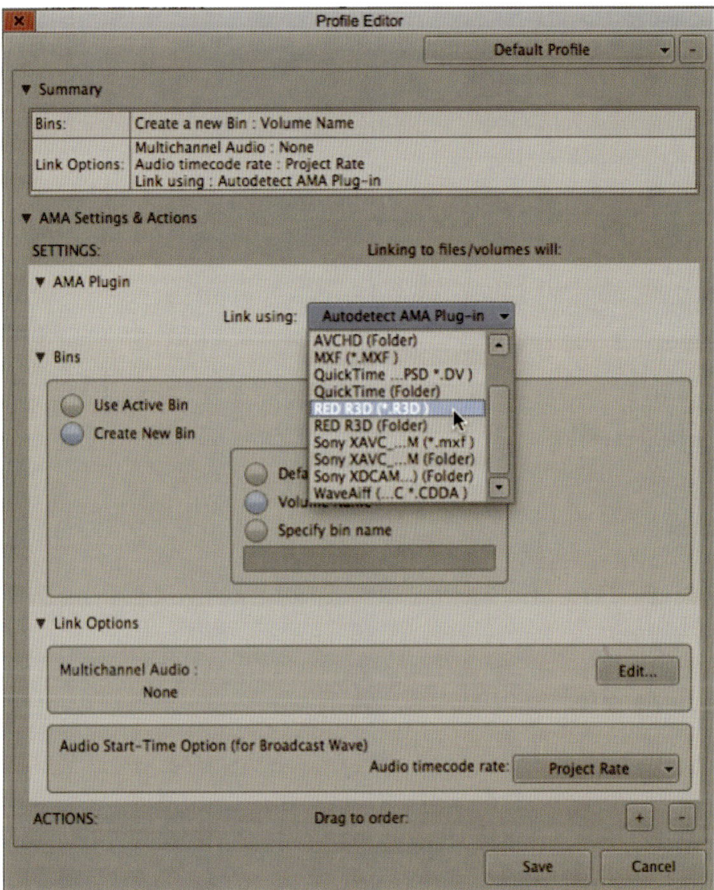

Fig. 1.56 The Profile Editor for Dynamic Media Folders allows you to customize the settings that dictate how file-based media is treated when it is placed in the Dynamic Media Folder.

Under Link Options, there are options for how to deal with multichannel sound (see Figure 1.57). Using the Edit button you can assign how those tracks are mapped to Avid audio source channels. There's also an option for how Broadcast Wave files are handled. I'd stick with the default of Project Rate, unless you have an important reason to choose a specific frame rate.

Finally, and most importantly, at the very bottom of the Profile Editor window is the Actions section. Click on the little "+" icon to the right to add an action. The choices for adding an action come from the spin-down menu and include Copy To Folder, Consolidate, and Transcode.

The Copy to Folder option simply copies the media to a specific location. This is useful if the computer running Media Composer is the first computer to receive the camera card. That way, you can automatically create backup copies of the camera card onto different drives, for safety.

The Transcode and Consolidate options are basically the same as described above in the Moving AMA to transcoded media section. Indeed, the Dynamic Media Folder is basically just trying to do all of that work automatically and in the background, so the options are essentially the same. The spin-down arrow next to the Action button allows you to see the options for the Transcode or Consolidation. When you have customized this, choose the Save button. This launches a small dialog window allowing you to name your customized profile. Give it a descriptive name. This profile can be used for many other purposes (see Figure 1.58).

Back in the Dynamic Media Folders window, you can now choose this profile from the Profile pulldown menu. When you choose it, you can see that that profile is added to the Profile column of the DMF you selected (see Figure 1.59). You can then click on the Enable button of that DMF in

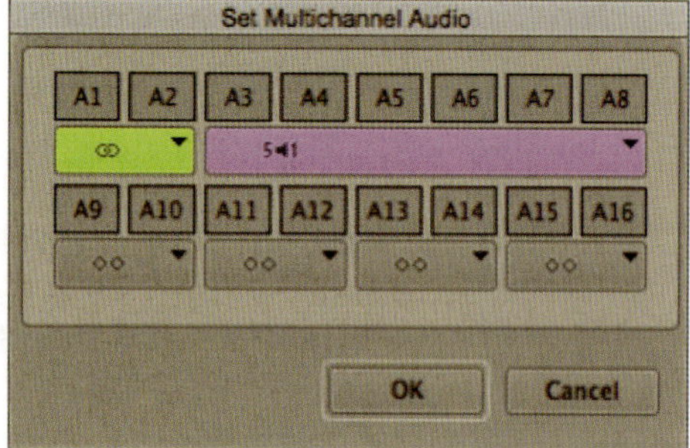

Fig. 1.57 Under the link options, you can determine how multichannel sound is linked. In this example, channels 1 and 2 are brought in as stereo, the next five are 5.1 surround, and the rest are mono. These multichannel designations are indicated by the linked or separated circles or the 5.1 below the track numbers.

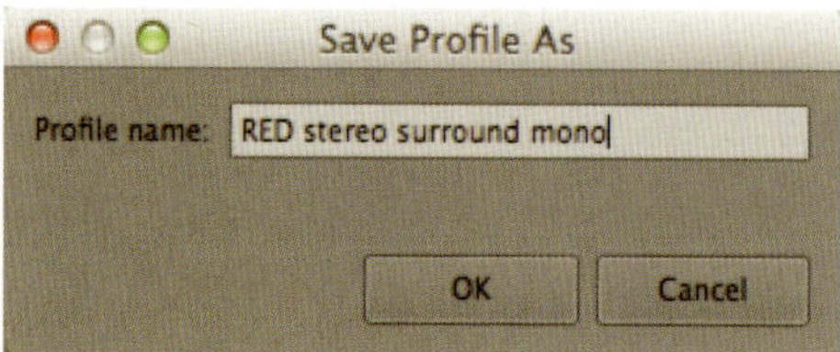

Fig. 1.58 Save this profile with a descriptive name because it can be used for many purposes.

the list and close the window. Now the Dynamic Media Folder is ready to have materials dropped into it to process in the background.

When a proper file or folder type is dropped into the DMF, an icon to the right of the audio meters in the Timeline indicates the status of the footage. A green logo of a circle forming around a dot indicates that the Profile is executing and footage is being transcoded (or whatever the Profile describes; see Figure 1.60). When the circle goes away and leaves just the green dot, the footage is ready (see Figure 1.61). If the circle turns amber instead of green, there was an error in executing the profile with the footage that was dropped into the folder. Clicking on this icon allows you to acquire the footage or investigate the error (see Figure 1.62).

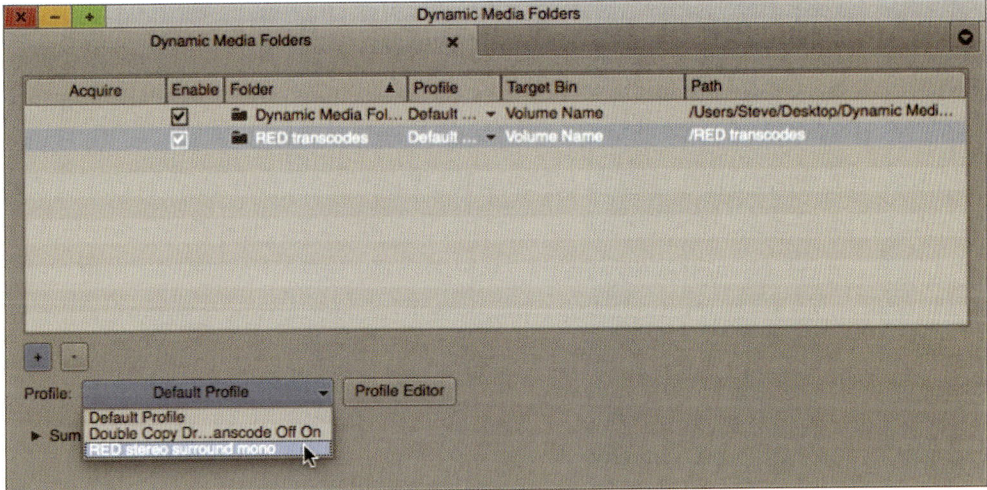

Fig. 1.59 The newly created "RED Stereo surround mono" profile selected in the Dynamic Media Folder.

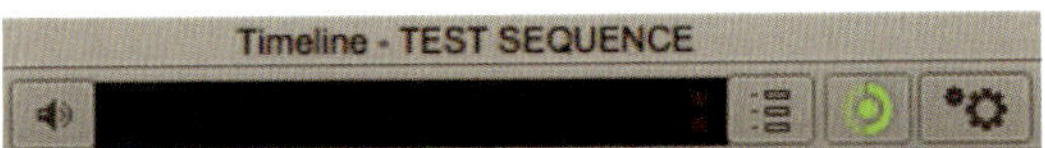

Fig. 1.60 In the Avid timeline window, to the right of the audio meters, the green circle icon indicates the processing of footage in the Dynamic Media Folder. In this image, the files are partially transcoded.

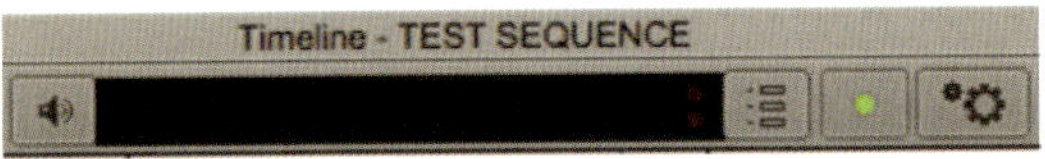

Fig. 1.61 In this image, the transcode is completed and ready for editing.

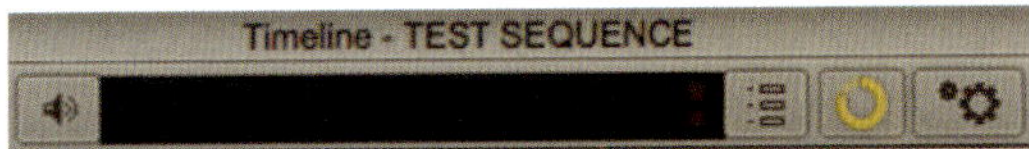

Fig. 1.62 In this image, there has been a processing error of some sort in the footage placed in the Dynamic Media Folder.

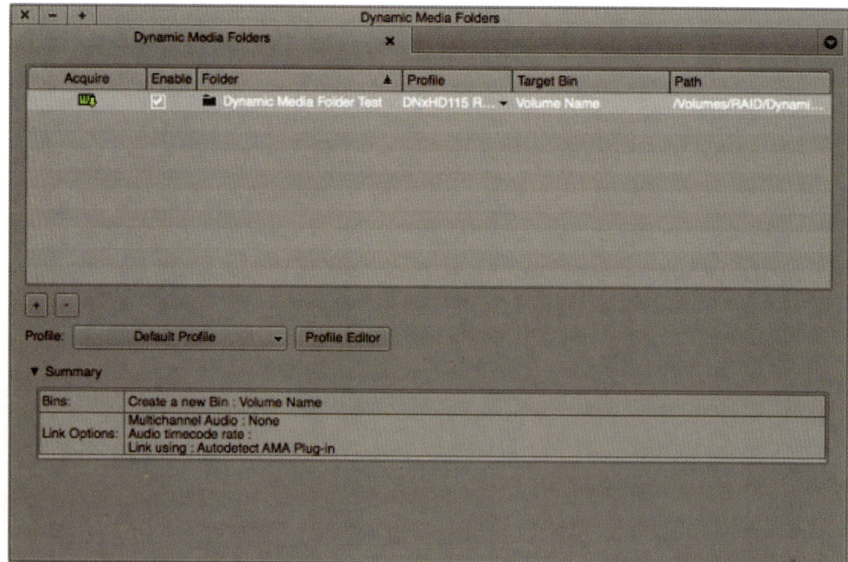

Fig. 1.63 The Acquire icon (green) shows that there is media ready for acquisition from the Dynamic Media Folder.

TIP

As soon as the green dot appears inside the moving green circle, you can click on it and begin acquiring the footage. The AMA material will be instantly accessible and the transcoded media will be off-line until each transcode is done. But the transcoded files will come on-line as soon as they're done.

If there is an error, right-click on the icon and choose Generate Error Report in Console. To see this report, Command-6 (Mac) or Control-6 (PC) launches the Console. Scroll back using the Console window's scroll bar to see a list of errors. These are going to be relatively cryptic and unhelpful, but they may help you diagnose the error or at least pass some information along to Avid Support. If you're having errors, drop a small test QT movie into the DMF. That almost always works.

If the icon is a green dot, the DMF window appears. In the DMF list, your DMF should be at the top with a small green Acquire logo next to it (see Figure 1.63). Clicking that icon will bring the footage from the DMF into a bin. The bin name is assigned by your Profile selections.

High-Resolution Media

The two big things to remember about being able to fully exploit high-resolution media (2K and 4K and beyond) is that it has to come in AMA and you need to use the Source Settings, which are a contextual menu available by right-clicking the Source monitor or by right-clicking on the clip in the Bin. It's a little easier to find from the Source monitor, because the Source Setting is at the top of that list. From the bin, it's kind of buried about halfway down. Source Settings is going to be one of those important things that will make people REALLY want Avid to allow contextual menu options to be mappable. (MC7.01, released as this book was being delivered to the publisher, actually added the Source Settings option to a standard menu, allowing it to be mapped to the UI or a keyboard, thanks to a suggestion I made following the release of 7.0. It was nice to see such quick action to a feature request! They added this less than a month after I first suggested it.)

If you import a high-resolution clip, it automatically "down-reses" the clip to your project settings (HD of some flavor). The problem with this is that if you want to resize the imported footage, you will have access only to the HD resolution of the footage, instead of the full 2K or 4K material.

However, if you AMA the clip into Avid, you can load the clip into the Source monitor and right-click on the Source monitor to gain access to the full resolution of the media through the Source Settings menu. Using AMA with high resolution footage also allows you to utilize the additional resolution in the Effects mode which is especially useful for zooms and repositions. In Figure 1.64 you can see that a resize effect dropped on to an AMA 4K clip has the option for FrameFlex resizing. This will result in a much higher quality resize inside an HD project.

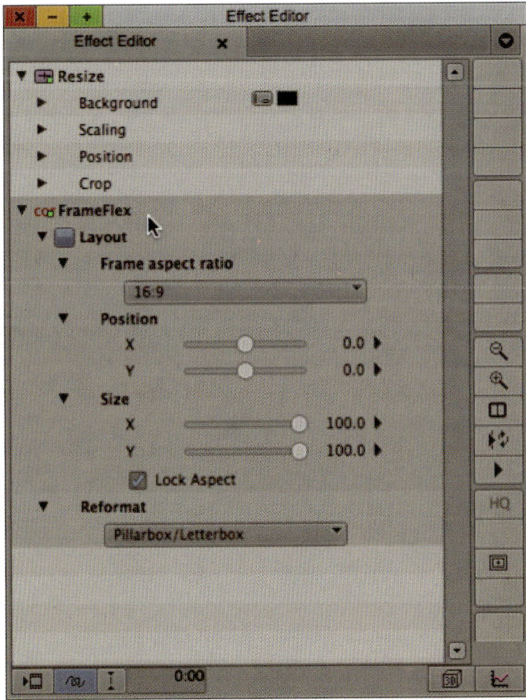

Fig. 1.64 FrameFlex resize effect.

There are two or three tabs in the Source Settings window (which is called up by right-clicking the Source Window, or, on a Mac, Control-clicking, then choosing Source Settings from the contextual menu): AMA Source Setting, Color Encoding, and FrameFlex. If the camera manufacturer doesn't provide special Source Settings, the AMA Source Setting tab will not be there. RED provides the AMA Source Settings (see Figure 1.65).

We'll discuss AMA Source Settings and Color Encoding in the next part of this chapter. The FrameFlex tab allows you to access the high resolution of the camera original material (see Figure 1.66).

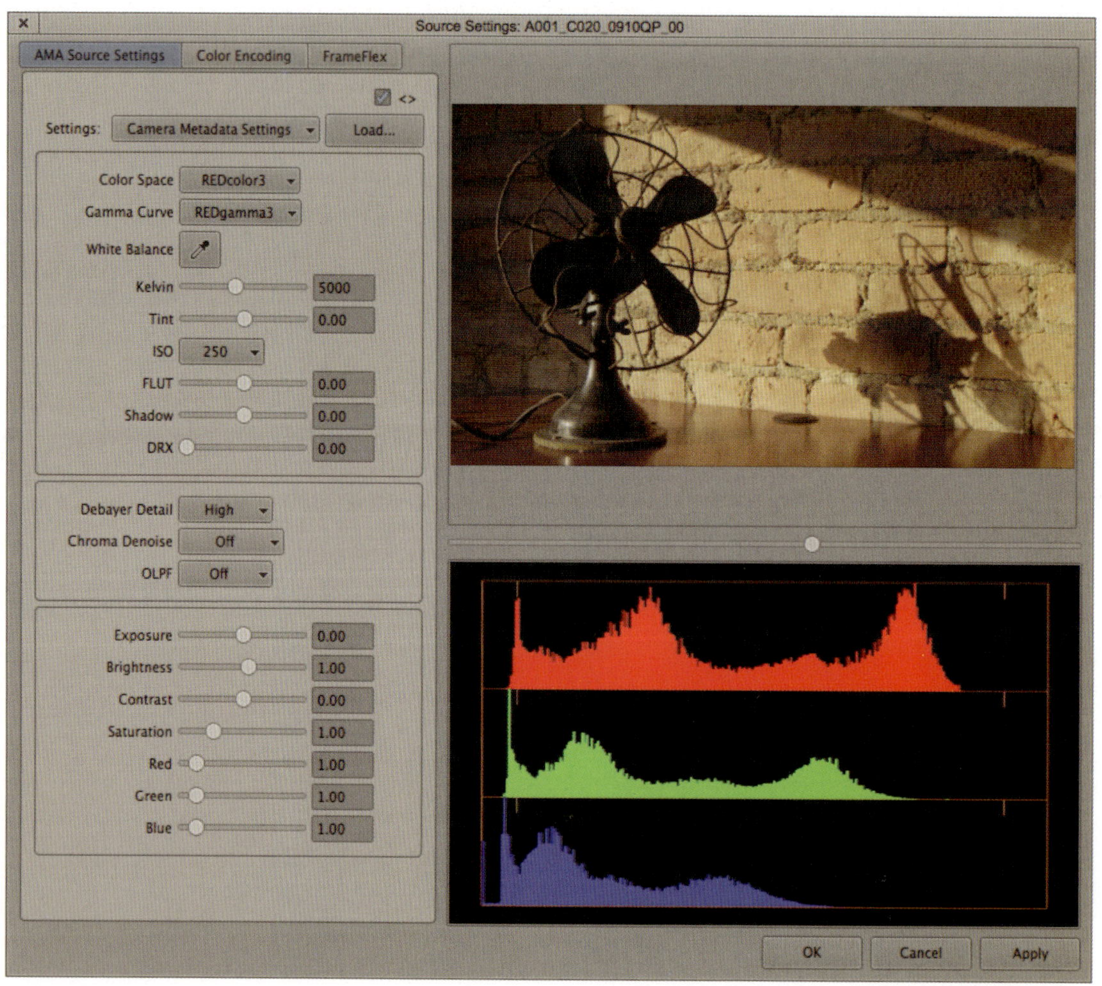

Fig. 1.65 RED footage accessed via AMA allows full access to the RAW nature of the footage, changing ISO and color balance, for example in post, prior to color correction.

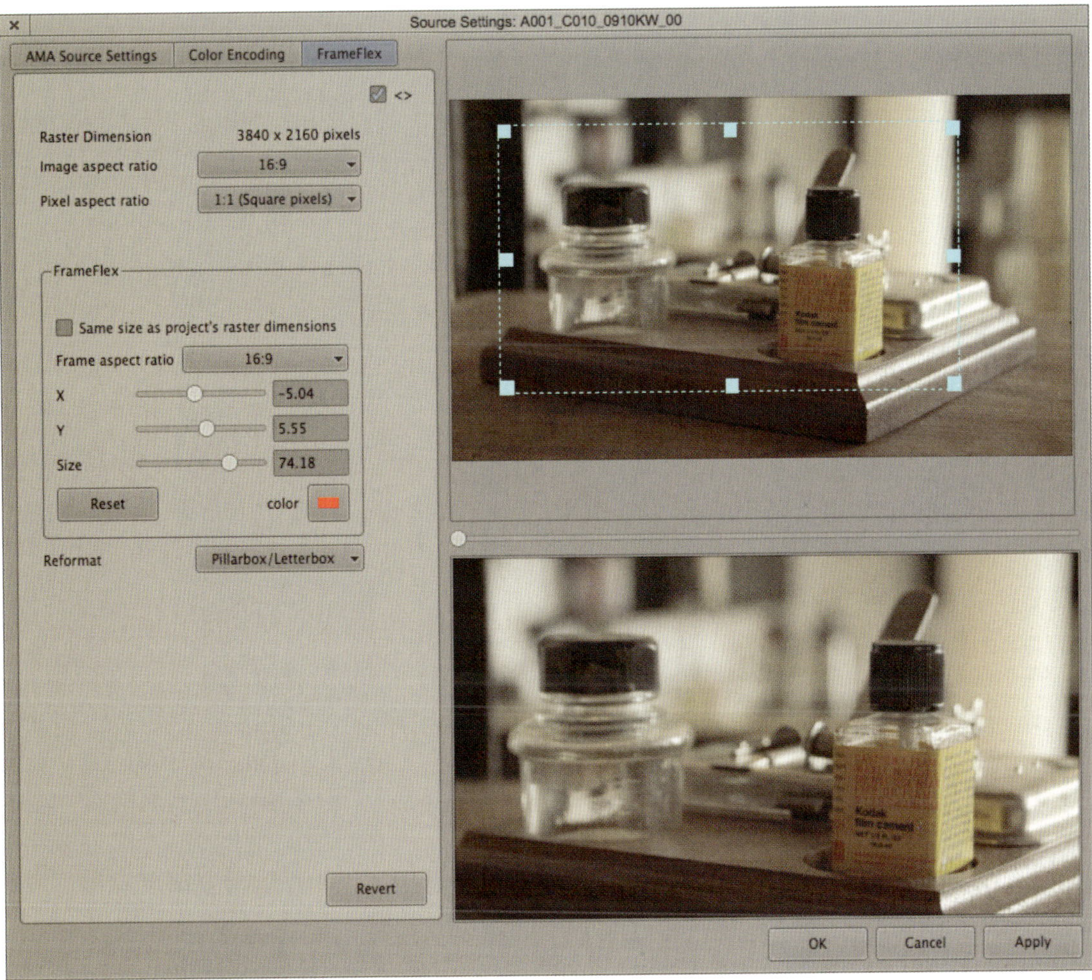

Fig. 1.66 FrameFlex UI allows Avid to access the full resolution of 2K and 4K media for repositioning and resizing.

One of the things that I wish this window had was a warning to indicate when you have zoomed in beyond the resolution of the current project. In other words, when do you pass the point where you have a 1:1 mapping of pixels? The workaround for this deficiency is to use the check box labeled "Same size as project's raster dimensions" (see Figure 1.67). With this box checked, you automatically zoom in to the 1:1 mapping of pixels of the source resolution to the resolution of the project. For a 4096 x 2304 UltraHD clip, this would be a zoom of 46.87, or a little better than half of both the vertical and horizontal resolution. But if you want to type in a specific size or drag the crop marks manually, there's no indication, so you'd

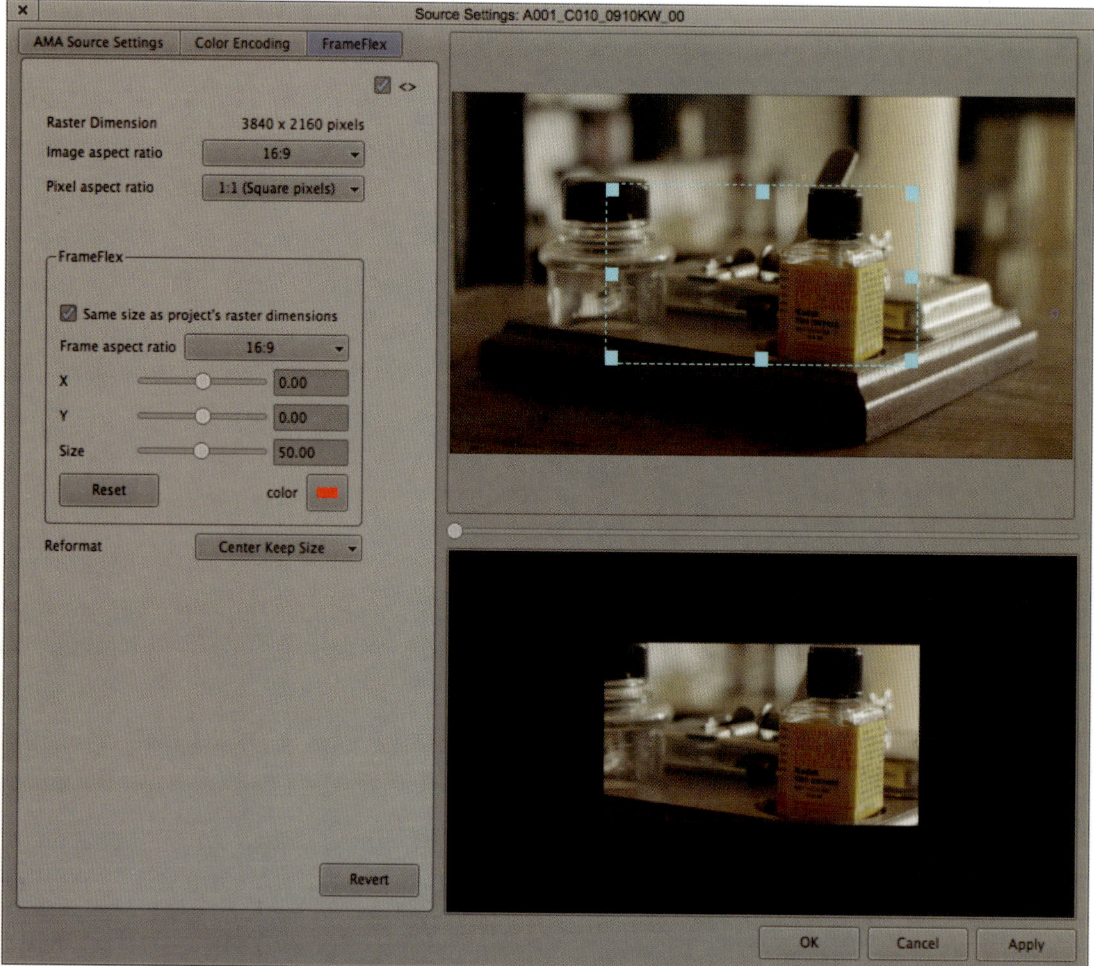

Fig. 1.67 Checking "Same size as project's raster dimensions" allows you to see how far in you can zoom and maintain a 1:1 pixel ratio. So the blue outline here indicates an HD pixel size inside a 4K image. Zooming in closer will result in image deterioration.

need to consider any crop that is tighter than 46.87. Obviously, you could zoom in more, but you would be zooming beyond the 1:1 pixel ratio, possibly adding artifacts or softening the image. On tests I've done with UltraHD footage, zooming beyond 1:1 does give very pleasing results. They are rarely pixelated or aliased in anything other than the most extreme blow-up. But still, it would be nice to know when you passed the point of the native resolution.

But to explore the rest of the UI for Source Settings, there are a number of controls, and most are pretty self-descriptive. One that's not is the little check box next to the <> indicators at the top of the window. This allows you to show/hide the windows to the right. The Raster Dimension describes the full size of the source media and is simply a display, not a control. Below that is the image aspect ratio of your output. The default is 16:9, but there are about a dozen options plus a custom option. Pixel aspect ratio really has only two choices: the default of square pixels (1:1 refers to the ratio of the height of the pixel compared to the width, hence square) or custom.

Then the real magic happens down in the FrameFlex section of the UI. One of the really important choices here, as I mentioned in two paragraphs ago, is the "Same size as project's raster dimensions" option. This check box will resize the crop box so that the pixels are mapped 1:1 to the project frame size (generally HD, or 1920x1080—possibly 720). The Frame aspect ratio gives you the same choices as the Image aspect ratio at the top of the window, but these choices are for the destination size of the media instead of to describe the source media ratio. X, Y, and Size sliders obviously give you control over the zoom and positioning of the destination framing inside the source. The size button does not work well when sizing back to 100 percent if the media is not perfectly centered. It's better to use the Reset button to get the positioning and resize back to their default of displaying the entire source frame.

The Reformat button describes how the source should map to the destination if the aspect ratios are not the same. I'm surprised that Stretch is the default. I would choose Center Crop or Center Keep Size. To get a better idea of what these do, try changing the aspect ratio of your Source to something like 1.85:1, which is similar to 16:9 but not quite the same. You'll see that instead of stretching or squishing the image to make the two aspect ratios fit, it will crop a little off the sides. I would rather that the shape of the image be kept the way it is in the source footage. The Color option, according to the MC7 Workflow documentation, lets you "Set color of framing box outline in the viewer." But this doesn't change the crop color for me and has no effect (thankfully) on the background color of the pillarbox/letterbox. It also doesn't change the color of the clip in the timeline.

Use the Apply button to see how your changes will look and the OK button to leave the Source Settings mode. Cancel and Revert buttons will get you out without any changes. Obviously it's nondestructive, so you can't do any permanent damage here.

The even more awesome capabilities occur when you drop certain effects onto the clip and go into Effects Mode. Now, in many effects, there is a FrameFlex option (see Figure 1.64). These can be found in the Resize Effect, Avid Pan and Zoom, 3D Warp and others. The FrameFlex parameter gives the effect access to the full resolution of the image, instead of just to the HD size. This also allows you to do keyframed moves, like zooming or panning around a 2K or 4K image.

Color Management

Color Management isn't quite as sexy as the FrameFlex stuff, unless you're really into color. Basically this feature allows you to apply LUTs and to preserve any color management that is embedded into the file from the camera manufacturer. Any color management that's embedded will automatically be extracted and applied to the clip. LUTs and CDLs (color decision lists) are applied in the Source Settings in the same way that FrameFlex worked.

From the Bin or the Source monitor, right-click on the clip to call up the contextual menu and choose Source Setting from the menu choices (see Figure 1.68). Color is pretty simple. There's the source color space

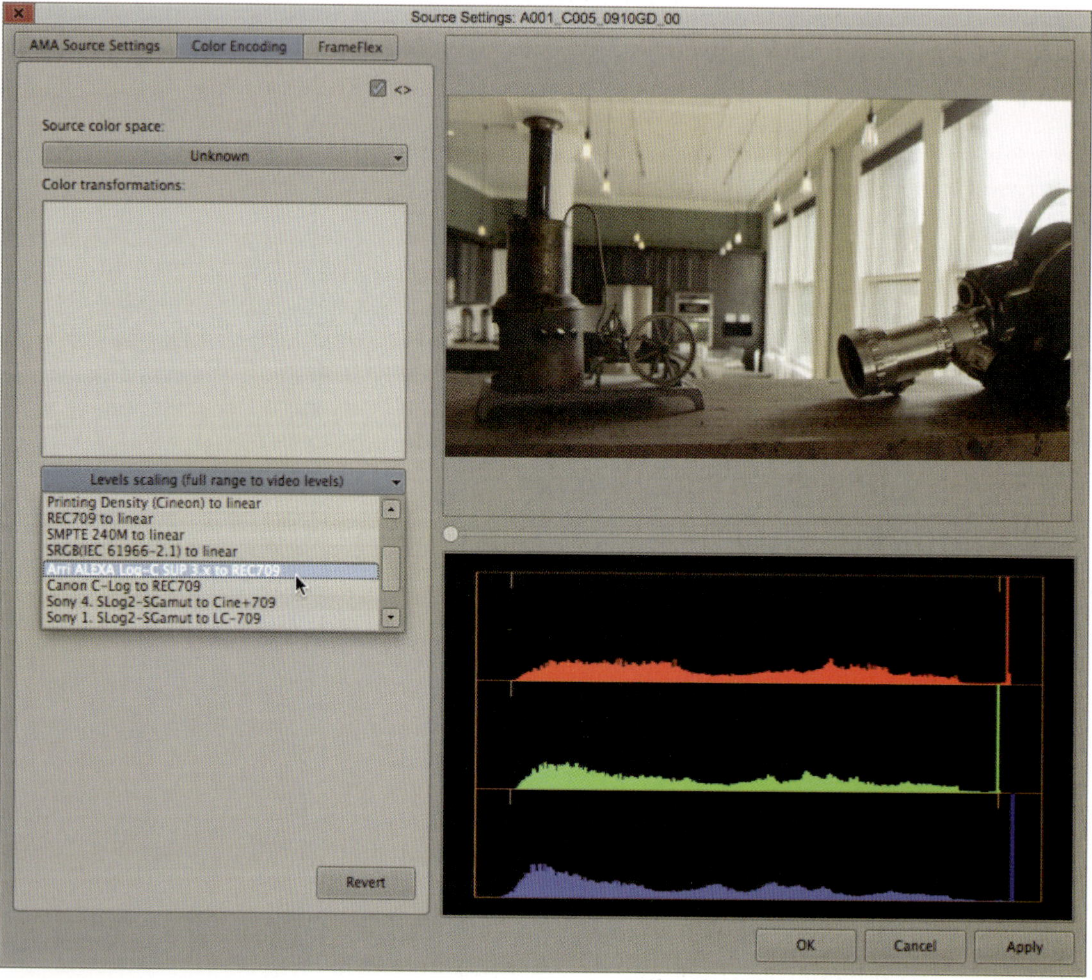

Fig. 1.68 The Color Encoding tab of the Source Settings. An Arri Alexa Log-C LUT is being applied to the footage.

and there's the transformation that converts it to the output color. There are options to set these automatically, but you should find out from the DP, the DIT, or the footage provider what the color space is for the source.

The transformation can be from one of the built-in settings or from an external LUT or CDL. Supported LUTs include 1D and 3D LUTs. Three-dimensional LUT types include IRIDAS, KDM, Kodak, LUTher, and Mistika. Other specific supported LUTs include Avid DS and Autodesk.

The first step is to install the LUT, which must be either downloaded or provided by the production. Once installed, it will be included in the list of Color Transformations in the Source Settings. Any changes made in Source Settings will show up when clips are edited into the timeline. If you had already started editing with these clips before installing and applying the LUT, you'll need to refresh the sequence by right-clicking the sequence and choosing Refresh Sequence > Color Adapters.

Installing an External LUT

From the Project window, in the Settings List, select the Color Management setting and choose Install LUT from the subsequent window. Browse for your file, select it and click Open to install it. The LUT becomes available in the pulldown list of color transformations.

Applying an External LUT

From the timeline, or a bin, right-click on a clip and choose Source Settings and the Color Encoding tab. The drop-down menu for the Color transformations will have installed LUTs at the bottom, with "External" before the name. Apply this by itself or in concert with other CDLs or LUTs.

Using Color Decision Lists (CDLs)

These CDL values are imported via an EDL or ALE. When you do an export as an AAF, ALE, or EDL, the values are included along with the associated media.

Any clips with CDLs to which you link will automatically apply the CDL values.

AMA Source Settings

Certain file formats, like RED, that have an AMA plug-in installed on your system also provide access to their color information in the AMA Source Settings tab in the Source Settings window. Color space adjustments should be made on this tab for cameras like RED. This provides a lot of power. You can alter the exposure, color temperature, ISO settings, gamma curve, and other settings (see Figure 1.56).

Audio from Double-System Sound

> **TIP**
>
> AMA should never be used for double-system sound sync in Media Composer. If you AMA audio, Avid will default to using the QT AMA plug-in which doesn't allow for pull-up or pulldown or for proper decoding of timecode. The standard file format for double-system sound is BWF (Broadcast Wave File), and there is no BWF AMA plug-in.

The standard dailies workflow for bringing in double-system sound is to have the syncing between the camera footage and the audio recording done by the company providing the dailies. That audio is linked to the video files and ingested into the Avid system if the audio had always been incorporated with the video. Avid Media Composer is also capable of syncing sound recorded separately from the video using the AutoSync tool. Avid actually considers this the best solution because it allows for one-step sync with not just the mix track, but with all of the isolated sound tracks as well. Avid also suggests syncing in Avid because it is possible to slip sync from within Avid with AutoSync'd footage and track the amount of that slip throughout post production. Slipping sync can only work in a "film" project of 23.976, 24 or 25fps. AutoSync'd footage also obviates the need to track audio timecode in the SoundTC column.

It's important to realize that in 1080HD projects, the SoundTC column only tracks a rate of 30fps. If you work at any other rate, use one of the Auxiliary TC columns for sound because they always match the project rate.

BWF files carry a huge amount of valuable metadata which can include specific mics used on each track or even character names as well as Scene and Take information. If this information hasn't already been incorporated by the dailies team when syncing the audio, you can utilize it with AutoSync. When you AutoSync, Media Composer uses the .sync default naming appended to the name of the video clip, but with cryptic file-naming conventions coming from camera manufacturers, this isn't always helpful. You can easily use the information from the BWF file, including scene and take, to rename clips. Before you AutoSync, create a custom column and duplicate the contents of the Name column—Command-D (Mac) or Control-D (PC)—select SoundRoll from the pop-up menu, and paste the file name information there. With the clips synced, select the Custom column with the Scene and Take information and duplicate it back into the Name column. The BWF data (file name info) will be placed in the TapeID column. Unfortunately, the EDL Manager can't read this info when creating EDLs, but, if you want, you can duplicate the column and add it to the SoundRoll column (from the pop-up menu), and the SoundRoll info is available to EDL Manager.

Prepping Feature Films

One of the premiere editorial assistants in the world of high-end film editorial is JC Bond. Bond is also an accomplished film editor in his own right, but he's assisted on some of the biggest movies of the last decade including *Men In Black II, Harry Potter and the Prisoner of Azkaban, Charlie and the Chocolate Factory, The Taking of Pelham 1-2-3, Alice in Wonderland,* and *Unstoppable*.

Bond elaborates, "I've been fortunate enough to work on a lot of large projects in the recent years. These are projects with 4,000–6,000 bins. Since then I've done a few that were smaller. Currently I'm cutting as an editor a much smaller project, a $17 million project."

"I basically organize every project the same way whether it's a $5 million movie or a $200 million movie," Bond states, "The organizational methods that I use come from having to deal with the 5,000 bin projects. I follow the same pattern, the same ideas whether I'm dealing with a smaller project or a larger one. Once you have to deal with that many bins and that many clips and that much footage, if you follow the same ideas, they scale down pretty easily." The larger project taught Bond the procedures he uses to insure that the editorial department doesn't get "painted into a corner" at the end of the production.

"The philosophy I try to follow is to keep everything as simple as possible," Bond explains. "I don't do anything because anyone taught me 'That's the way you should do it.' I do something because I have found out that in my experience that I do A, B, and C because of D. Obviously having taken many projects over the last twenty years from beginning to end, I've gotten to the point where I've seen the A, B, C, or Ds of why you have to do certain things on preproduction because it pays off when you're delivering the movie a year later."

The first layer of organization for any movie cut on Avid is the Project level, which organizes using folders, subfolders, and bins. Bond explains that he organizes these projects the same way virtually every time. "I always receive dailies from the lab or the set or from wherever they're coming from and I have a folder that I call PREP and within that folder, I have a folder called PICTURE and a folder for SOUND and one for SYNC material. So I have a set of bins for Day 1, 2, 3, 4, 5 that are picture only; Day 1, 2, 3, 4, 5 that are SOUND only; and I have the result of adding those two together, which is my subclips of all the materials SYNC, Day 1, 2, 3, 4, 5, etc. It is my preference when the size of the crew or the size of the movie allows it to sync the picture and sound in the Avid because it allows for better accounting of where things came from. It provides for better accuracy on the sync. And the reality is that companies out there that do this on the fly in the middle of the night they don't do it accurate enough and I end up having to fix more of their mistakes. I'd rather do it myself and do it right the first time. Or not myself, but somebody in my crew."

Sync is critical to Bond, so he has multiple methods for insuring that it is as accurate as possible. "If I am doing a movie where the lab is syncing the material, I never give the editor a master clip. If I'm editing myself I don't want to be editing with a master clip. I always want to be working with subclips. And the reasons for that: First, the master clip cannot be perf-slipped. Subclips can. And second, when you get to the point towards the end of the project where you're doing DI pulls or visual effects pulls and you want to keep track of what has been pulled and what hasn't been pulled, it's easier if you're dealing with subclips than if you're dealing with master clips. Again, this is something that you wouldn't hit until you're dealing with the DI. But if you're working with subclips, that process is extremely easy if you're dealing with master clips then you have to go through hoops to get there." Perf-slipping is available only in film projects (those projects where film was checked when creating a new project) and allows audio to be slipped in subframe amounts, depending on the gauge of the film chosen when the project was created (see Figure 1.69).

"It has become sort of the standard in the DI process nowadays that the cutting room just turns over an EDL of the cut and lets the DI department deal with the entire nightmare," Bond confides. "In the days that we were dealing with film, I would provide them with scan lists and say, 'This is what I want you to scan. If you scan this, when I give you an EDL of the cut, you'll be able to assemble that.' So to get to that process, I needed to be able to keep track in my project of what I had given the scanning department, orders to scan. And to do that I would create subclips of the scan orders. If I have a take and that take is three minutes long, but I'm only scanning a minute in the middle I would create a subclip for that minute

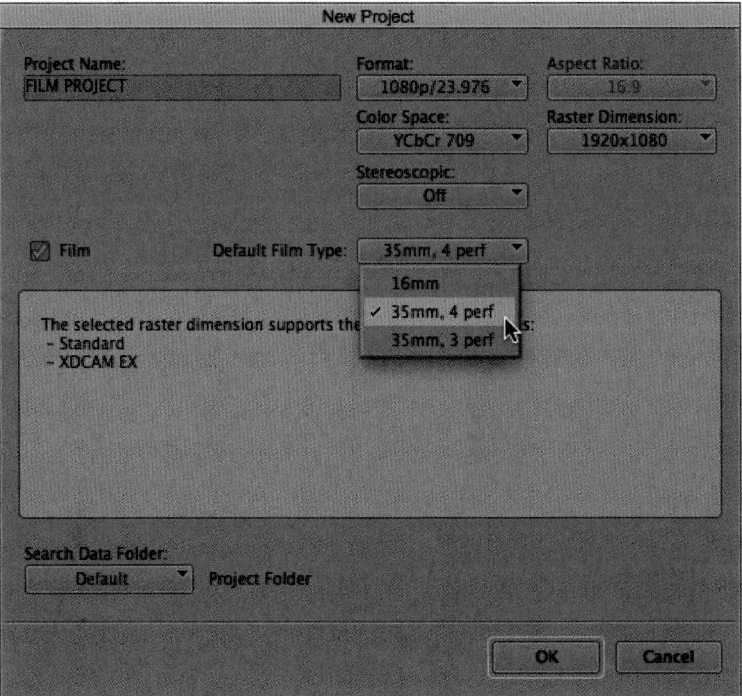

Fig. 1.69 Film project setting.

I was scanning. It's very easy to create those subclips if you do a decompose of your sequence and you tell the decompose command, give me a decompose of this sequence with eight frame handles. If you're doing that with a sequence that was created with master clips, you're dealing with decomposes that are master clips and those master clips now are useless to you. You can't relink them to media. You can't do anything. But if you do a decompose of a sequence that was created using subclips, the decompose result is subclips. And you can then relink those subclips to media, and those subclips represent whatever you ordered as a scan order."

Most large projects still deal with file-based media as if it was shot on film. In other words, it goes through what is basically a "dailies" process in which the original camera file is treated like a camera negative. Instead of being telecined to tape, it's converted, prepped with a one-light color correction—probably using a LUT or two—and synced with audio. "In the project that I'm cutting on now, it was shot using ARRI RAW. It is my plan that the moment that the Avid can process ARRI RAW directly, like it can now with ARRI ProRes, then I will do away with the lab altogether."

The files come from Technicolor as Avid media files with a bin. Bond elaborates, "From there in this particular show, Technicolor did pre-sync the material, but I'm taking that material and subclipping it and perf-slipping it to get the sync more accurate. So initially I end up with—in a folder—a picture subfolder, a sound subfolder, and a sync subfolder. And inside each of those, I have a Day 1, which includes everything from day one for picture, and then anything for day one for sound and then anything from day one sync. Then once that sync is prepped, I Option-drag those into scene bins. That way they stay in the original sync bin and are copied to the scene bin."

Bond continues explaining the process, "At that point that ends up with two copies of the same subclip. One inside the prep folder, sync bin, and one inside the dailies bin. Currently that does create a problem with Avid down the line with the search function, because when you match back, it's possible that the system will match back to either the copy of the subclip in the dailies folder or the same subclip in the prep folder. So what I do in the meantime is I take my sync bins out of the prep folder and put them outside of the project (at the finder level) so that it doesn't match back to that."

Bond then explains the rest of his project organization past his Prep folder (see Figure 1.70): "From that point on, I have a folder with my current cut. The top structure of my project is a current cut is the first folder, then it's dailies, then it's Incoming Sound, with folders for ADR, a folder for music, a folder for what I call elements, which is anything other than ADR or music, and then I have a folder for outputs and in there are subfolders for who the output is going to and then within the

GLOSSARY

ADR: Automated Dialog Replacement. Or Additional Dialog Recording, Also known as Looping. This is dialog recorded at a recording studio after the scene is shot to replace dialog that may have been difficult to hear because of technical problems, noisy recording environments, a desire for a different performance or even a change in the script.

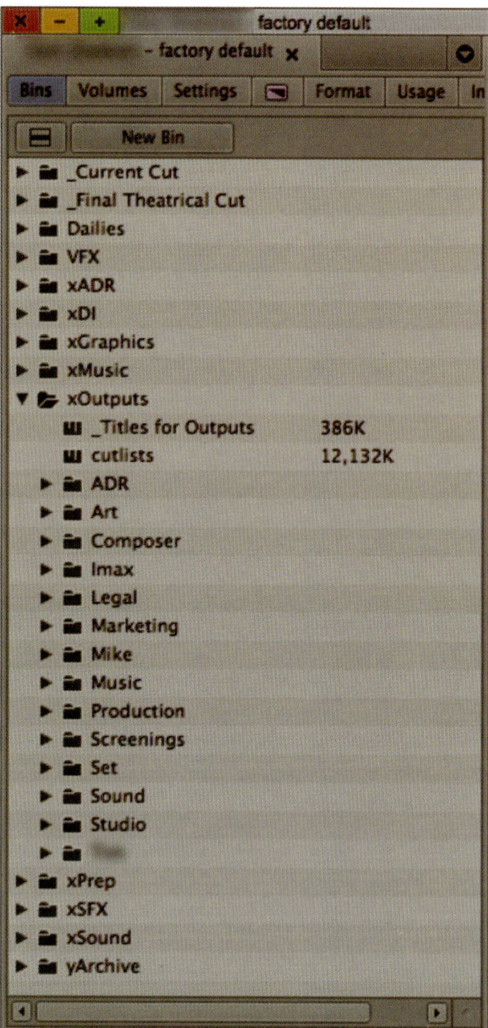

Fig. 1.70 This is the Project window for one of Bond's blockbusters. There are 60 folders and subfolders in the project, each with as many as 150 bins.

output folder of who it's going to I put bins every month. If I'm doing outputs to the director, I will create a folder for the director and inside that will be a bin for 2012 June, 2012 July, and all of the outputs for that director for that month go into the same bin."

Bond also creates a folder to put all the ADR with separate bins for each character. His music folders are organized with a separate bin for each composer. These are the composers of the temp music used by the editor before the composer for the film delivers the final music for the film. An additional bin is eventually created within the music folder when the final music arrives. Bond also has a folder for temp sound effects that is organized with separate bins for each type of sounds, like doors, clocks, and cars.

Bond picks up by describing the VFX folder for the project: "There's usually a visual effects folder and that contains multiple subfolders and within that it's typical with visual effects to create sequences which may or may not include more than one scene, so I have a folder for every sequence. I have a folder for turnovers, which is giving stuff to the visual effects facility. I have a folder for receiving stuff from the visual effects facility and then within each of those folders I have bins per sequence and I have that duplicated both in the outgoing, which is turnovers, and the incoming shots received."

Bond continues, "The only other thing is that I keep a folder in the Project that I call Archive and in that folder I keep copies of every reel—as the project moves into reel structures—and I keep a versioning system as the editor's assembly is finished we have version 1, then version 2, so every time there's a major event, let's say that we're done with the editor's assembly. That becomes a turning point and at that moment I archive the sequences of the project and put it into the Archive folder. And then we keep moving towards the director's cut. The first time we show the director's cut to one of the producers or any kind of significant event in the process or time of the postproduction, it goes into the Archive folder.

"Inside the Current Edit folder there's a bin for each reel. And inside those bins there's usually only a single sequence for that reel and then alts for scenes that belong to that reel. And alternate copies of the reel would go into the archive. If the movie gets rebalanced or whatever, we shift where the scenes fall so if there's alternates for those scenes, they live within the same bin."

We'll hear more from JC Bond in chapter 3.

GLOSSARY

Rebalancing: With feature films, as the length of the film changes, the break between reels may also change. "Rebalancing" is the act of altering where the reel changes happen in a feature film.

ScriptSync

Avid proves its pedigree when it comes to feature films. About 90 percent, if not more, of studio theatrical films are cut on Avid. From 2009 to 2012, fifteen of the twenty films nominated for the Best Editing Oscar were cut on Avid (*Inglorious Basterds, Life of Pi,* and *The King's Speech* were cut on LightWorks; *The Social Network* and *The Girl with the Dragon Tattoo* were cut on FCP by the same editor and director.) *All* of the prime-time Emmy-nominated dramatic and comedy shows for those years were edited on Avid. One of the awesome features that put Avid at the top of the hill when it comes to feature film editing is ScriptSync.

If you have not played with ScriptSync or haven't even heard of it, you should give it a try with some scripted drama dialog. It's also been used in nonscripted documentary editing. One of the main proponents of nonscripted use of ScriptSync has been Steve Audette, who has cut for *NOVA* and *Frontline* on PBS. Before we get into this slightly unorthodox use of ScriptSync, let me describe its most common use and how it radically changes the approach to editing scripted material.

ScriptSync allows you to import a feature film script and link takes (clips) to specific lines in the script that the take covers. The methodology is similar to the way that a script supervisor marks up a script during shooting. (Tip: I thought "scripty" was affectionate, friendly lingo for script supervisor but learned the hard way that this is a very *un*politically correct term.) Because ScriptSync reflects the script supervisor's method of documenting the production on-set, let me start the explanation of ScriptSync with a brief diversion into the methods employed by the script supervisor.

Script Supervisors and Lining a Script

I had the privilege of co-editing the Sony/TriStar feature film *Courageous*. I can tell you from experience that the script supervisor is the editor's best friend. One of the script supervisor's main responsibilities is to insure that the editor has all of the materials to edit the film together. How?

The script supervisor has numerous critical functions and responsibilities, including continuity, but one of the most useful to the director and editor is to make sure that each scene has the proper "coverage." In other words, has every scene and line of dialog been covered by enough camera angles and editing possibilities to allow the scene to be properly cut together in the edit room? It's certainly possible that a director could decide that some bit of dialog is done only in a wide shot or a two shot and that individual over-the-shoulder or close-ups aren't shot at all, but for most scenes, they are shot with multiple options for editing possibilities. And the person who keeps track of whether a line of dialog has been successfully covered is the script supervisor. The way they notate this coverage is by "lining" the script.

> For more in-depth understanding of the script supervisor's duties, Focal Press (Taylor & Francis) publishes *Beyond Continuity: Script Supervision for the Modern Filmmaker* by Mary Cybulski.

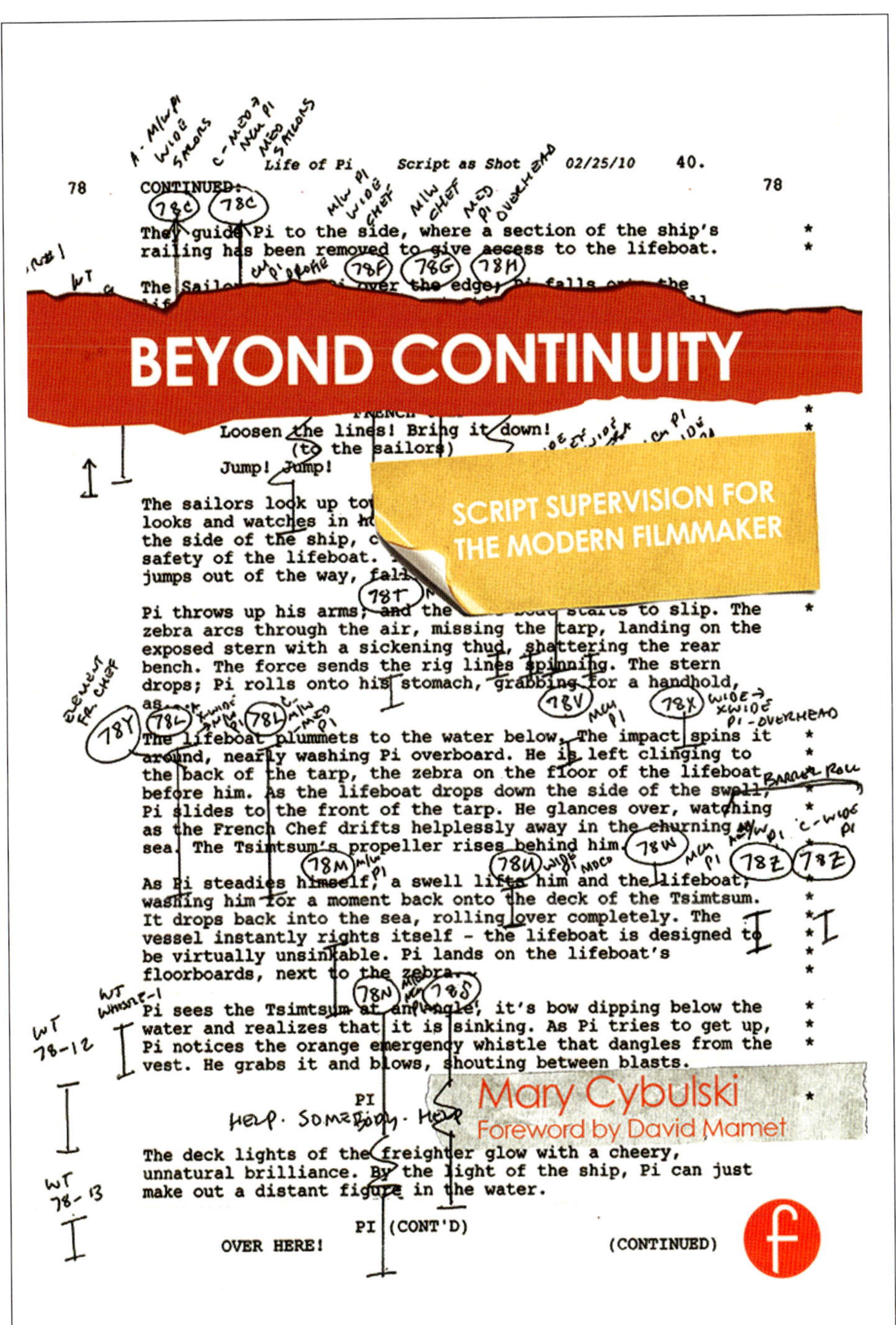

Fig. 1.71 Mary Cybulski's "Beyond Continuity" book explains the role and duties of the script supervisor and has in-depth discussions of marking scripts for feature film and TV. The script supervisor is a close ally to the editor.

The script supervisor sits on set with the shooting version of the script. As each set-up or take of a scene is completed, they note the angle, subject, and length of the take, including who is on-camera and who is off-camera. Each set-up (series of takes) is represented by a vertical line extending down the length of the script page, including the circled take or set-up number at the top (see Figure 1.72). Some takes may extend for several pages. Some takes may only cover a single line of the script. As wide shots and close-ups and over-the-shoulder shots and two shots are done, the script supervisor draws a line indicating where the take started and where it ended. When a scene is complete, it is easy to see, based on the lines extending down the page, which dialog lines have been covered and how many options the director and editor will have for a specific line of dialog.

I spoke to script supervisor Mary Cybulski, a fellow Focal Press book author, about script supervision. Cybulski's script supervision credits include *Life of Pi, Eternal Sunshine of the Spotless Mind, Eat Pray Love, Michael Clayton*, and *Maid in Manhattan*, among numerous others.

Cybulski has a very different approach to script supervision from some of the old-school supervisors, who emphasized how hyper detail oriented the best script supervisors have to be. Mary feels that this not only makes the job seem virtually impossible but also demeans the true contributions that the script supervisor makes. She does appreciate one old-school methodology: eschewing modern script supervision computer programs for the traditional method of using a pen to line the script in a three-ring binder. And though the most fundamental part of the job is lining the script, there are numerous other responsibilities that are centered on continuity that require the script supervisor to interface with nearly every department.

Cybulski says of her approach to script supervision, "The technical stuff seems overwhelming, but the new way to look at it is that the script supervisor is a film maker and a collaborator. And the script supervisor needs room to breathe."

When the lined script and the footage finally get to the editorial department and the editor starts to edit a scene, he or she can refer to the script supervisor's lined script to get a great visual reference of the options for editing a scene. It's also a way to double-check the various takes in the bin to make sure that all of the options that were shot—according to the script supervisor—are present in the bin and haven't been misplaced by an assistant or failed to make it through the dailies process at all.

You can see (Figure 1.72) that the lined script can provide a wonderful game plan for approaching the edit.

Avid took this idea and incorporated it directly into its NLE. Once the script is imported, each take or set-up can be lined in exactly the same way that the script supervisor used to do, but with the difference that now those lines can be used in much the same way as an Avid bin,

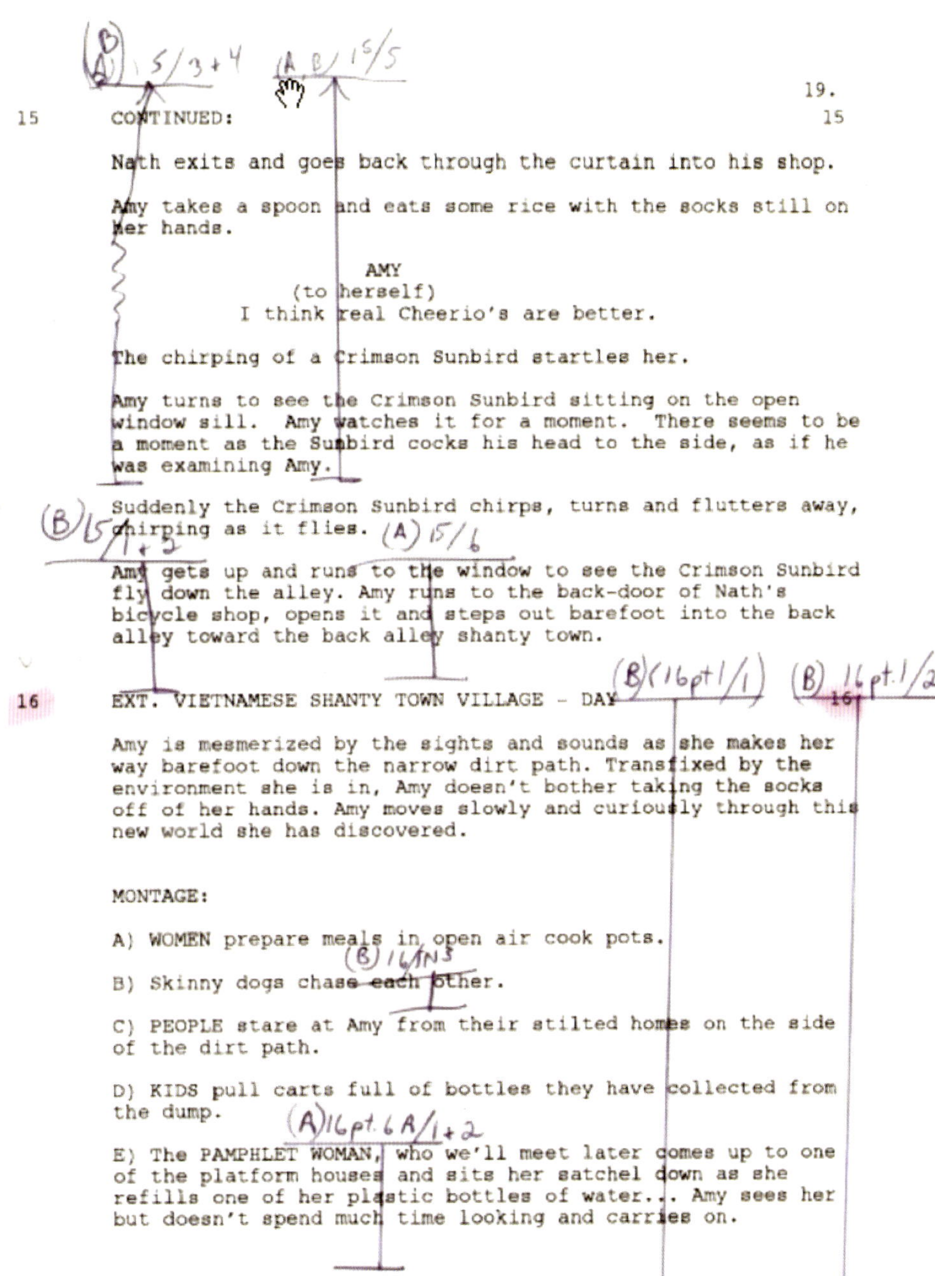

19.

15 CONTINUED: 15

Nath exits and goes back through the curtain into his shop.

Amy takes a spoon and eats some rice with the socks still on her hands.

AMY
(to herself)
I think real Cheerio's are better.

The chirping of a Crimson Sunbird startles her.

Amy turns to see the Crimson Sunbird sitting on the open window sill. Amy watches it for a moment. There seems to be a moment as the Sunbird cocks his head to the side, as if he was examining Amy.

Suddenly the Crimson Sunbird chirps, turns and flutters away, chirping as it flies.

Amy gets up and runs to the window to see the Crimson Sunbird fly down the alley. Amy runs to the back-door of Nath's bicycle shop, opens it and steps out barefoot into the back alley toward the back alley shanty town.

16 EXT. VIETNAMESE SHANTY TOWN VILLAGE - DAY 16

Amy is mesmerized by the sights and sounds as she makes her way barefoot down the narrow dirt path. Transfixed by the environment she is in, Amy doesn't bother taking the socks off of her hands. Amy moves slowly and curiously through this new world she has discovered.

MONTAGE:

A) WOMEN prepare meals in open air cook pots.

B) Skinny dogs chase each other.

C) PEOPLE stare at Amy from their stilted homes on the side of the dirt path.

D) KIDS pull carts full of bottles they have collected from the dump.

E) The PAMPHLET WOMAN, who we'll meet later comes up to one of the platform houses and sits her satchel down as she refills one of her plastic bottles of water... Amy sees her but doesn't spend much time looking and carries on.

Fig. 1.72 A lined page of the same script that was referenced earlier in this chapter, courtesy of producer, Stephen Marinaccio.

allowing the editor to instantly call up the lined shot into the Source monitor directly from the imported script.

To improve the tool even more, Avid uses phonetic technology to precisely match the exact audio of every word in the script to the actual text of the script, so clicking on a specific word in the imported script will instantly jump the timeline locator in the Source window to that exact word. The precision of this audio matching (ScriptSyncing) is amazing. It allows you to quickly compare all of your takes at a specific line or word of the script.

The ScriptSync feature includes licensed technology from a company named Nexidia, so ScriptSync is an additional option on Media Composer instead of being included with the purchase of your Media Composer license. If you are interested in trying this, download the trial and use the tutorial footage and sample script to try out this feature for yourself.

Using ScriptSync is really a two part process. In high-end workflows, the first part would be done by an assistant editor, who does all of the preparation. The second step is to actually use the prepared script inside the Avid editing environment to efficiently assess the options available for cutting the scenes together.

If you don't have ScriptSync or a trial of ScriptSync installed, do a search on your editing system computer for AvidLicenseControl, which should be in the Avid Utilities folder in the Application or Program folder. A possible file tree (for a Mac) might be Applications > Avid Utilities > LicenseControl > AvidLicenseControl. Launching this app allows you to activate (install and register or run trial) for ScriptSync, PhraseFind, PhraseFind Language Packs, and other options. Just choose Activate and run in trial mode or enter a temporary or permanent license key. There are other ways to install and activate PhraseFind and ScriptSync as well, depending on how you purchased or plan to purchase the option. See the manual for details that pertain to your specific situation.

Preparing to Edit with ScriptSync

The first step is to prep the script. Avid needs a simple text document to import, so if you're using something like FinalDraft (the industry standard script software), you'll need to export the script as a text document or ask someone in production to generate this file for you.

> To follow along with the tutorial, simply access the "Avid Uncut—The Short Film.txt" file on the companion website, www.focalpress.com/cw/hullfish.

With the text document (see Figure 1.73) in the Avid editing computer, or at least accessible by it in Media Composer, select File > New Script

A guy sits in an office on a leather coach. Another guy walks in and shakes his hand. First guy does not get up.

STEVE
Dave, right?

DAVE
Yeah. Steve?

STEVE
That's me. So you're a producer?

DAVE
Obviously. As you can see, I have a leather coach and a jukebox.

STEVE
You're clearly both creative and successful.

DAVE
Clearly.

STEVE
So you're looking for an editor.

DAVE
Yeah. For a romcom. Cutting on Avid

STEVE
Have you heard of ScriptSync?

DAVE
No. Sounds boring.

STEVE
It could save you a lot of money.

DAVE
You had me at money. How's it work?

STEVE
Works like a charm. Do you know the one about how many producers it takes to screw in a lightbulb?

DAVE
No.

STEVE
2 P-As.

DAVE
I don't get it.

Fig. 1.73 This is the script of my not-quite-award-winning comedy *Avid Uncut*.

(see Figure 1.74). This brings up the Select a Text File dialog. Locate your script and open it. Your script will be imported into the Project window, just like a new bin, but with a "script" icon next to it (see Figure 1.75).

Open the script from the Project folder (double-click), and it opens the Script window (see Figure 1.76), which basically looks like a text document with a series of icons along the top. To prepare the script for editing from this point, you need to have the clips that cover the script in a bin.

> If you're going to follow along with the tutorial material, import or AMA the *Avid Uncut* clips: Uncut the Movie WS, Steve and Dave 2 shot, Steve XCU, Steve MS, Dave Jukebox PU, and Dave CU. These files are on the companion website (www.focalpress.com/cw/hullfish).

Watch each clip and note the starting and ending lines of dialog in the take. In the tutorial, the WS, two shot, and two shot cover the entire

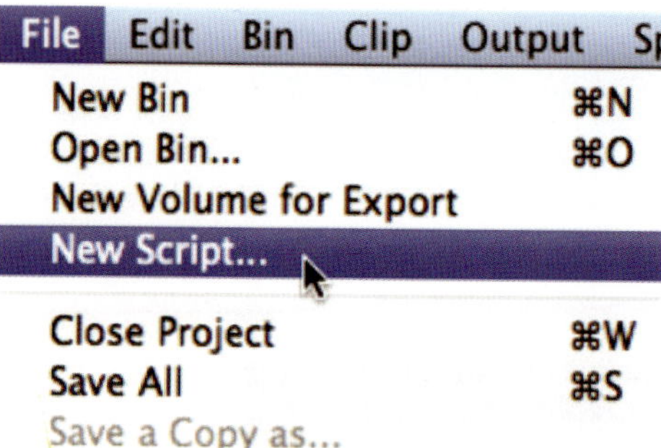

Fig. 1.74 Scripts are imported from the New Script option in the File menu.

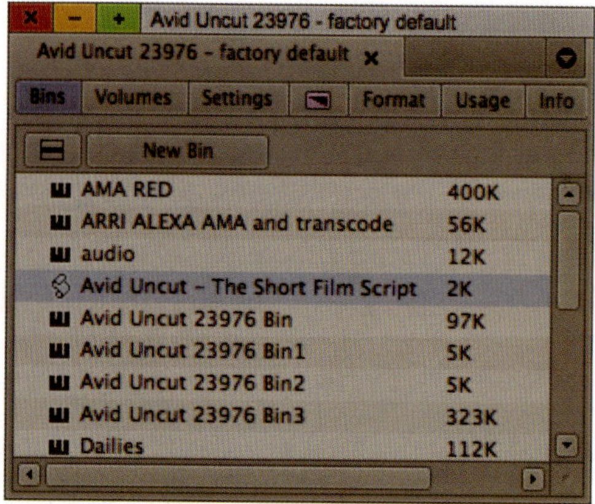

Fig. 1.75 This is how the script looks in the Project window.

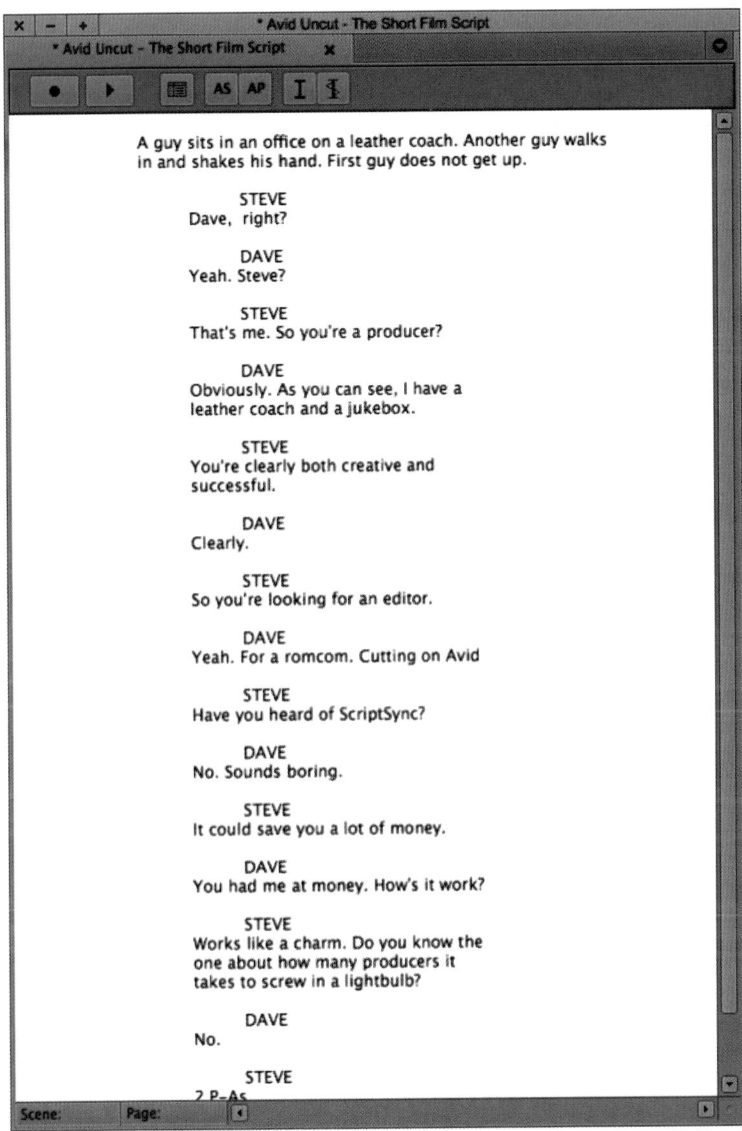

Fig. 1.76 This is what the script looks like when opened in Media Composer.

script. To link a clip to specific lines of the script, open the script from the Project window and click and drag or Shift-select the lines of script that are included in the take. With the lines selected on the script (indicated in gray; see Figure 1.77), drag the clip from the bin or Source monitor and drop it on the selected lines (see Figure 1.78). For each take, select the correct lines (indicated in gray) and drag the take from the bin to the selected lines (see Figures 1.79 and 1.80).

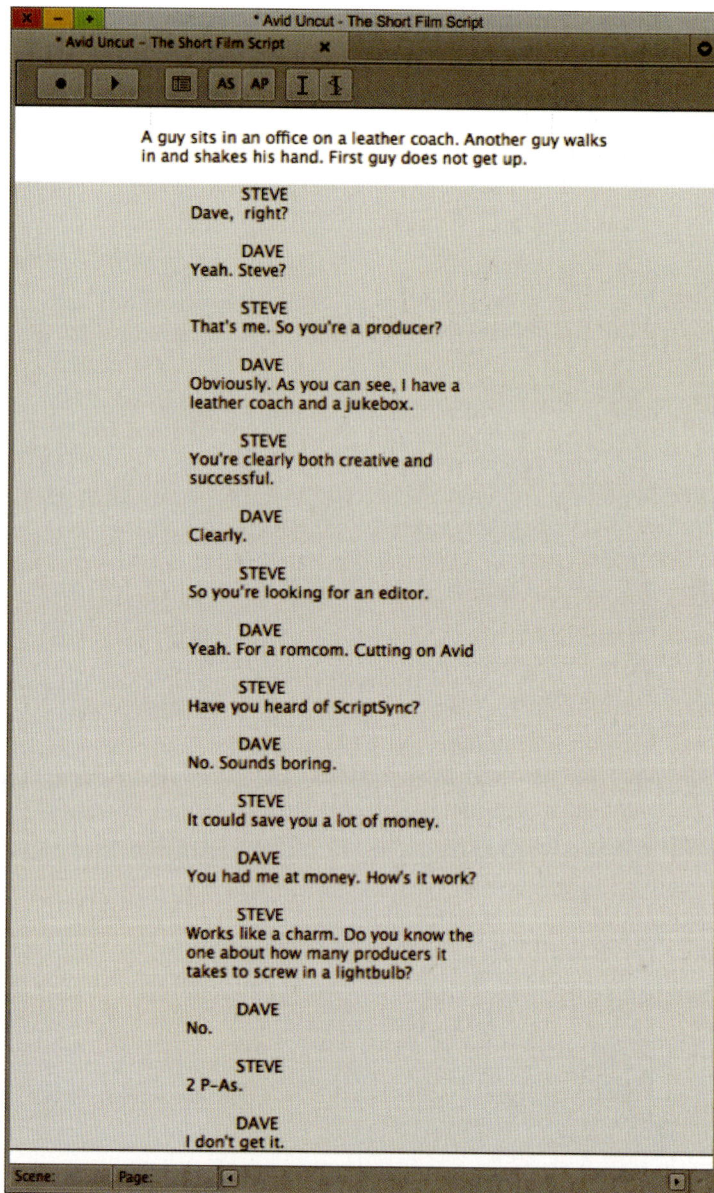

Fig. 1.77 Note the gray color of most of the script. The wide shot (WS) covers all of the lines in the script, so I clicked the first line and Shift-clicked the last line, selecting them all.

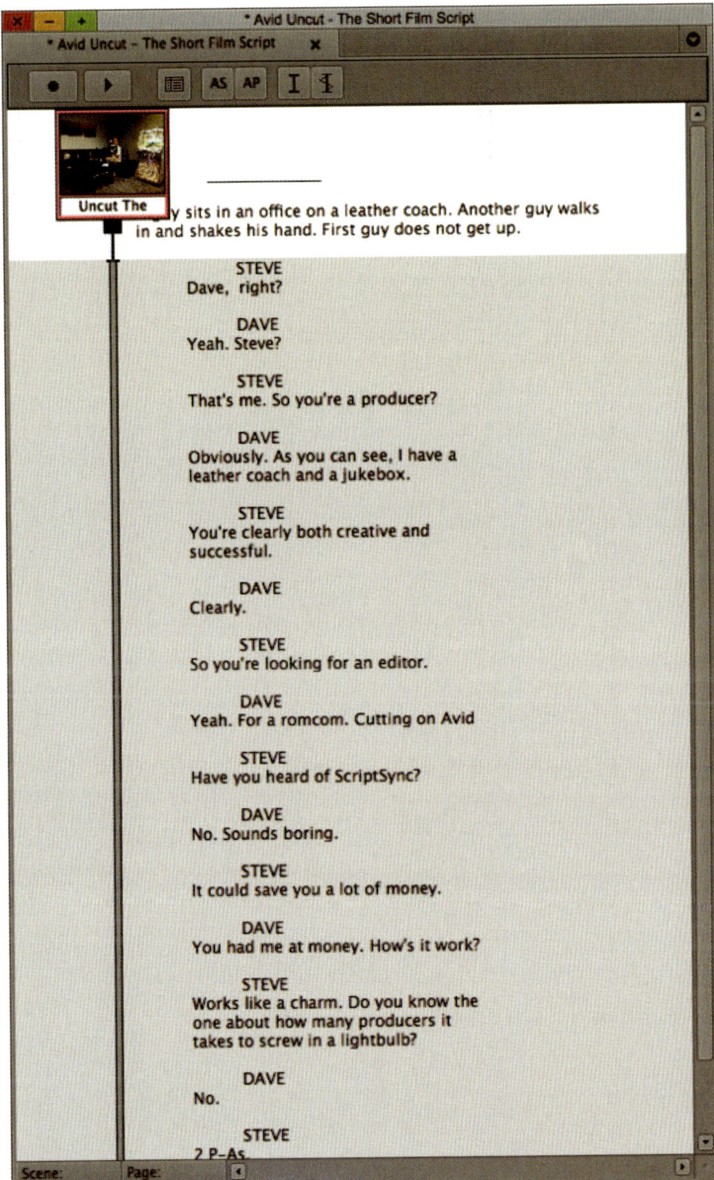

Fig. 1.78 Dragging the WS from the bin to the gray area, the selected lines, links the shot to those lines. The two shot covers those same lines, so the clip was also dragged onto these same lines.

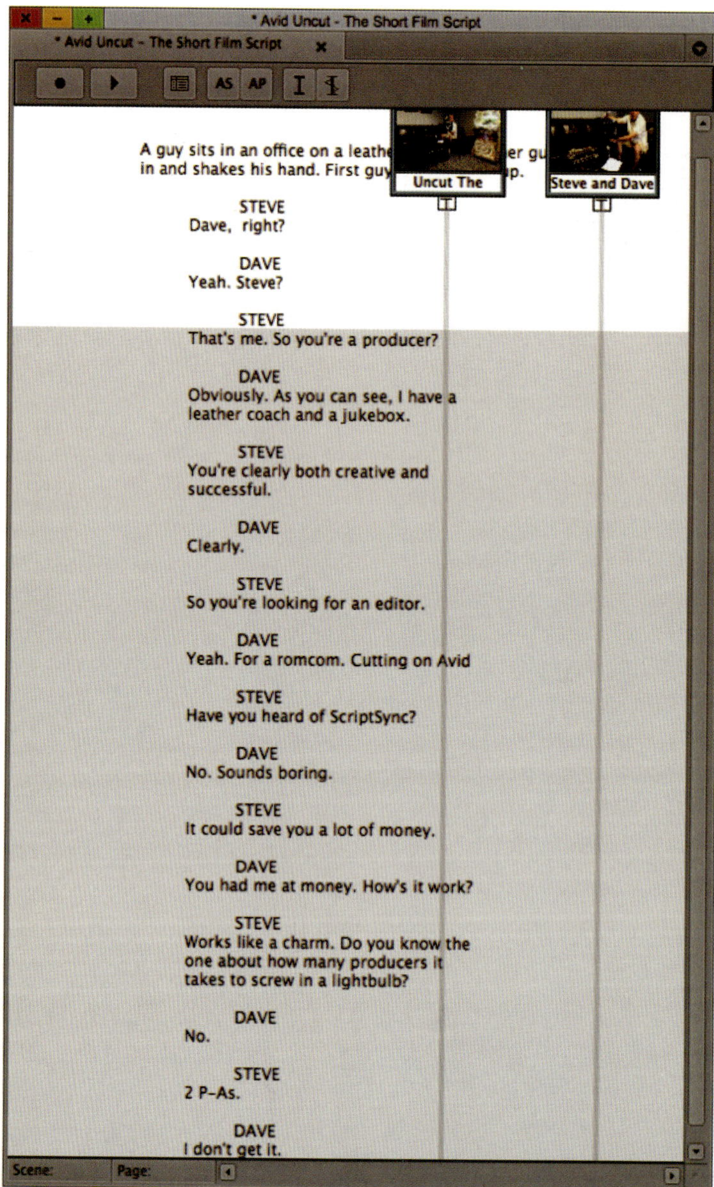

Fig. 1.79 The "Steve XCU" shot starts on "So you're a producer?" and continues to the end of the page, so a different set of lines was selected and that shot was dragged onto them.

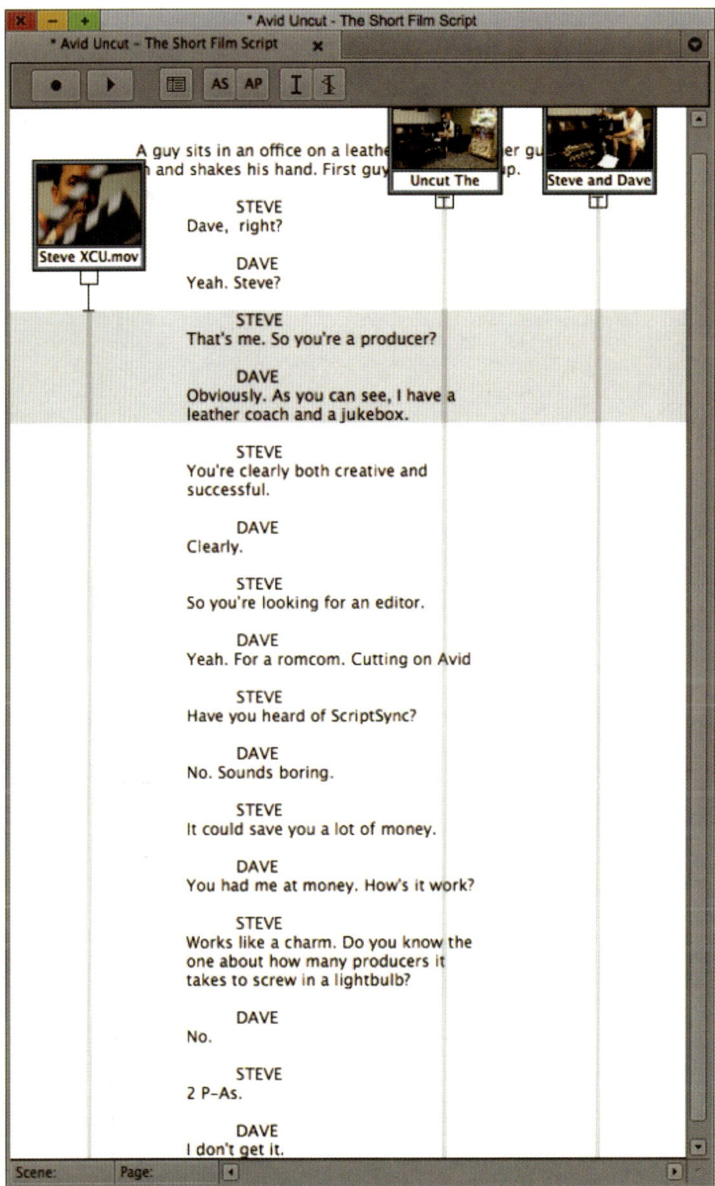

Fig. 1.80 The "Dave Jukebox" shot starts at "So you're a producer?" and continues though "leather coach and a jukebox," so only those lines were selected before that clip was dragged onto the selected lines.

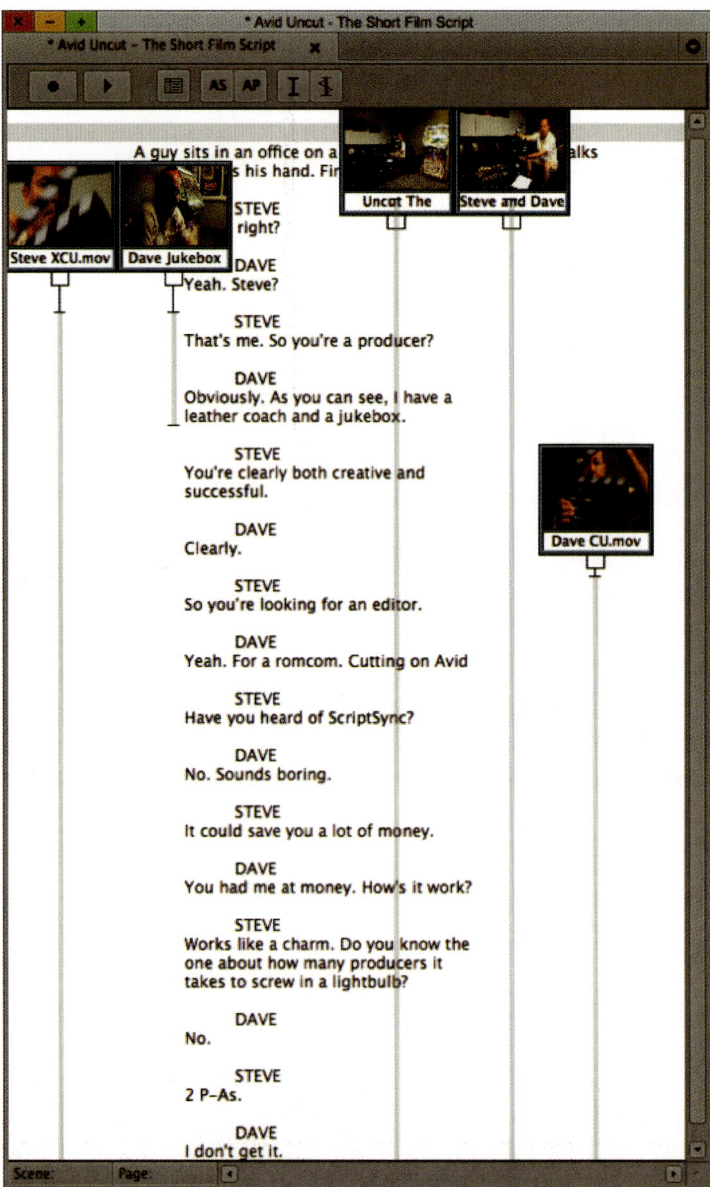

Fig. 1.81 This is the script with all takes linked to the correct lines. Note the gray vertical lines that extend down from the thumbnails showing which shots provide coverage for each line.

When you have an entire scene linked to the coverage, Shift-select all of the clips in the script (see Figure 1.81) and, from the main Media Composer menus, select Script > ScriptSync (see Figure 1.82). This uses the phonetics of the audio to match with the script in the text. Note that there are a bunch of options in this dialog window (see Figure 1.83). You can choose a language—probably North American English, but you have six other choices, or as many as thirty-one if you buy them in a ScriptSync language pack. Choose the audio track or tracks that have the best quality dialog recording, Skip lines that have capital letters (so ScriptSync doesn't try to sync to your character names), Skip text in parentheses (so ScriptSync doesn't try to sync to the "wrylies," or parenthetical direction), and Skip lines indented less than dialog. This last option allows you to analyze the text document and figure out how far in the lines of dialog are indented. This will help ScriptSync ignore the scene description. With these options engaged, choose OK to start the syncing process, which will happen quite quickly.

When the process is complete, you'll see that each line of dialog now has a small double-facing arrow along each line that extends down from the clip thumbnail (see Figure 1.84). This indicates that ScriptSync has matched the line of script with the audio in the clip. Now it is possible to jump to an exact line of dialog in any of the clips by simply clicking at the juncture where the descending line from the thumbnail meets the line of dialog in the script.

TIP

And to have your mind thoroughly blown, try this: Click on every vertical line that goes through a given line, clicking on the little sideways-facing triangles, but instead of double-clicking to hear that one line, select ALL of the lines all the way across a line of dialog by Shift-clicking. Then hit the Play button at the top of the script page. What happens? It auditions EVERY take of that line back to back to back in a loop! Very, very cool! (Thanks for this tip, David Helfand!)

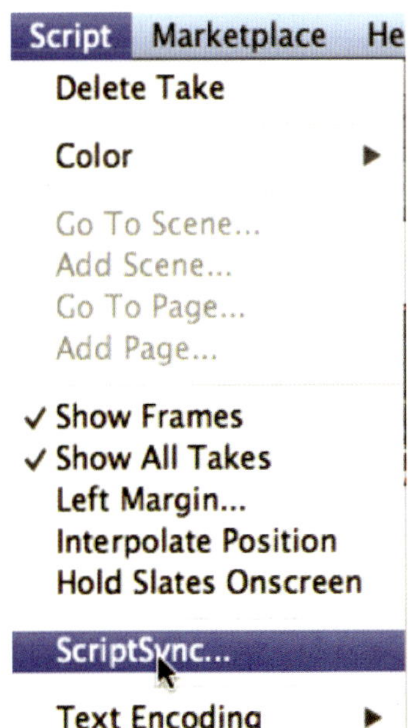

Fig. 1.82 To phonetically link the words in the script to the audio spoken in the clips, use the ScriptSync option in the Script menu.

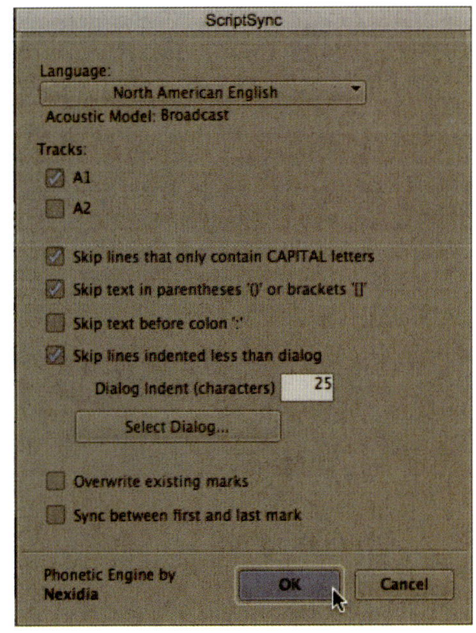

Fig. 1.83 After choosing ScriptSync, the ScriptSync window appears providing several options and settings for getting the best sync between script and audio.

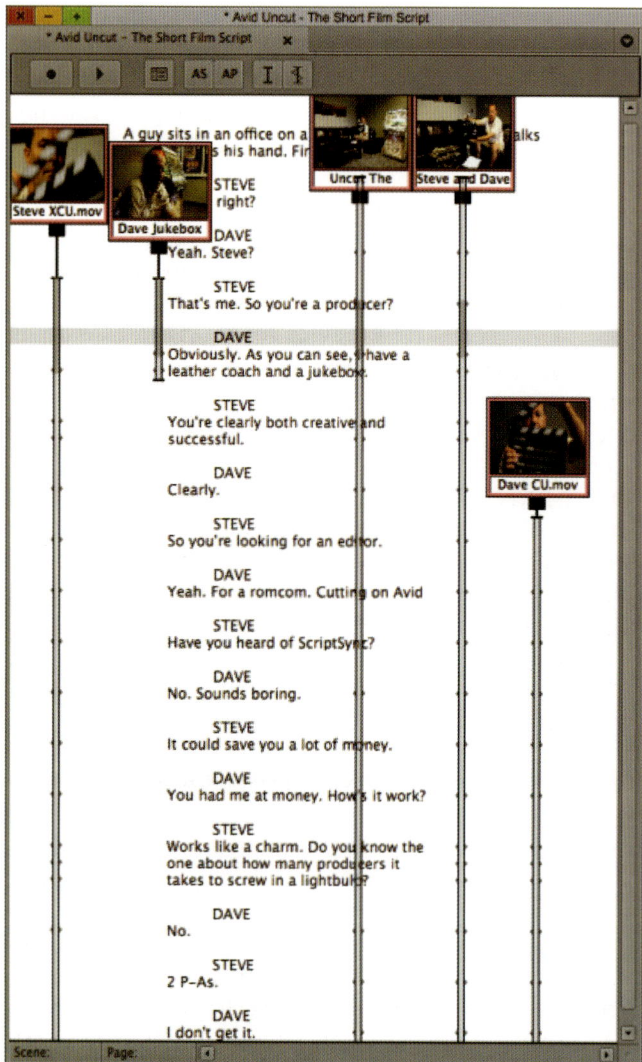

Fig. 1.84 This is a synced script. The small outward facing "carrots" or arrows studding the vertical lines of the takes indicate that the line has been synced.

Using the ScriptSync Script to Edit

The previous steps are usually done by an assistant editor. Here's where the editor steps in to utilize all of that hard work. By clicking across on each descending line at a specific dialog line, you can quickly preview (and load in the monitor) each covering clip for that line, loading the clip and marking an in point automatically, at that exact line of dialog. This is even fairly accurate if the actors slightly modify or ad lib lines, though if extensive ad libbing was done, the script supervisor should have rewritten the script to include the ad libs. But, even with the tutorial materials, you'll see that some lines were not read exactly as written and ScriptSync still syncs them up with the script fairly well. Earlier in this chapter, and later in chapter 3, David Helfand describes how he deals with improvisation and adlibbing in his use of ScriptSync.

Obviously, you can adjust the in point that is automatically set, either before or after you edit the clip into the timeline. Setting the out point needs to be done manually, as many editors prefer to edit with "sloppy tails."

GLOSSARY

Sloppy Tails: An edit without a predetermined outpoint. This allows the end of the scene to be determined more organically and intuitively.

Using the script like a bin, the edit continues until the scene or scenes are cut together. At any time, you can close the script window and continue editing normally using bins and the timeline.

Using ScriptSync with a Nonscripted Shoot

One of the main proponents of using ScriptSync for nonscripted (documentary-style) editing is Steve Audette. Audette has been using this method since ScriptSync was introduced. While this method has definite advantages, it has also been supplanted somewhat by the introduction of the PhraseFind option, which allows Avid to do many of the same things without the need for transcription.

I was one of the first people to "steal" Audette's technique and have used it on many productions since the beta of ScriptSync came out, but Audette has refined the use of ScriptSync for this purpose, and his methods really make a difference in the usefulness of the tool.

The main use of ScriptSync in nonscripted editing is to locate specific pieces of sound bites in interview footage that has been transcribed. For the example below, I used the transcription of JC Bond's interview with me. It should be noted that the interview was conducted over the phone and recorded with a ZOOM recorder sitting next to a speakerphone. Mr. Bond has a lovely accent and neither the accent nor the horrible speakerphone quality was any problem for Avid. And, actually, the transcription wasn't perfect in many places. Still ScriptSync accurately placed 99.5 percent of the lines to the syllable. To complete my ScriptSync of the Bond interview, I followed Steve Audette's time-tested recipe:

Instead of importing a script from FinalDraft, I took the Word document (see Figure 1.85) of the transcript and prepared it in a very specific way. The preparation of that text file is the trick that Audette discovered. Because Audette's use is not the *intended* use of ScriptSync, there are several issues with simply importing a raw transcript into the Avid for use with ScriptSync. The main one is that because of the way Hollywood scripts are formatted each line of dialog is fairly short, so each

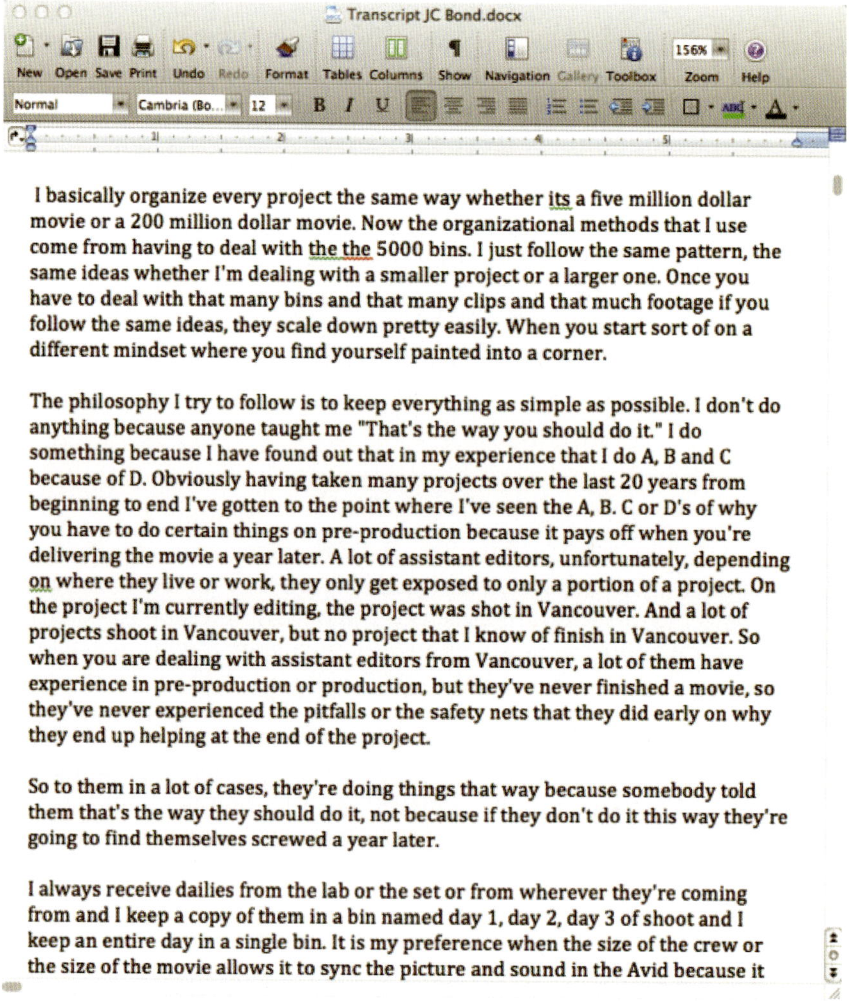

Transcript JC Bond.docx

I basically organize every project the same way whether its a five million dollar movie or a 200 million dollar movie. Now the organizational methods that I use come from having to deal with the the 5000 bins. I just follow the same pattern, the same ideas whether I'm dealing with a smaller project or a larger one. Once you have to deal with that many bins and that many clips and that much footage if you follow the same ideas, they scale down pretty easily. When you start sort of on a different mindset where you find yourself painted into a corner.

The philosophy I try to follow is to keep everything as simple as possible. I don't do anything because anyone taught me "That's the way you should do it." I do something because I have found out that in my experience that I do A, B and C because of D. Obviously having taken many projects over the last 20 years from beginning to end I've gotten to the point where I've seen the A, B. C or D's of why you have to do certain things on pre-production because it pays off when you're delivering the movie a year later. A lot of assistant editors, unfortunately, depending on where they live or work, they only get exposed to only a portion of a project. On the project I'm currently editing, the project was shot in Vancouver. And a lot of projects shoot in Vancouver, but no project that I know of finish in Vancouver. So when you are dealing with assistant editors from Vancouver, a lot of them have experience in pre-production or production, but they've never finished a movie, so they've never experienced the pitfalls or the safety nets that they did early on why they end up helping at the end of the project.

So to them in a lot of cases, they're doing things that way because somebody told them that's the way they should do it, not because if they don't do it this way they're going to find themselves screwed a year later.

I always receive dailies from the lab or the set or from wherever they're coming from and I keep a copy of them in a bin named day 1, day 2, day 3 of shoot and I keep an entire day in a single bin. It is my preference when the size of the crew or the size of the movie allows it to sync the picture and sound in the Avid because it

Fig. 1.85 Unformatted Word document of a transcript.

line of script includes only about six words at the most (see Figure 1.73). ScriptSync can only "sync" with the granularity of a single line. But with unformatted transcripts, a line of dialog might be dozens or hundreds of words long. To make searches on words more specific, Audette takes the transcripts and formats them with very narrow margins—about three inches, and a monospaced font, like Courier (see Figure 1.86). Then he exports them (or "Saves As" in some text editors) as Text Only with line breaks (see Figure 1.87). This means that, when the transcript is imported into Avid as a script, the accuracy of finding specific words quickly is

GLOSSARY

Granularity: an adjective describing the smallest unit of size measured.

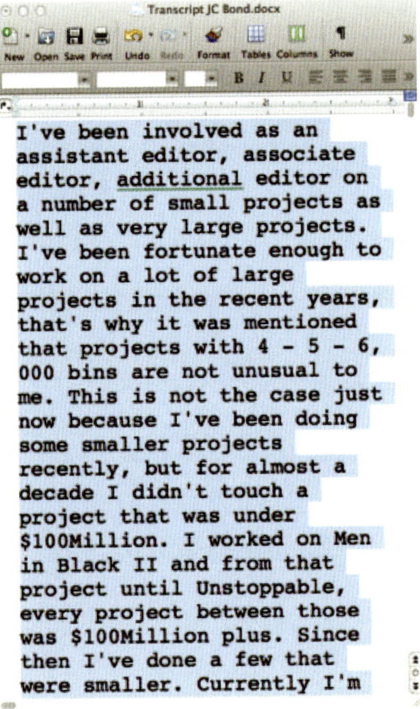

Fig. 1.86 Transcript with narrowed margins and a monospace font, like Courier.

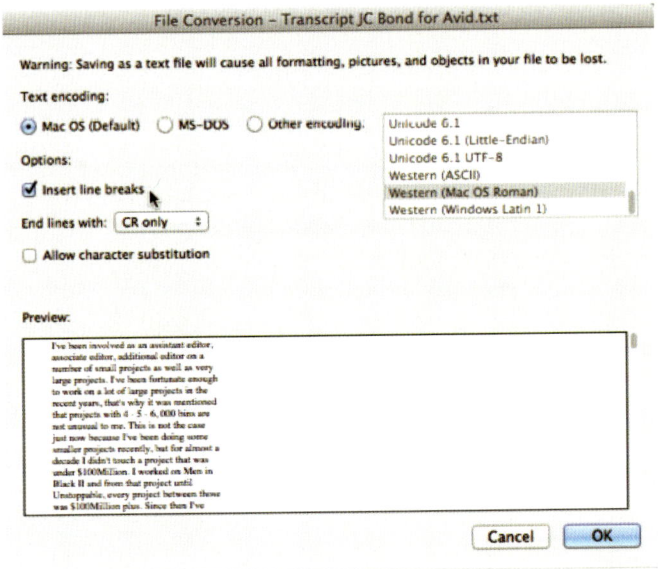

File Conversion – Transcript JC Bond for Avid.txt

Warning: Saving as a text file will cause all formatting, pictures, and objects in your file to be lost.

Text encoding:

Mac OS (Default) MS-DOS Other encoding:

Unicode 6.1
Unicode 6.1 (Little-Endian)
Unicode 6.1 UTF-8
Western (ASCII)
Western (Mac OS Roman)
Western (Windows Latin 1)

Options:

Insert line breaks

End lines with: CR only

Allow character substitution

Preview:

Cancel OK

Fig. 1.87 Saving the document as plain text (.txt) *with* line breaks is critical.

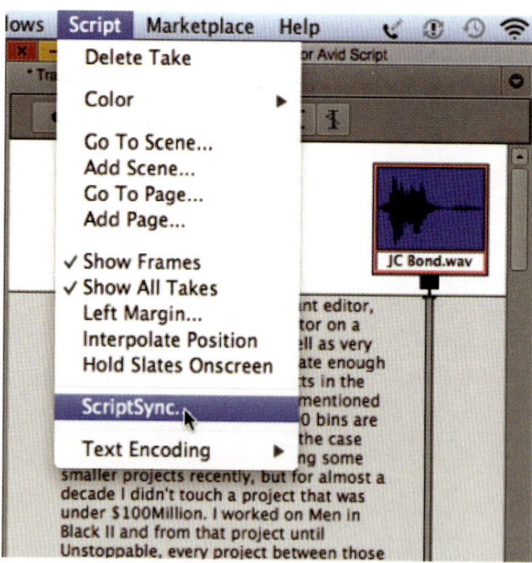

Fig. 1.88 From the Script pulldown menu, select the ScriptSync option.

greatly improved. (I don't find the need for changing the font necessary, but Audette has done this a lot, so we'll go with his sage advice.)

That text file is imported in the same way as the text file of the script (described in the section, "Preparing to Edit with ScriptSync"). Go to File > New Script and browse to the location of the specially prepared .txt file of the transcript. Open it. Then select all of the lines of the interview in the imported script and drag the interview clip from a bin or the Source monitor onto the selected lines. For me, there was a lot of "slop" before and after the interview, and I found that subclipping the interview clip so that it started and ended where the transcript started and ended helped a lot, especially at the beginning and ending of the ScriptSync. With the long, vertical line descending from the thumbnail in the transcript selected, choose ScriptSync from the Script pulldown menu (Figure 1.88). The interview, which was almost an hour long, was synced in less than thirty seconds (Figure 1.89). It is possible to hand-sync scripts and transcripts, but you can see that you don't have to do this too many times to have ScriptSync pay for itself in time saved.

Using this imported transcription for editing is useful if your producer isn't very good about providing specific timecodes for the lines of dialog in a "paper cut." If this is the case, you can locate the words in the producer's "paper cut" by searching the imported transcriptions. You can do this search with Command-F (Mac) or Control-F (PC). Note in Figure 1.90 that the search can be done on a *selected* transcript (script) or on *all* scripts. If the transcript was prepared with tightened margins, then

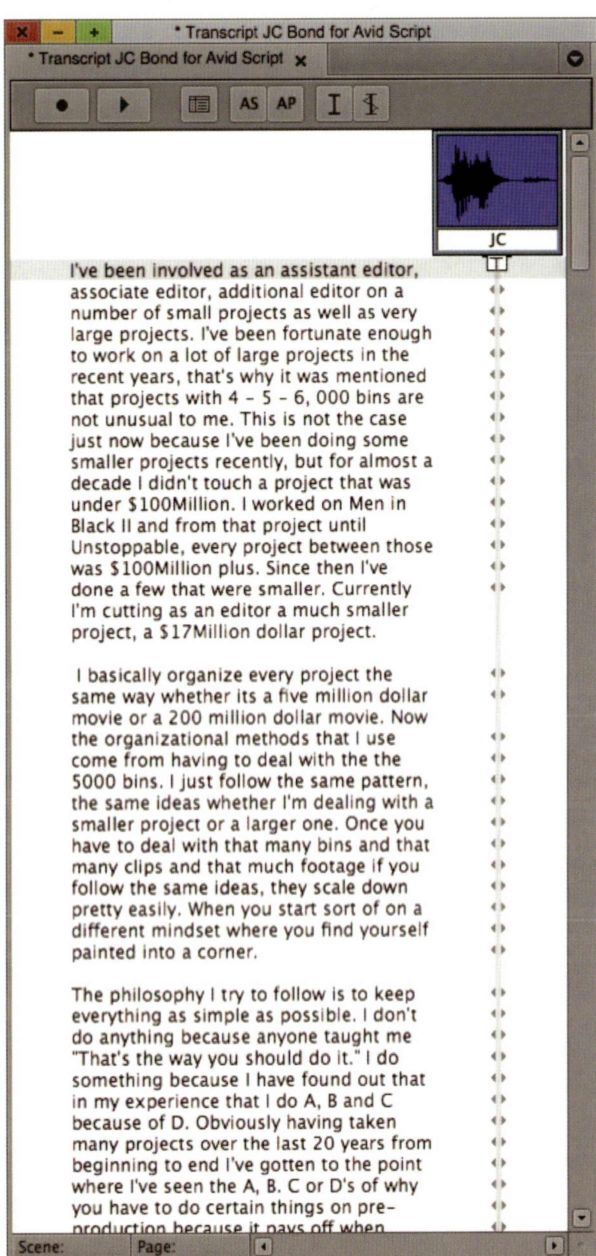

Fig. 1.89 After ScriptSync, each line will have a small indicator of two outward-facing triangles at each line of dialog along the take's vertical line. Each of these indicators shows that the line has been synced successfully. Note that there are a few lines that don't have indicators on them. This was not Avid's fault. These were lines that I did not transcribe word for word.

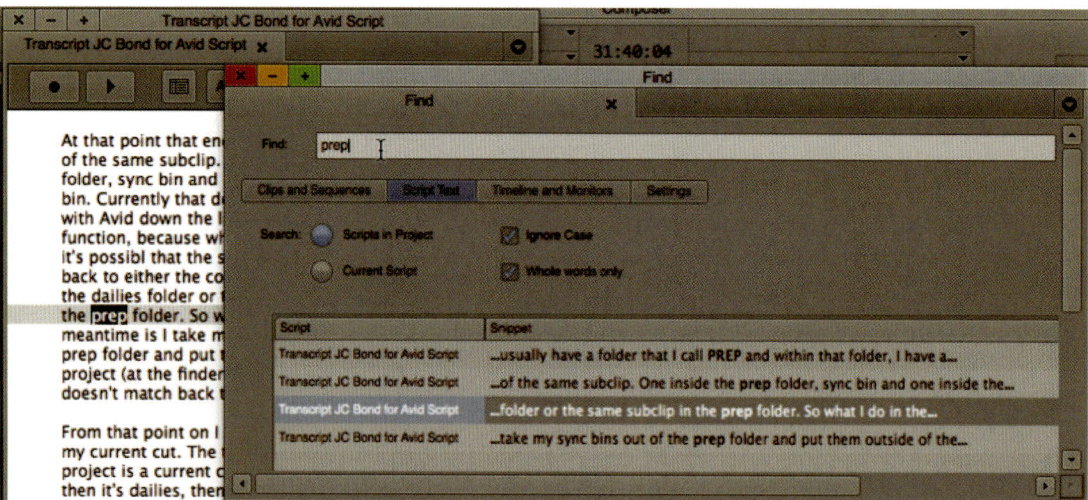

Fig. 1.90 With ScriptSync, the Find function in Avid allows you to search scripts. In this example, I searched for the word *prep*. There were four successful hits of this word in the transcript.

the "find" on the transcript will locate your bite within just a few words of your search. You can see that searching the JC Bond transcript for the word *prep* yielded several results. If I click on them, they would take me straight to the line in the transcript, but by clicking on the line in the transcript, it would take me directly to the proper line in the clip itself. Clearly, this is of enormous benefit to nonscripted editing.

But, for Audette and me, the real benefit of being able to search on these transcripts and immediately jump to the possible resulting words is when you need to "Frankenbite" a sound bite. Normally this is done because, in the sound bite from the interview, the speaker may not actually finish a thought at the same point that the producer or writer wants him or her to finish the thought, so the voice inflection may not be correct. ScriptSync allows you to search for that same word or even syllable throughout the rest of the script, hopefully finding another reading of the same word or syllable with the correct inflection. One of the easiest ways to do this is to do a search on the final word, including a period after it in the search window. If your transcriptionist was paying attention to punctuation, that should be all you need to cut in a version of the word that "buttons up" the thought. Because you are editing in only the audio of a single word, the lip-sync will probably be perfect and unnoticeable if done correctly.

> At the time of publication of this book, Audette had a Vimeo clip describing his process: http://vimeo.com/17502817.

Editing Proxy/Low-Res/Up-Rez to Final Quality

One of the important things for Avid is to maintain the ability to relink between various transcodes, the off-line quality footage, and the on-line quality footage. To do this, there are a few important caveats.

Make sure that the when you create your project, you match your project frame rate to the frame rate of your camera original footage.

You'll need to create some custom columns in the bins and duplicate information between columns. The clip name starts as the name of the camera file. Once you've copied this information to a few other locations, this can be renamed to something descriptive, like a scene and take number. Unlike FCP, the name of the file is not sacrosanct. When you transcode the files, the Drive will change, so you will want to duplicate the data in the Drive column of the original footage and paste it into a custom column, possibly named Original Footage Drive. See if the Tape ID column is populated with data. If it is, leave it alone. If not, paste the Drive column data here. The Take column is one of the places where you should copy and paste the name data. Camroll can be used to include a barcode or some asset or library information to the asset.

One of the problems with describing workflows is that there are a ton of them based on the camera, the dailies methodology, the media used during off-line, the system used for color correction, the media used for color correction, and the system used for finishing. Even if you had only two choices for each of the above scenarios, I think that would mean detailing sixty-four different workflows.

Overall AAD and Transfer Tips

If you link a sequence back to the AMA original sources, refresh the sequence before you transcode (Clip > Refresh Sequence > All).

When relinking to camera original media or when moving this media between drives, it's sometimes useful to have a third-party program installed that allows for fast, indexed file searches. These programs can provide quick access to the full finder-level, folder, or directory paths for files that you're looking for.

With linked AMA sources where you want to just create an AAF that just links to the media without transcoding, enable AAF Edit Protocol in the AAF export settings in Avid.

When exporting in Resolve 9, in the Delivery tab, under the Output Options, select "Render each clip with a unique file name" so that Resolve doesn't overwrite clips in instances where multiple clips are used from the same master clip.

A RED Preproduction Workflow

Let's take an example: A RED Epic, dailies in RED CINE-X PRO, DNxHD36 media for off-line, sent to Resolve, grading using the original R3D files, with the finish back on Media Composer.

Using the above choices, here's the short and sweet methodology.

REDCINE-X can be used to create DNxHD 36 (off-line or proxy resolution) MXF files (see Figure 1.91).

> If you want to attempt this method, the companion website (www.focalpress.com/cw/hullfish) has a folder of RED footage provided by Michael Cioni of Light Iron. Download that footage and then download REDCINE-X from the RED website (www.red.com/products/redcine-x).

Launch REDCINE-X PRO, and navigate in the Browser (top left) to the .RDM level of the folder with the Light Iron footage. Thumbnails of each shot will appear in the Browser, to the right of the folder you chose. Drag them to the Project area at the bottom of the REDCINE-X window. Then choose File > Export from the REDCINE-X menu bar and set your Export

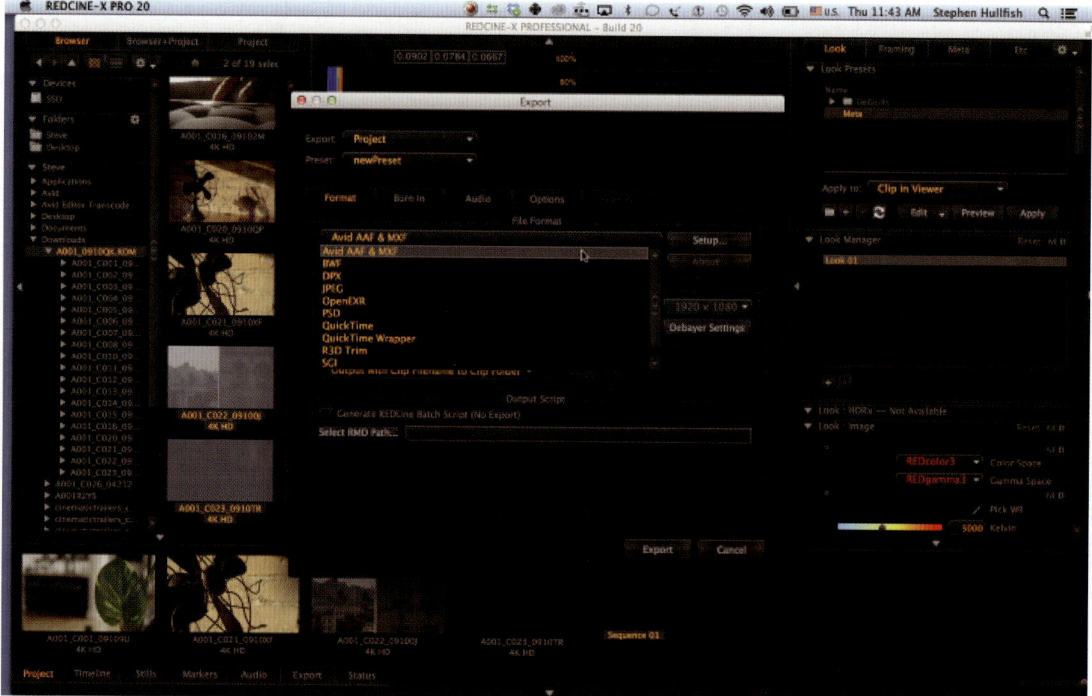

Fig. 1.91 This is the REDCINE-X UI (circa summer 2013). It is not an Avid product and can be downloaded free from RED.

Format settings to Avid AAF & MXF (see Figure 1.92). Then, to the right of that pulldown window is a Setup button. Click on that and choose a specific frame rate, resolution (1080/720/etc.) and DNxHD resolution (see Figure 1.93). When you choose Export, it will ask you to specify or create a new folder for the export to go in to. I suggest creating a new folder and naming it for the resolution you created and the date it was done.

When this is done, each batch of footage ends up as a folder full of MXF files, a folder full of AAF files, and a single ALE file. The number of MFX files will equal the number of clips that were shot times the number of tracks of audio that were synced to the video tracks. Drag all of these files

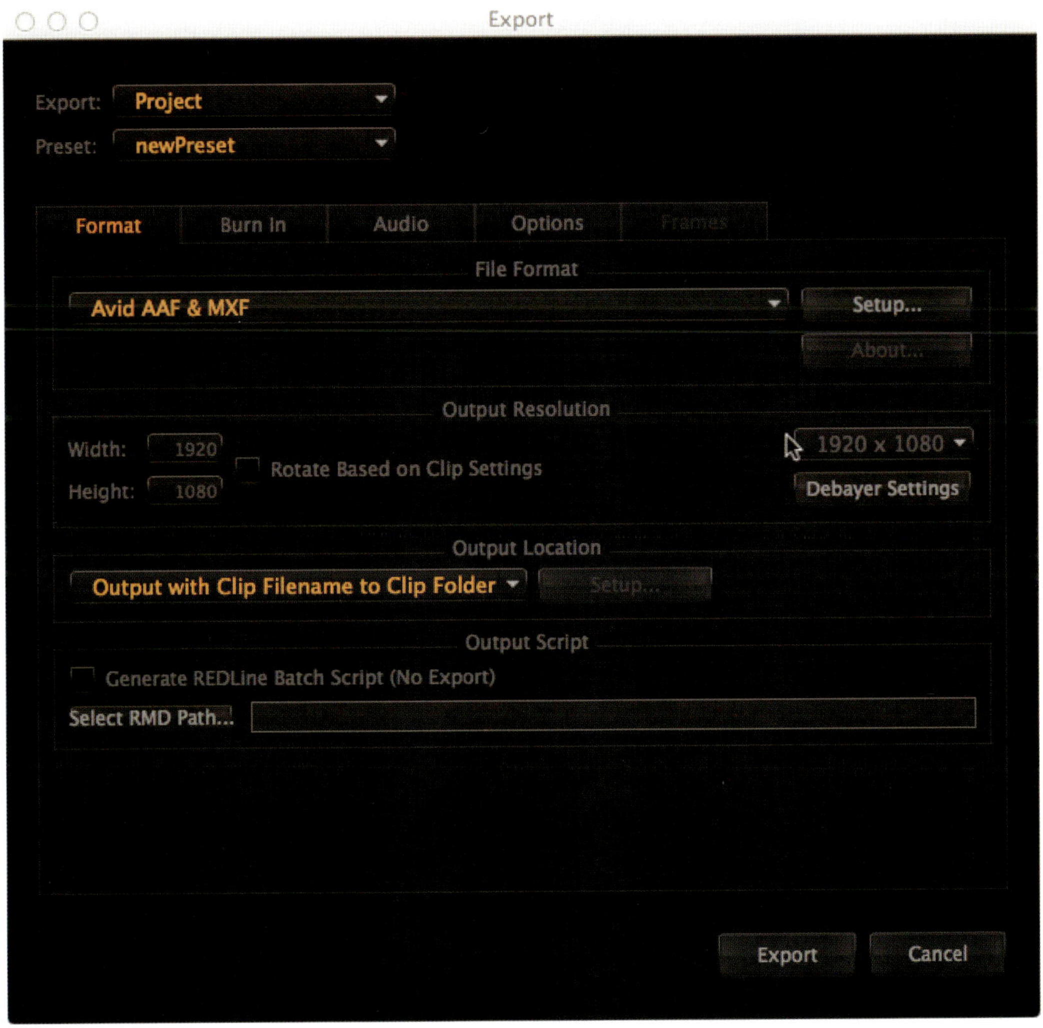

Export
Export: Project
Preset: newPreset
Format | Burn In | Audio | Options | Frames
File Format
Avid AAF & MXF | Setup... | About...
Output Resolution
Width: 1920 | Height: 1080 | Rotate Based on Clip Settings | 1920 x 1080 | Debayer Settings
Output Location
Output with Clip Filename to Clip Folder | Setup...
Output Script
Generate REDLine Batch Script (No Export)
Select RMD Path...
Export | Cancel

Fig. 1.92 The Export window of RED CINE-X. Note that the Avid AAF & MXF option has been chosen under Format.

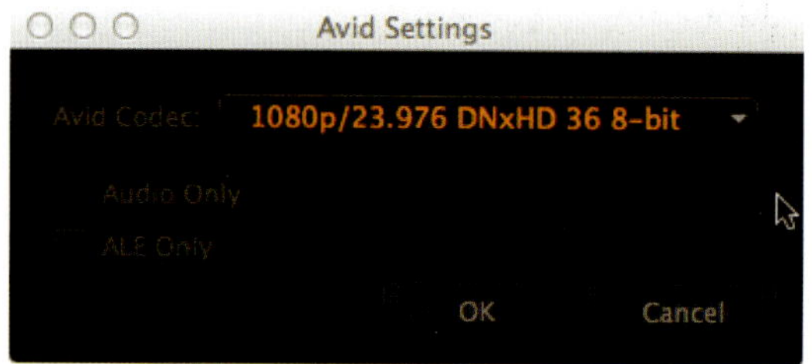

Fig. 1.93 In Figure 1.92, because the Avid option was chosen, an additional dialog comes up to set the specific Avid codec and resolution.

onto your inside the Avid Mediafiles folder on your Avid Media storage. The path will probably be something like this: RAID or media drive > Avid MediaFiles folder > MXF folder > # folder. (The symbol # stands for a numbered file inside the MXF file. You can actually number these folders within the MXF folder with any numbering scheme you want, as long as it is just a number. Maybe use the number of the shoot day and card number.) Then, in Media Composer, import all of the AAF files from the AAF folder generated by the REDCINE-X. By importing the AAFs, the Avid associates them properly with the current project. It also immediately links to the MXF files that were just dragged into the Avid MediaFiles folder and it supports audio and video. Then import the ALE file using the "Merge based on known sources" option. Avid recommends importing AAF first and then merging that with the ALE as a second step. If you like to edit from a selects reel or want to create one for other members of the production team, select all of the clips in the bin and, from the menu bar, select Bin > Autosequence.

This methodology obviously precludes being able to use the FrameFlex and Color Management aspects of the Avid because REDCINE-X has already transcoded down to a straight HD file and applied its own color management.

If you would prefer to AMA the RED footage, the RED plug-ins obviously have to be installed. Then it's simply a matter of creating a new bin, using File > AMA Link, and pointing to the RMD folder level. This quickly displays all of the clips in the folder. You can then use the Color Management tools and FrameFlex within the Source Settings to manipulate the footage (or not) and then transcode it all to a more easy-to-use Avid DNxHD format or even compressed Standard Def for editing, depending on your storage requirements.

As JC Bond suggested, you can then sync your clips in Avid, if desired, then subclip every shot to allow for perf slipping sync (in film projects) and to enable any future decomposed footage to remain a subclip instead of becoming strictly a new master clip.

We'll follow this media through the editing and on-line process in chapters 3 and 5.

An ALEXA Preproduction Workflow

There are a number of ways ARRI Alexa footage can be recorded. The footage provided on the companion site of this book was recorded on the camera card as a simple QT ProRes movie. It can be downloaded and used to practice with.

> In the case of the *Avid Uncut* companion website project files, this is a folder called A001R2YS (www.focalpress.com/cw/hullfish).

In order to maintain the ability to use Avid's FrameFlex and Color Management capabilities, the files from the Alexa must be brought in via AMA directly. This is simple to accomplish in the case of the QT versions of the Alexa files. Start in Avid's menu bar at File > AMA Link and simply point at the level of the folder directly above the .MOV files.

You can transcode all of the files by selecting them and going to Clip > Consolidate/Transcode. That calls up the Consolidate/Transcode window. The rest of this has been covered several times throughout this chapter, but the basics include selecting a Drive and a resolution and clicking the Transcode button.

Similar to the above example with the RED footage, it is suggested to subclip the footage and to edit from the subclips. This is not necessary, but it helps in the case of film projects to do perf-slipping of sync and also to allow for maintaining links or relinking to the original off-line media after the footage has been consolidated or decomposed.

As with the RED footage, we'll follow this media through off-line and on-line in chapters 3 and 5.

Metadata in Action

Light Iron is one of several companies providing on-set dailies (among other services) to the TV and feature film production community. Light Iron's Michael Cioni relies heavily on a deep understanding of metadata to deliver efficiencies in the dailies and finishing on both sides of off-line. Michael describes the rich metadata that his company culls, curates, and follows through the entire post process: "We worked closely with Colorfront to invent a protocol called EMD, which is an Express Metadata File, and this is actually a process that RED initially conceived of, so we took the properties of the RMD that work really well and applied them to something even bigger." The EMD file was expanded from RMD so that it would work for every other camera—right down to GoPros—and includes archiving and sound. Cioni continues, "That EMD contains not only file information, but corresponding sound, timecode, timecode offsets, camera names, file names, Avid bins, Avid sync points, CDL information, LUT information, even down to camera magazine serial numbers. So it's this really fat file that once you plug anything in, this little EMD is able to start gathering information from every direction. So this becomes the key that holds everything together."

Cioni continues, "Then to get the EMD file into Avid, we spit out an ALE file and we give Avid the appropriate things that it wants. The imported ALE file links to the already transcoded, off-line quality DNxHD files and populates the bin columns of everything from Kelvin to tint to ISO to CDL values to sound offsets, sync points, slate points. All that stuff's inside the ALE."

To create the EMD files and to generate the off-line media for Avid, there are two portable computer systems that Light Iron packages into a convenient, portable cart that Light Iron rents to its clients. The first system is called Lily Pad (see Figure 1.94), which is used on-set, and the second is called Outpost (see Figure 1.95) and is in the offices where the off-line edit is being completed.

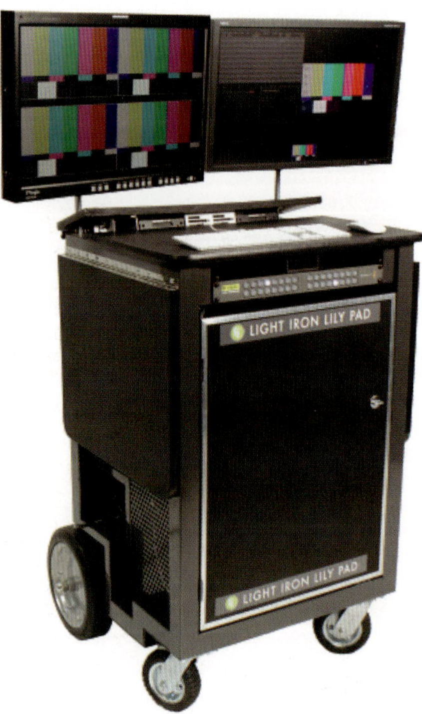

Fig. 1.94 Light Iron's Lily Pad on-set dailies cart.

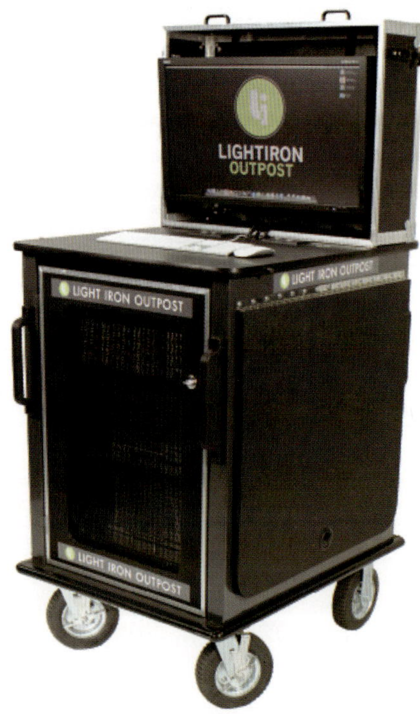

Fig. 1.95 Light Iron's OutPost "Mobile Processing Lab".

Michael describes one of the productions that uses their system: "So on *Criminal Minds*, the Lily Pad (see Figure 1.94) is on-set doing all those processes. When the files go to Outpost (see Figure 1.95) for the archiving and additional web uploads for CBS, the last thing Outpost does is it renders to Avid directly. It's connected through gigabit ethernet to the Unity and every camera magazine gets transcoded and loaded automatically onto the Unity and then the assistant editor gets a bin and they just start setting things up."

CHAPTER 2

Settings

Settings are easy to ignore as a video editor or as a book author. Either way, that would be a major mistake. Settings help editors keep from making mistakes and avoid unnecessary and repetitive work. The time you spend customizing settings is rewarded by making the Media Composer work with you instead of seemingly against you. Settings would be fairly easy to ignore in a book aimed at more advanced users, but even the most experienced editor has gaps in his or her knowledge, and if that gap includes missing out on exploiting a beneficial setting, it could be a costly gap. And even if you know how to do some of this stuff, use it as a prompt to finally getting around to setting up your Avid for peak efficiency. As Abraham Lincoln once supposedly said, "If I had six hours to chop down a tree, I'd spend four hours sharpening my ax." Time spent doing this will save you pain and effort for years to come.

For Final Cut Pro editors and editors from other NLE backgrounds, Avid settings are analogous to Preferences. Personally, I like Avid settings. At least, compared to Final Cut Pro, the Avid settings seem to consolidate almost all of the preferences into a single place. For me, when I was a rookie Final Cut Pro editor, there were too many places I had to check for preferences of how things worked. Many were in the Preferences pulldown menu, but there were a lot of other places where the various preferences or settings could be found. For Avid, there is really only a single list of preferences and that is the Settings tab of the Project window (see Figure 2.1).

For those who are used to finding these kinds of things in the Preferences menu, you can go to the Media Composer menu and pull down to Preferences, or use the keyboard shortcut, Command-comma for Mac or Control-comma for PC.

Project, User, and Site Settings

If you open the Settings tab, you can see three columns of information. The first column is the name of the Setting (e.g., AMA, Audio, Audio

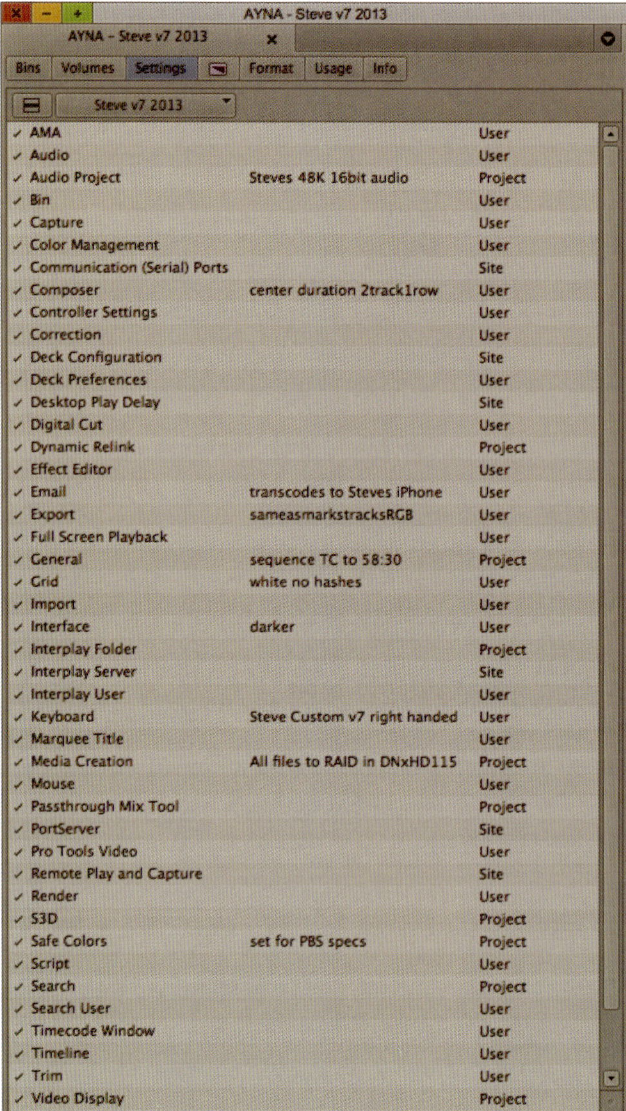

Fig. 2.1 Settings tab in the Project window.

Project). The second column allows for custom naming of customized Settings (more on that later). The third column is the type of setting.

There are three types of Avid Settings: Project, User, and Site. The reason for these three types of settings is that project settings are applied on a project basis, no matter which user is working on the project. Project settings are valuable so that all editors working on a specific project are using the same settings. A good example is that Audio Project settings are project based because it makes sense that all users of the project have the same audio sample rate. User settings are applied only to certain users, no matter which project is open. A typical user setting is Workspace Views. The Workspace View customizes the layout of tools in the Avid user interface (UI). For each editor, this is a very personal and unique thing, so it makes sense that each time a specific editor opens a project, the views are the same. The final type of settings is Site settings. These settings are unique because they are applied to all new projects and users on the site. *Site* is defined as a particular Avid Media Composer or Symphony system. Site settings are very useful to force certain behaviors across all projects and all users.

The kinds of things that can be useful as Site settings are things like Media Settings (the default resolution and where the media is recorded), Audio Settings, and Audio Project settings. A good example of this would be a postproduction shop or TV show that wants to have all projects and all users always import audio at 48K/16-bit. Normally, once the Audio Project setting is set to these preferences, they are set that way only for a single project. To conform to the house settings, an editor has to remember to set the Audio Settings to these preferences at the start of every project. But if this project setting is turned into a Site setting, then it forces all new projects to default to this house setting. If there is a specific reason to import audio at a different setting, it can still be changed, but it would have to be a conscious decision to change it.

Site settings can be created from user settings or project settings. Creating your own Site settings is easy. To force a specific settings option on all future projects or users, customize the setting to your preference (for example, 48K Audio on all projects). Then go to the Special menu and choose Site Settings. This opens a window showing the customized Site settings for your machine, or site. The first time you do this, the Site settings list will be empty. Now, just drag the setting you just customized from the settings list into the site window.

Some other uses of creating Site settings include saving Bin Views as Site settings. This is valuable because usually Bin Views are created by a specific editor for his or her own use, but in a group or shared editing environment, multiple editors may want access to those same views and information that can be accessed only from custom columns. The Media Creation setting is also a good candidate for a Site setting. Even though projects can be digitized in a number of resolutions for various reasons, it is possible that a specific resolution could be desired as the default. By setting that resolution in the Media Creation setting and then saving the setting as a Site setting, all future projects will default to that setting, though it can still be easily changed for specific reasons. I've worked at houses where the Media Creation setting defaults to something like DV25 even though almost all projects are HD. Often, in these cases, a lot of work is lost because hours of footage are captured or imported at DV25 by mistake when an HD resolution is desired. Creating an HD Media Creation setting can prevent these mistakes. Similarly, if a post house wants all of their editors to be able to import or export files in a standard way, import and export settings can be customized and saved as Site settings so that all editors have access to these presets. This helps to standardize import and export methods.

Settings Fast Menu

The Settings Fast Menu, sometimes called the "hamburger menu," is a way to filter the settings list, providing the editor with a view of only the settings he or she needs. There are a few options that you may find valuable: Active Settings, All Settings, and Base Settings. In addition, there are settings to filter for Bin Views, Export Settings, Import Settings, Timeline Views, Title Styles, Workspaces, and Workspace Linked (see Figure 2.2).

Most of these are pretty easy to understand. Active Settings is one of the most valuable. It simply shows all of the active settings. In other words, these are the settings that are having some effect at the moment. When you have the Active Settings selected, you'll notice that all of the settings in the Settings tab have check marks to the left. This check mark indicates that the settings are active.

All Settings shows every setting that has been created by default, in the project, by the current user, or has been set as a site setting on the machine. This can be a huge list, especially after a user has been editing for a while. But if you need to see every setting, this is the one to choose. One of the reasons that this list is so long is that each time a Bin View or

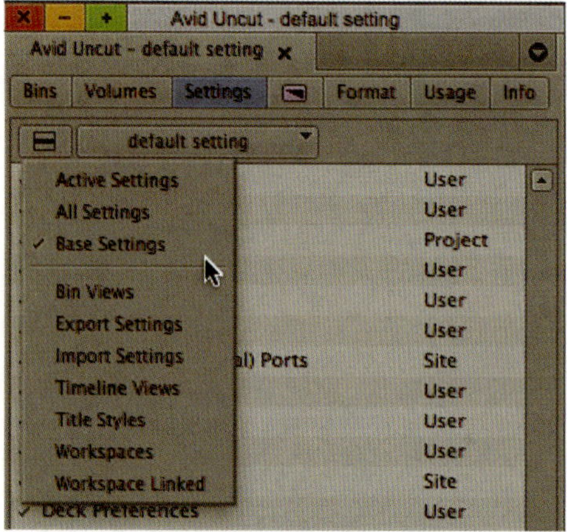

Fig. 2.2 Hamburger menu icon.

Timeline View is created, it is added to the list along with all of the many import and export settings, of which there can be dozens.

Base Settings is very similar to Active Settings, and in some cases, they may be nearly identical. The difference is that Base Settings includes Project, User, and Site settings, but not display views. The other thing (not mentioned in the Avid documentation) is that it also hides all of the multiple customized settings of the Base Settings. For example, if the import setting is not customized, it will show up in the Base Settings list, but if it is customized and given a customized name to describe the way it has been customized, then that customized import setting will no longer show up under Base Settings.

For me, for most purposes, I stick with Active Settings. And I rarely use the options that include Bin Views, Export Settings, Imports, Timeline Views, Title Styles, Workspaces, and Workspace Linked. But obviously, you may find it useful to limit the settings list to include only one of these groups.

TIP

Import settings and export settings seem to multiply like rabbits. First, there are many import and export settings that are preloaded "from the factory" for every user. Most of these, you'll never use. You may also accidentally create an import or export setting that doesn't work the way you want it to and then have to revise it. These settings don't really take up much space on the drive, but they create a lot of visual clutter in the settings list and when you open the Import Settings or Export Settings dialog windows. It is possible to delete these unwanted settings by simply clicking on the setting and hitting the delete key.

Important Settings

I often joke that I could write an entire book about just Avid Settings, so in order to leave space for something other than settings, I'm going to have to limit my tips and advice to a few specific aspects of the most important settings. Let's run through the Settings list in alphabetical order.

AMA has several settings (see Figure 2.3), but one that can have the greatest impact is the ability to turn off AMA Volume Mounting. The main reason for considering this would be that—for the many times you want to launch Media Composer and you're fairly certain you won't be needing AMA linking—it speeds up the launch of Media Composer

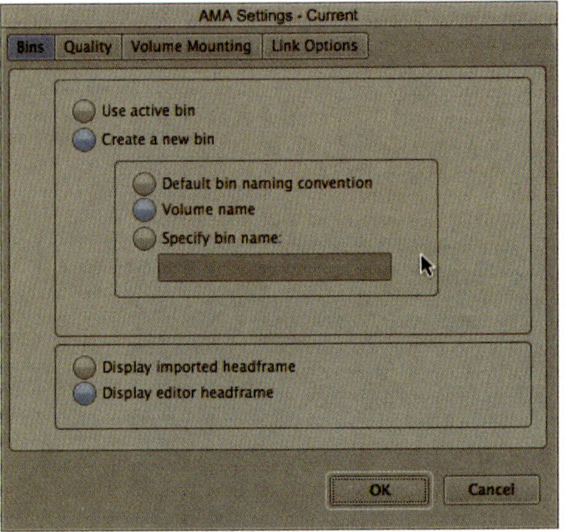

Fig. 2.3 AMA Settings

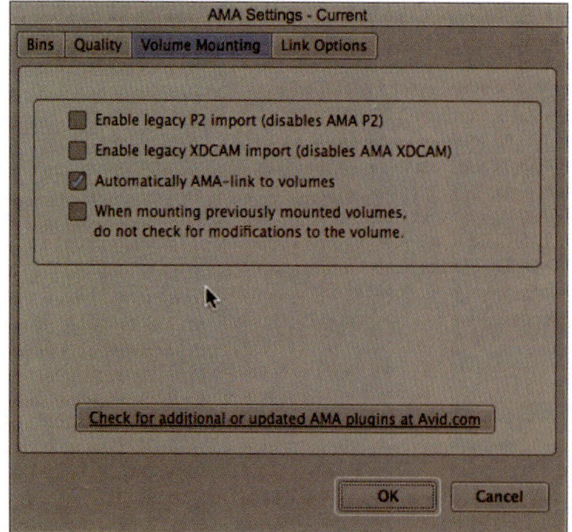

Fig. 2.4 Enable AMA Volume Mounting.

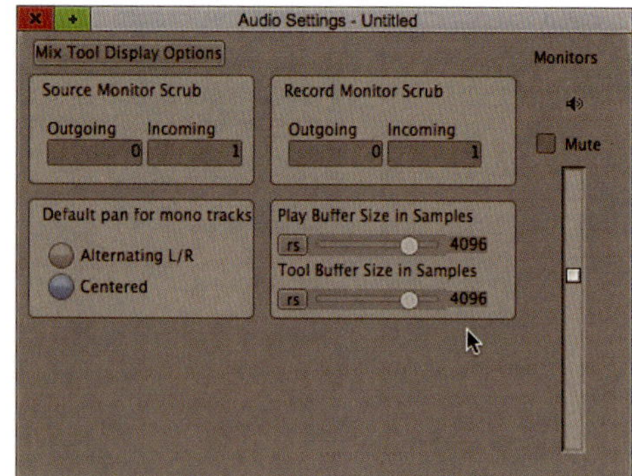

Fig. 2.5 Audio Settings

considerably. If you find you need to use AMA, simply recheck the option and relaunch. There are four tabs in the AMA Settings (see Figure 2.4). Select the Volume Mounting tab and uncheck the Enable AMA Volume Mounting check box.

Audio Setting: Only one selection is important. Do you like your audio channels to be panned left and right as a default or centered? Just remember that this setting is *only* about the default pan settings, and each track can easily changed in the Audio Mixer. The alternating pan works with odd numbers panned left and even numbers panned right. For me, music should usually be in stereo tracks anyway and, other than that, almost all of my audio from cameras or imports is mono, so I'd use centered instead of left/right (see Figure 2.5). Also, since MC7, there's a Mix Tool Display Options button that allows for the customization of multiple controls and displays in the Mixer Tool.

Audio Project Settings: The main settings in Audio Project Settings are Sample Rate and Bit Depth, which are in the Main tab (see Figure 2.6). Audio Sample Rate options are 32, 44, and 48. The latter, 48, is really the sample rate for professional digital video. That's what I prefer. Obviously 44 is the sample rate of CDs, and 32 is basically consumer video. You don't really save yourself much by choosing the lesser options. The Bit Depth options are 16-bit and 24-bit. The 24-bit audio does take a bit more space. I personally can't hear a difference between 16-bit and 24-bit, but for those who do, I'd go with that option. For me, I'm a 48kHz/16-bit guy. You can choose whatever you prefer, but I suggest making that choice and then pretty much sticking with it forever. An easy way to do that is to set the settings the way you want and then save the Audio Project Setting as a site setting, so that this will be the default for all future projects. (See "Project, User, and Site Settings" in this chapter.) Audio Settings also allows you to set input and output settings, including (if you have the

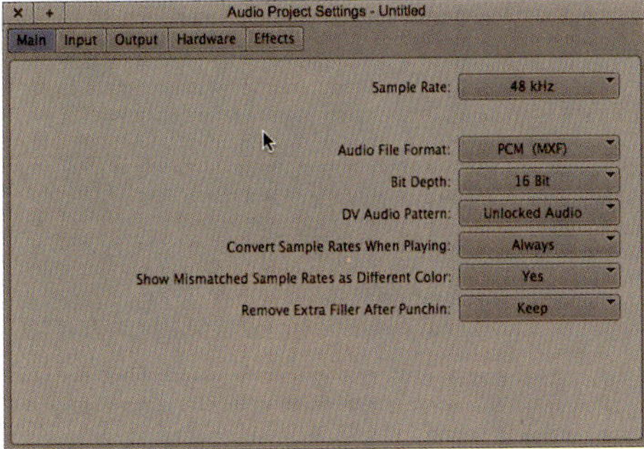

Fig. 2.6 Audio Project Settings.

hardware for it) 5.1/7.1 surround options. Additional tabs also let you access the audio hardware, which is often part of the video card, and the Output tab which allows you to set whether the output is mono, stereo, or several flavors of surround.

Bin Settings

Bin Settings are so in-depth and wide-ranging that they need their own heading in the book. We'll continue with the rest of the settings in the Settings list in a few pages. Bin settings are all about access to information. Having the proper bin settings lets you easily see what you want to see, making the right shot available to you at the right moment.

The bin settings in the Bin Settings list basically define how often things are saved. To me, this setting should be called the Attic Setting. The dangerous thing with this setting in shared environments is that it's possible for one person's Bin Settings to wipe out the Attic for other people on the same machine. If you have Maximum Bins in a project's Attic set for 1,000 and mine is set for 10, then every time I log in, Media Composer looks at the Attic and says, "Oh, there should be only 10 bins!" and it will delete the other 990. This is why it's very smart—especially in a shared environment—to make the Bin Settings a site setting. Site settings allow you to force all projects or users to use the same options for specific settings.

Auto-Save Interval: The interval will depend on how stable your edit system is and how fast you work. If you make ten or twenty edit decisions in five minutes, you might want to set this to five minutes. The default is fifteen minutes, which is where I leave mine. I figure I can always remember what I did in the last fifteen minutes. If you go longer than that, you have the potential to lose a lot of work. If your system is very "crash-y" (I'd suggest that something is very wrong), then you may want to have it set to a shorter interval. My systems are very stable, and I go days without a crash, so I could probably move mine to thirty minutes, but I like the insurance. The best rule of thumb is to decide how much work you are willing to lose, which varies from project to project, though personally I rarely alter this setting on a per-project basis. It also allows me safety to go back into the Attic for different versions, even without an actual crash. Sometimes you accidentally delete a sequence that you needed and you can go into the attic to recover it. If you go to a short interval, you'll be interrupted regularly by the autosaves of the bins. Media Composer will try not to interrupt, but the Inactivity period option and the Force AutoSave interval will define the peskiness of those interruptions.

> To recover bins from the Avid Attic, the easiest method is to simply search for the Attic using Sherlock on a Mac or Find on a PC. The Avid Attic is saved in a particular location on each computer.
>
> On a Mac:
>
> Macintosh HD/Users/Shared/Media Composer/Avid Attic folder
>
> On a PC:
>
> (Windows XP) drive:\Documents and Settings\All Users\Shared Documents\Media Composer\Avid Attic folder
> (Windows Vista or Windows 7) drive:/Users/Public/Public Documents/Media Composer/Avid Attic folder
>
> Inside the Avid Attic folder is a folder for each project and inside that folder are folders for each bin. Inside those folders are the individually saved Avid bins. There is one version of the bin for each time that the Avid automatically saves or for which you manually save. Each version simply has a number, but each file has a created or modified time that can help you determine which bin you want to recover. If your most recent bin is either corrupted or has something that was deleted or changed, you obviously want to go back to the most recent version that has what you're looking for.
>
> As an example, a common use of the Attic is to recover a sequence bin where you accidentally changed an edit without making a duplicate of your sequence in case, affecting the original cut. In that case, you have to guess which bin (at which saved time) will have the original version of the sequence that you're looking for.
>
> When you find that bin, the suggested procedure is to duplicate and move the bin out of the Attic and put it somewhere else, like the desktop, and possibly rename it. You cannot open an Attic version of a bin while the current version is open in your project, so close the current version of the bin in your project, then use the File > Open Bin option to navigate to the Attic version and open it. Check to see that this bin has the information or version that you want and copy the sequence—or other asset—from the Attic version of the bin to another bin. Then close the Attic version and continue working. If you want to open the original bin and move the preferred version of the sequence or other asset back into that bin, you can.

Inactivity Period: Media Composer tries not to autosave while you're working hard and have a flow going. The inactivity period waits until you haven't hit the keyboard or moved the mouse for a while before saving. The period defaults to fifteen seconds. That's where I leave mine. But if you're feeling like you're being interrupted by auto-saves, consider changing this. But making it longer might actually cause you more interruptions because eventually the auto-save is forced, despite the inactivity period.

Force Auto-Save At: If you continuously edit longer than the regular autosave interval, eventually you want the Avid system to autosave. The Force Auto-Save interval defines the maximum time between saves, whether or not you are actively pushing buttons. I would never have this set longer than thirty minutes; twenty would be safer. Ten if you're an insurance freak.

Maximum Files in a Project's Attic: This interacts with the Max versions of a bin in the attic and the autosave interval. I consider how far back I might want to go to find a bin. For me, that's at least twenty-four hours, maybe even forty-eight. So, if you have a short autosave interval, many bins are being saved each hour and you'd need a higher maximum bins number. For me, my interval is fifteen minutes, so I'm saving about four versions of every bin every hour. If I'm editing twelve hours a day and go back two days, I would need to have about one hundred versions of each bin. Many times I have about ten different bins in a project—sometimes more, sometimes less—so that means one thousand saved bins in a project. Remember, though that you're only autosaving the bins that are open, so that's pretty generous. So divide sixty by your autosave interval, multiply by the number of hours you want to be able to rescue, and multiply that by the average number of bins you have in a project to get a good number for this. This also depends on how big the hard drive is that stores your Attic. In the old days, you didn't want this set too high or you'd fill up your 100 GB hard drive!

Max Versions of a File in the Attic: See above. How far back do you want to be able to go for a specific bin? For me this almost always means, "How many versions of my SEQUENCE bin do I want?" I rarely need to go into the Attic for any bin other than the bin I keep my active sequence in. If I autosave 4 versions of my sequence every hour, and I sometimes edit 16 hours a day, that means I need to save 64 versions for each day I want to go back in time. If I want to go back two days, that means 128 versions of a bin. Really, though, I keep this set at 80.

Double-click loads object in: Source/Record monitor and New Pop up Monitor are the options. I have never liked new pop-up windows, so I keep this at the default.

Enable Edit from Bin (Splice, Overwrite): Enabling this setting lets you highlight a clip or clips in the bin and hit the overwrite (or insert) command and they will cut into the Timeline directly from the bin.

TIP

It may freak you out to think about saving a thousand bins for each project, multiplied by the number of projects you do in a year. If so, consider a weekly or monthly maintenance schedule. Go into the Attic and save only the two most recent versions of each bin from any completed projects. I save the most recent two bins instead of only one because of the chance that one of them is corrupt. In reality, the only bin you really probably need to save is the bin with the current sequence. If you name your current sequence bin the same thing in every project, it makes it easier to locate and protect this bin.

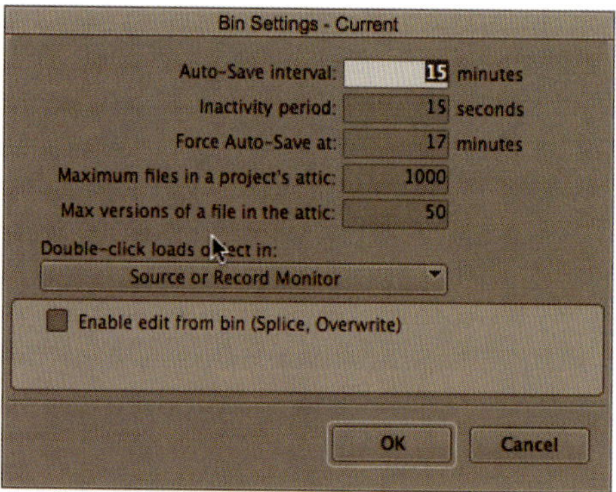

Fig. 2.7 Bin Settings.

Important Settings Continued

Bin Views: These bin views are really created in the bins themselves, mostly by choosing and setting which columns of data you want to look at. There are about a half-dozen factory default bin views provided (see Figure 2.7). Though you'd rarely choose or apply the settings from the settings list, having the Bin Views accessible here allows Bin Views to be copied and shared between editors.

Capture: OK, who captures from tape anymore? Seriously. If you want to customize how captures, especially batch recapturing for up-rezzing, this is the place to look.

Color Management: This is part of the latest changes in MC7. LUTs can be added here from the Install LUT button. By and large, if you are not dealing with footage from a production using dailies, you don't have to deal with this. Color management gives you access to LUTs, CDLs, and other ways to accept color data from the production.

Communication Ports: From here, you can set ports for communicating with VTRs. Again, not many people are capturing from tape anymore.

Composer Settings: This is one of the most important settings on an Avid system. I suggest that you spend ten or fifteen minutes sometime going through all of the settings to see if some of these things help your editing. There are five tabs in Composer Setting: Window, Edit, FF/REW, Multicam, and S3D View. In the Window tab, you can set the number of rows of data displayed along the top of the Source and Record monitors. You do NOT want to choose "off." This will disable your ability to use numeric keys to move around the Timeline, like jumping to a specific timecode or jumping forward +1:00:00. Two rows is sometimes helpful on the Sequence side because you can compare video track timecode with the timecodes on audio tracks. You can also set the numbers of rows of buttons under the monitors. You should really be pushing yourself to edit almost exclusively from the Timeline, so setting this to "off" is a great idea if you can do it because it saves screen real estate. Another highly useful setting is here: Center Duration. I highly recommend turning this on. It will show you the length of the marked section (between in and out points) of the Source or Sequence side, whichever one is active. This is a UI thing that I prefer over Final Cut. It's not necessarily better, but I find that it's easier to see at a glance than in FCP. In the Editing tab, there are some great options, most of which are pretty obvious. The two that may not be self-evident are Sync Point Editing and Phantom Marks. I think Phantom Marks is a must-have setting. Basically, if you have this turned on and you set an in- and a outpoint on either the Source or record side, Media Composer will create "phantom" marks showing where the in and out points will be on the other monitor. This is useful because if your in point is too close to the end of the media, you don't know that you don't have a long enough clip to "fill the edit" on the sequence side until you try to make the edit and get an error message. Phantom Marks let you know that there's not enough media before you make the edit, because the second edit point (usually the out point) will be missing. Sync Point Editing allows you to make marks on a segment and then place the Timeline locator at a specific point inside the marks, for example on a big visual "hit" on the source side. Then on the sequence side, place the Timeline locator on a big audio "hit" like a cymbal crash or drum hit, and the edit will be made matching the two synced points instead of basing it on starting with the in points. This functionality can also be had without enabling the setting by using the Replace Edit button. The FF/REW tab is really misleading. What this really does is determine what points the FF and REW buttons jump to. The options are Head Frames, Tail Frames, and Markers and to ignore track selectors. If you don't ignore the track selectors, then only edit points or markers in the active tracks matter. In the Multicam tab, you can choose whether to show multicam splits on the "client monitor" or whether you want to just show full-screen display. We'll cover the S3D View options when we discuss stereoscopic editing, in Chapter 3 starting on page 238.

Controller Settings: This is where you set up specialized controllers. If you don't have an external mixer or similar device, you don't need to worry about this. If you do own one, then the device should come with instructions on how to set it up.

Correction Settings: These are for color correction. Definitely check the boxes for Eyedropper 3 × 3 averaging and Show Eyedropper Info. In its default, the eyedropper samples only the single pixel under the tip of the eyedropper. But this pixel may not be representative of the actual area you want to sample. It's possible that video noise means that the chosen pixel is very different from the values immediately surrounding it. By choosing 3 × 3 averaging, you are selecting the pixel under the eyedropper and the eight other pixels around it and averaging them.

Deck Configuration: Another setting for capturing from tape. This used to be important. Now, not so much. When I did training sessions

Video Tutorial

The Composer Setting is so important that I created a video walkthrough of the options. Check it out in the video tutorials section of the project files, www.focalpress.com/cw/hullfish.

Avid has a helpful Timecode window that can also display any of these timecode related numbers. Go to Tools > Timecode Window to open up the window. It can be resized to almost any size (great for producers sitting behind you). It can also be customized with what times, durations, sources, or tracks are being displayed (see Figure 2.8).

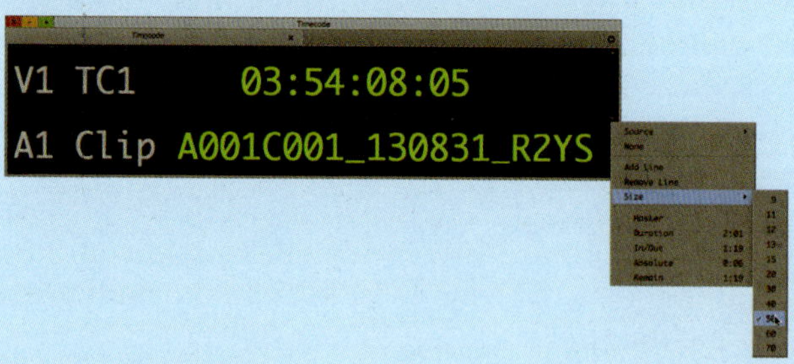

Fig. 2.8 Timecode window.

as recently 2008, I'd spend quite a bit of time explaining the best ways to set this up for multiple decks, so if you're still capturing from tape, check this out to be able to create specific named decks with specific settings that make it easy to use multiple decks without having to customize each time you swap decks. Name the Configuration at the top, click Add Channel, set the type of port, and add a deck by manufacturer name and model. Now that deck configuration can be easily called up by name.

Deck Preferences: Another setting for capturing from tape. Most settings are pretty obvious.

Desktop Play Delay: This is for setting the delay between a FireWire playback to a monitor (through a camera or transcoder) so that the playback on the desktop matches the playback on the external monitor. The easiest way to see this delay is when playing back a series of edits This setting will take some experimentation. However, this method of Firewire playback is not used much anymore.

Digital Cut Settings: Layoff to tape is becoming increasingly rare. The settings are not actually accessible from the settings list. To adjust these Digital Cut settings, choose Output > Digital Cut. If you find you have to lay-off to tape and don't want the hassle of renting a deck, check out DigitalServiceStation.com.

Dynamic Relink: This is an Isis/Interplay thing allowing Dynamic Relinking between the initial AMA clips and the versions being transcoded. If you don't have an Isis, you can ignore this.

Effect Editor: Customizes the way the Effects Editor looks and acts. This stuff is pretty self-explanatory. I don't touch any of it, but I'm old school.

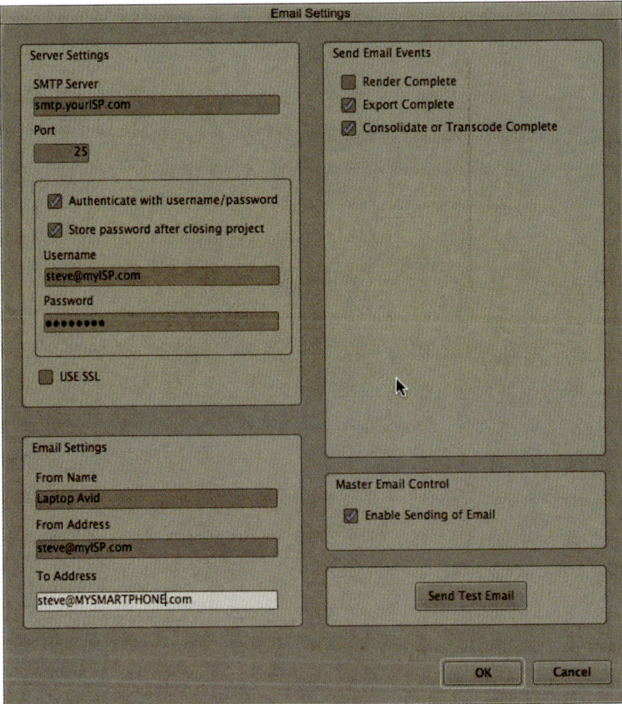

Fig. 2.9 E-mail Settings.

E-mail Settings: These are pretty awesome. This setting allows you to set up your Media Composer to e-mail you when a Render, Export, or Consolidate/Transcode is complete (see Figure 2.9). So you can unchain yourself from the edit desk while waiting for these procedures to complete. Set your phone e-mail to alert you when you receive an e-mail from the system and then take a nap or go get coffee.

Full Screen Playback: There are a lot of fairly complex settings in here, including for stereoscopic. I'm not going to go into these here, but if you want to use one of your dual screen computer monitors to display full-screen video, the settings you need are here. If you experience "tearing" of the monitor in full-screen mode, check the Flip on Vertical Sync button.

General Settings: The Avid engineers clearly didn't know what they should do with this hodgepodge of settings. The biggest one is the default starting timecode of sequences. Most people prefer this at 1:00:00:00, but I worked at a place where the default was 58:30:00. Use a semi-colon between the numbers to switch to drop frame timecode. You can also prepare for Windows compatible file names and names that can exceed twenty-seven characters (see Figure 2.10).

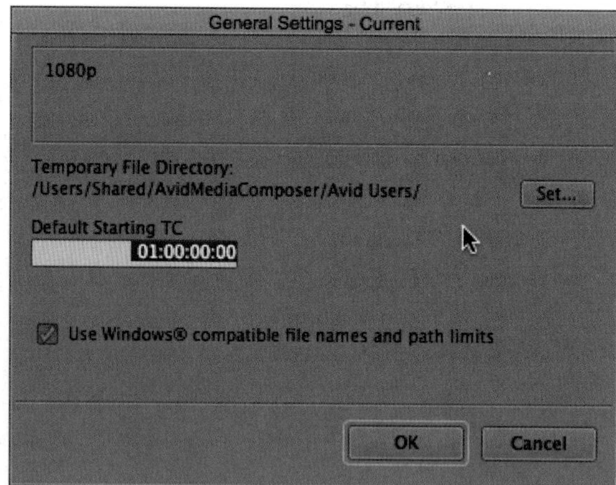

Fig. 2.10 General Settings.

Grid Settings: These settings let you customize the safe title and safe action overlays. Lots of options. To me, the most important setting is the Scale mode, which allows you to combine various aspect ratio displays inside or outside each other. One of the most common of these for me is 4:3 inside a 16:9 monitor, which allows me to see a 4:3 protected composition or graphic inside the 16:9 monitor, which is a fairly regular delivery requirement for broadcast (see Figure 2.11).

Interface: Use it to change the look of the entire UI—make it darker or brighter, change the highlight colors, show tooltips, etc. And again, the UI designers toss in a completely random unrelated setting, Automatically Launch Last Project at Startup. This is similar to the standard FCP functionality of relaunching the last closed project. I personally work on a number of projects at the same time, so I prefer to keep this unchecked (see Figure 2.12).

Interplay: The three Interplay settings are applicable only if you are operating in a shared environment with an Avid ISIS and Interplay. If you're on an Interplay system, these settings were probably set by the installer or system administrator.

Keyboard: We will discuss the keyboard shortly, when we map the keyboard.

Marquee Title: These settings allow you to automatically create or promote Titles to Marquee instead of using the standard Avid Title Tool. Marquee is a much more feature-rich titler than the Avid Title Tool, but its use requires more time and a greater learning curve. For most of my day-to-day titling needs, I stick with the Avid Title Tool, so I prefer not to default to Marquee. If you find yourself using Marquee extensively, revisit these settings.

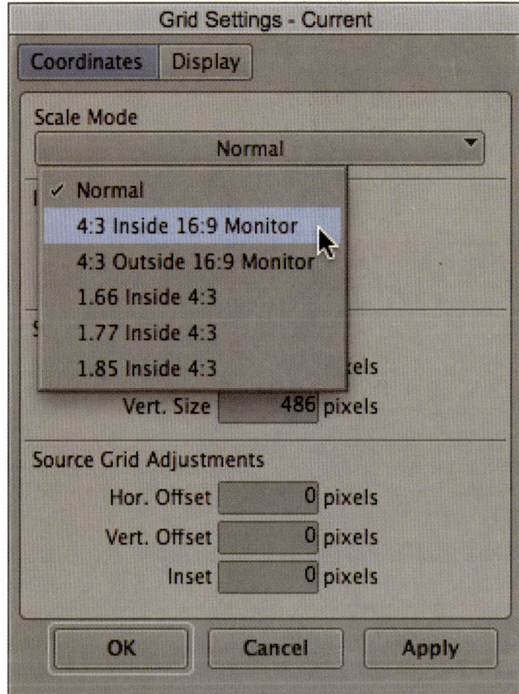

Fig. 2.11 Grid Settings.

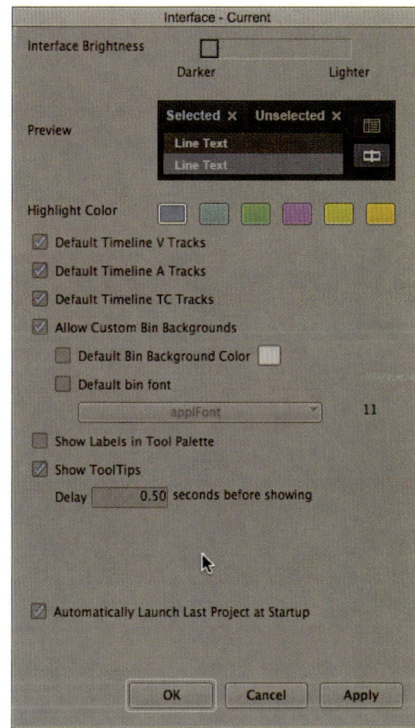

Fig. 2.12 Interface.

Media Creation: These settings are very important. Essentially they are the equivalent of the Scratch Disk settings in FCP. There are eight tabs in the Media Creation setting, but really the first is the most important (see Figure 2.13). If you have any issue with a drive not showing up in your Capture Tool or if you are Importing or Transcoding, the reason could be in the Drive Filtering and Indexing tab. This tab is designed to insure that media does not get recorded onto drives that you don't want used for media. The options are Filter or Not Allow Media on Network Drives Based on Resolution, the System Drive, or the Launch Drive. The second two are checked by default and probably shouldn't be checked unless you're on a laptop and want to record to the internal drive, but that's not usually a good idea. Storing media on network drives that don't have the throughput to play back media reliably is also a bad idea and having this option checked is also usually a good idea. The Capture, Titles, Import, Mixdown & Transcode, Motion Effects, and Render options help direct media to the proper drives and record the media in the proper video format. Media Management is covered quite extensively in chapter 1.

Mouse Settings: Allows you to change the scroll speed of the mouse, but more importantly it allows you to map commands to extra buttons on

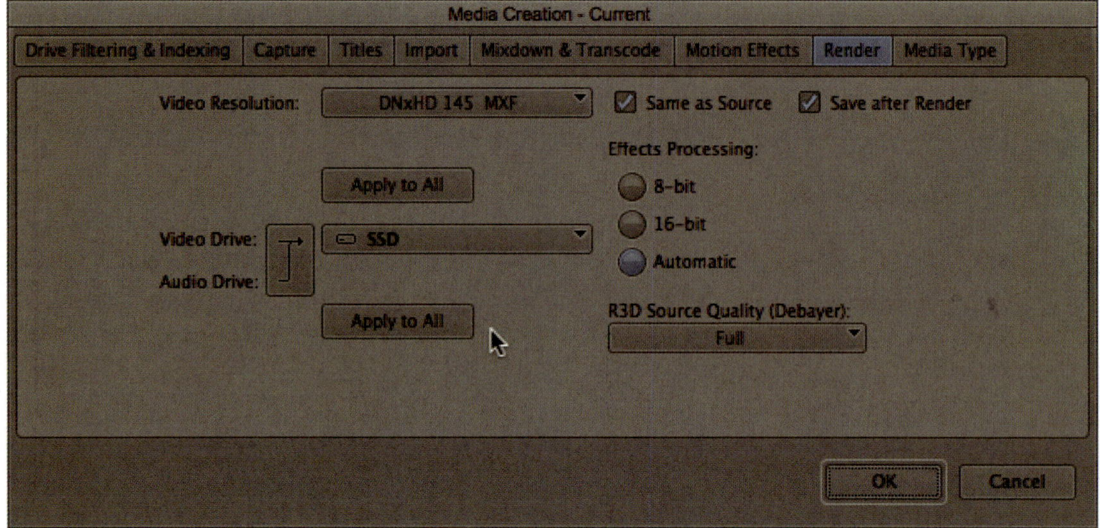

Fig. 2.13 Media Creation Setting.

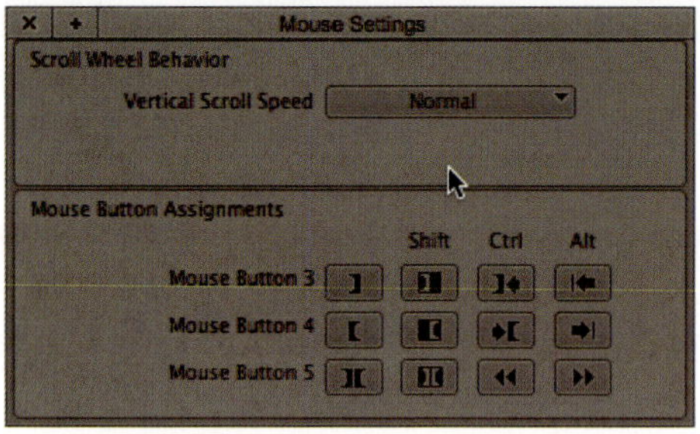

Fig. 2.14 Mouse Settings.

the mouse, if you have them. Doing this is just a matter of calling up the Command Palette (Command-3 Mac, Control-3, PC). Then dragging the commands you want from the Command Palette onto the buttons in the Mouse Setting dialog box (see Figure 2.14).

Passthrough Mix: This is really more of a tool than a setting, but you can definitely get to it through the Settings list. The Passthrough Mix Setting allows the editor to adjust the volume of the monitor levels. The levels set in the Passthrough Mix Setting do not affect recorded or exported levels, only monitored levels.

Portserver: This setting doesn't apply to many people. If you are on a shared LANshare workspace, this setting can start or terminate that connection.

ProTools Video: This is actually controlled from the ProTools Video option dialog.

Remote Play and Capture: These settings let your Media Composer act like a video tape deck. Remote Capture lets the MC start recording immediately when the linked deck starts to play without having to set ins and outs. You can also control the Media Composer as if it's a play deck. But obviously, most houses nowadays are using less and less tape-based record or playout.

Render Setting: This has a series of settings on how to render effects and what sound should be played when that render is complete (see Figure 2.15). Explaining all of the options of Image Interpolation, Timewarps, and Motion Effects could take several pages. Best to check the Avid documentation for more on these.

S3D: These settings are best discussed in the context of understanding stereoscopic, so if this matters to you, check out the stereoscopic editing section in chapter 3.

Safe Color Settings: To be honest, I haven't had much luck with Safe Color. It allows specific parameters to be set to indicate limits for composite video, luma, and red, green, and blue gamut (see Figure 2.16).

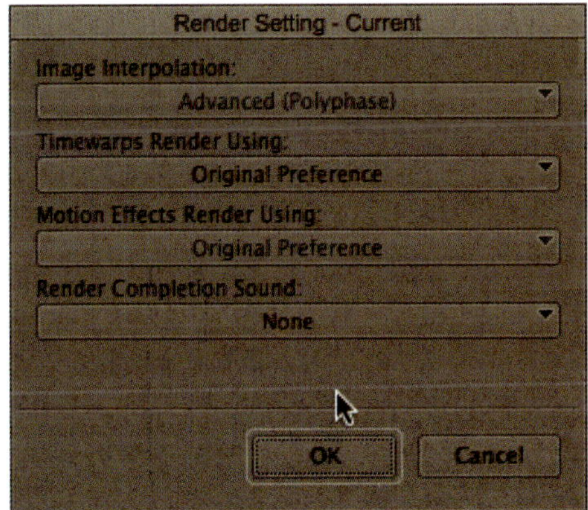

Fig. 2.15 Render Setting.

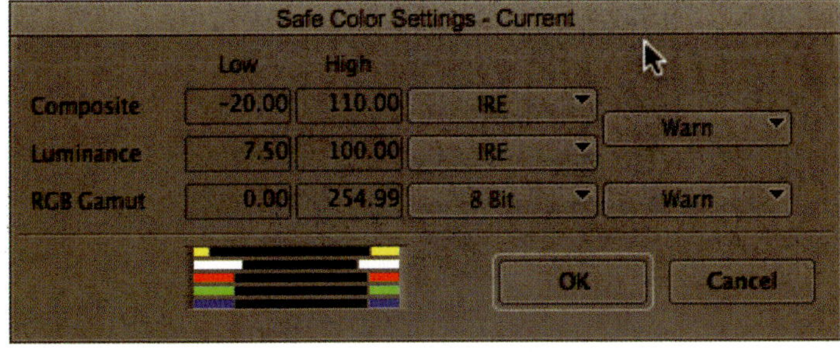

Fig. 2.16a Safe Color Settings.

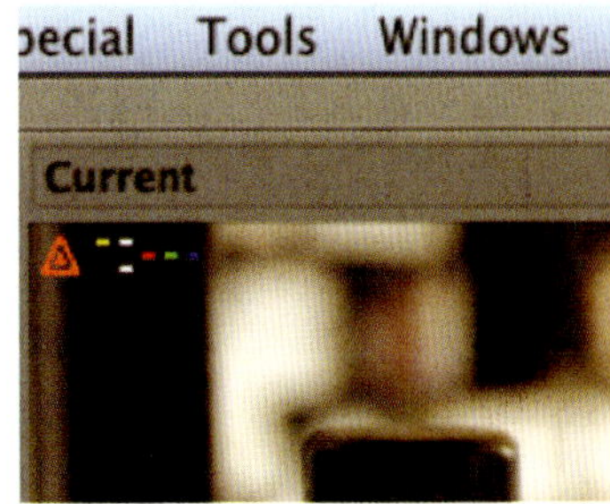

Fig. 2.16b The white indicators show that the luminance levels have "illegal" blacks *and* whites (top and bottom levels are out of spec). The yellow indicator shows that the composite levels are too high. The red, green and blue indicators in the middle show that the gamut levels are all fine.

When these limits are set, Media Composer provides specific color indications of any violations of these limits on the monitor. The indicators show whether each of the violation types is above or below the customized levels. The only issue is that often the low indicators are triggered by slight blanking issues, or differences in the size of the frame, where a small black border or edge can be seen around the frame or on one side. I'm a big fan of external scopes, and with those, Safe Color isn't really necessary or helpful, but without the external scopes it can be useful. When this setting is set to Warn, there will be a small indicator in the top corner of any clip that has illegal video levels of any type. The indicator has five colors: yellow, white, red, green, and blue. Yellow means composite levels. White means luma levels. And red, green, and blue are indicators of color channel gamuts. Each indicator has three positions: high, center, and low. If the indicator is in the top position, the level is too high. If the indicator is in the low position, the level is too low (see Figure 2.16b).

Script Settings: These settings are useful only if you have purchased and installed the ScriptSync option. With ScriptSync, it allows you to set fonts, font sizes, margins, colors of takes, use of thumbnails and other items (see Figure 2.17).

Search and Search User: These settings are not accessible from the Settings list. But there are numerous options for doing searches if you call up a search by using Command-F (Mac) and Control-F (PC). These allow for searches on Clips, Sequences, Script Text, Timelines and Monitors. They allow searches on specific columns and using specific Boolean search parameters. If you have the PhraseFind option, this also allows setting PhraseFind Settings including languages.

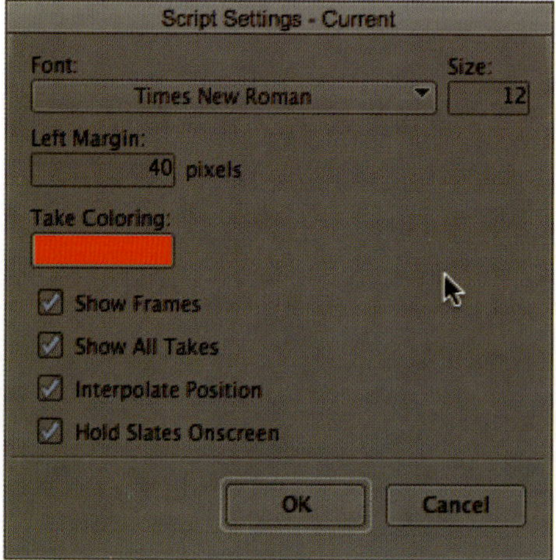

Fig. 2.17 Script Settings.

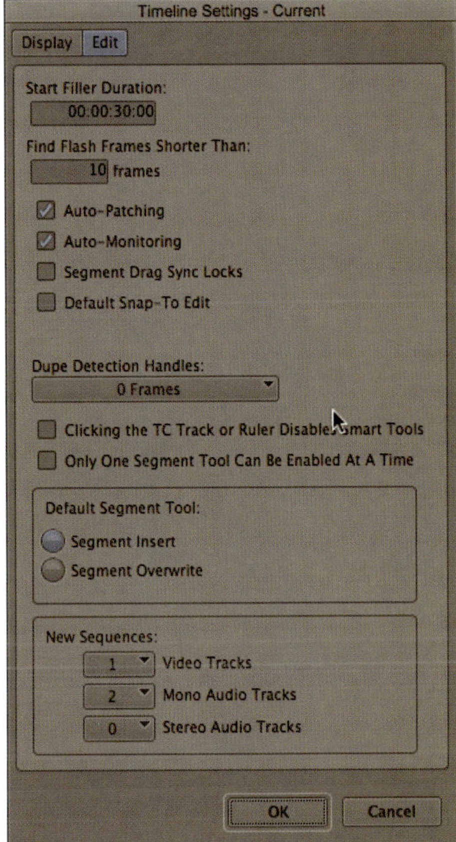

Fig. 2.18 Timeline Settings.

Timecode Window: This window is not accessible from the Settings list. However, it allows you to save and recall customized versions of the Timecode window. In the window, you can customize the number of lines of timecode, the source of that timecode, and the size of the font that displays the information. Better to just launch the Timecode window tool and set the settings from within the tool.

Timeline: A lot of work gets done in the Timeline, so the importance of this setting can't be overstated. Most of the settings are done simply by checking check boxes on the two tabs (see Figure 2.18). If you are devoted to editing with the keyboard instead of the UI buttons, you can uncheck the Show Toolbar menu. This will open up some screen real estate. Most editors enable Show Marked Region, which colors the section of the Timeline between marks on enabled tracks. This will also be comfortable for editors coming over from FCP, though the marked region displays a bit differently. Show Marked Waveforms used to be valuable if you liked to see audio waveforms in the Timeline but don't like to wait for the waveforms to redraw. In MC7 (released in the summer of 2013), waveforms became cached, so there's not really any advantage to this as far as redraw times are concerned, but if you'd just rather not have the "messiness" of seeing audio waveforms except in a selected area, then this is your setting. By turning this option on, the waveforms draw only between marks. Double-click to Show Nesting is something that some people hate and some people love. Put me in the hate category. But that's what the setting options are for—so some people can turn them on and others can turn them off. If you like to see layers of nested effects, turn this on. This is useful when you want to adjust a nested effect while seeing the finished result. Stepping in to adjust means you don't see the finished result. This can also be done in the effect editor now that all nested effects are accessible there. If you tend to accidentally double-click (as many FCP editors do), then showing the nesting is just annoying and means more clicks and aggravation to collapse the nest again. You can also set the way the Timeline scrolls while video is playing through the Timeline. Scrolling takes some computational power. Many veteran Avid editors will leave this set to none. Auto-Patching and Auto-Monitoring are usually left checked, but if you don't like the behavior, it can be turned off. Auto-Patching makes it easier to patch tracks between the Source monitor and the Sequence monitor, depending on what tracks are active. Auto-Monitoring is something I often turn off. It makes sure that the active video track is monitored (displayed). But sometimes, when stepping into and out of effects, this can cause the monitoring to change to a track lower than one you expect to monitor. One thing that FCP users will be grateful for in Avid is that changing the monitoring of a track doesn't undo rendering on the upper layers. Snap-To Edit can be turned on for FCP editors who like "magnetic" Timeline features. Of course, just holding down Command (Mac) or Control (PC) while dragging along the Timeline will do the same thing. For those who don't like the behavior of the SmartTools, try customizing that behavior with Clicking the TC Track or Ruler Disables SmartTools or Only One Segment Tool Can Be Enabled At A Time.

Trim Settings: Trimming is one of the big differentiators of Avid editing systems. The way that it behaves while in trim mode can be controlled here (see Figure 2.19). There are two tabs. The Play Loop tab allows you to adjust the amount of preroll and postroll that are shown when Play Loop is enabled. I have never changed them from their default, but you may want a longer loop, starting earlier and perhaps ending later. There are several options concerning Small Trim mode. FCP editors may like this mode. I can't stand Small Trim mode and far prefer the standard Large Trim behavior. If you want to see which you like best, try both—Never Use Small Trim or Always Use Small Trim—and then go in to Trim

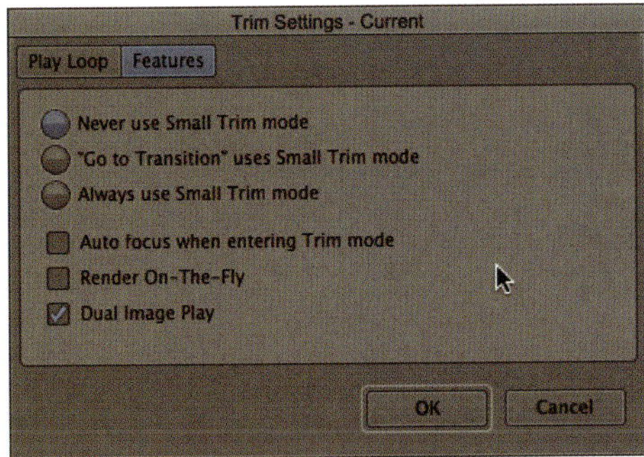

Fig. 2.19 Trim Settings.

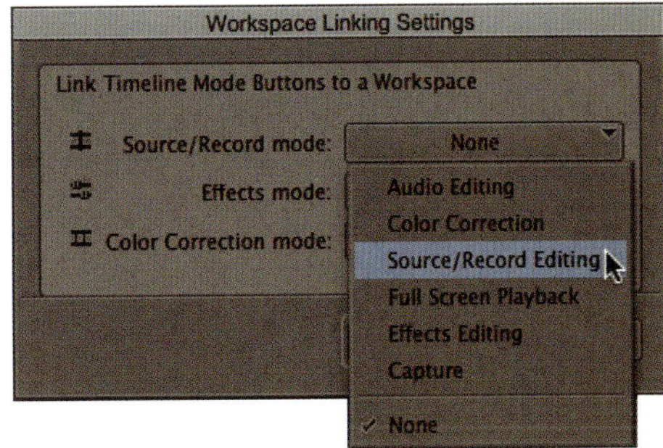

Fig. 2.20 Workspace Linking Settings.

mode. Arrows at Selected Transitions replaces the standard "roller" icons when in Trim mode with left and right arrows at the transition.

Video Output: If you are running a software-only version of Media Composer, you won't see this option. But if you have a video output card or Nitris DX/Mojo DX connected, you'll see it and it will include numerous parameters here for what signal is going out, the format, and so on. If you have a third-party card, like Kona, it will launch the third-party control panel.

Video Satellite: Not many people have the Satellite option. If you have it, you have people who know what to do with it.

Workspace Linking Settings: This is a cool setting. Usually, the mode buttons (Source/Record mode, Trim mode, Effects mode, Color Correction mode) take you to a specific mode, but this setting allows you to, instead, launch a specific Workspace when these mode buttons are used (excludes Trim mode, which doesn't have a Trim workspace; see Figure 2.20). This is a pretty useful thing to do, but you'll want to try this to see whether you like this behavior.

Keyboard Optimization

Of all the settings and possibilities to optimize and personalize your settings, the most powerful is to customize your keyboard, or at least learn and exploit the default settings.

Editing using the keyboard is the quickest route to efficiency and effortless, intuitive editing (see Figure 2.21). There are definitely times when I use the buttons and menus in the UI or edit with the mouse on the Timeline, but even there, the keyboard can come in handy. If you haven't been editing using the keyboard, I encourage you to slowly try to work more keyboard editing into your personal editing style. As with learning to play an instrument, this requires practice to develop "muscle memory." The easiest way to do this is to incorporate just a few new keys

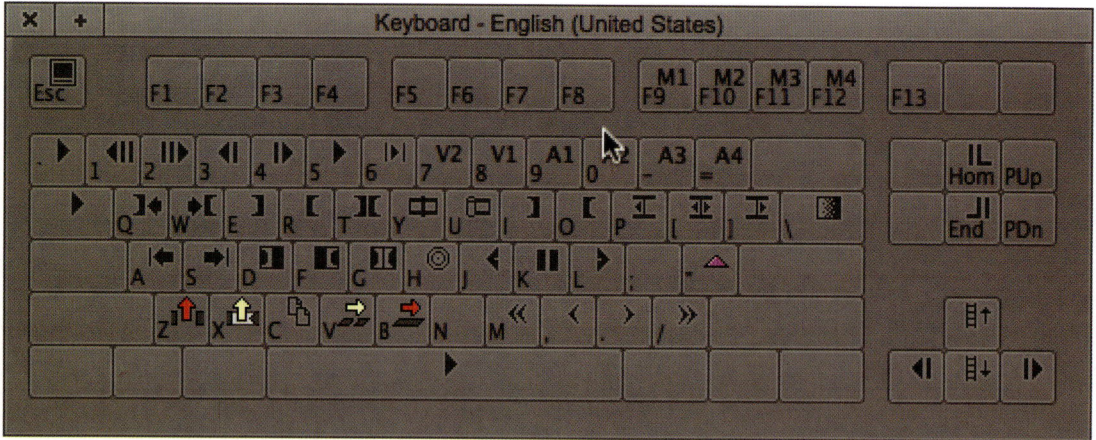

Fig. 2.21 Keyboard Settings.

and then work with just those keys for a week or even more before trying to add more keys to your arsenal. The list of truly critical keyboard functions is actually very short:

JKL—These are the reverse, pause, and play keys, but they also let you scrub and shuttle through footage. This is the same as FCP. Single tapping the J key goes backwards at full speed. Hitting it again doubles. Hitting it again triples the speed and so on up to 5x speed (the exact speed in frames-per-second depends on the playback frame rate of your sequence). The K key pauses. The L key plays. Hitting it again doubles. You can still hear audio play back at this speed. Hitting it again triples. Audio can still be heard, but it takes a practiced ear to understand speech at this speed. I can usually follow along at this speed if I am following a transcript. Above 3x speed, audio cuts out. Holding down JK together allows you to scrub backward. Holding down KL allows you to scrub forward. Holding your middle finger on the K key while rocking back and forth between J and L allows you to dial in a specific fine edit point. If you aren't familiar with using JKL, this is the first thing you should practice and implement.

IO—These are the In point and Out point buttons. These buttons are also the same as FCP. The advantage of marking in and out points with keys instead of on-screen UI buttons is that using keys doesn't stop playback.

The other four keys that are critical beyond JKLIO are the mapping of the Overwrite, Splice In, Lift, and Extract buttons. These are what allows you to selectively and exactly edit your footage into or out of the Timeline from the Source monitor.

But before we get to the specifics of using the keyboard, there are a few theories that the editors I've worked with through the years subscribe to when it comes to keyboard customization, and you need to decide which one makes the most sense to you.

The first theory is my own: customize from "scratch" based on your own personal needs and style, completely disregarding the default settings. I started out reading a great self-published book that was created back in the early 1990s by Steve Cohen and Basil Pappas. When I started editing on Avid in 1992, this book was my bible. I first mapped my keyboard based on the advice of their book. I customized every single key differently from the default mapping. There are about one hundred mappable keys on most keyboards. This is one of the places that Avid is lacking compared to the old Final Cut Pro system, which could map not only the basic keys and shift keys, but also Command, option, and control keys as well, just about doubling the number of potential keys. For me, though, knowing the location of one hundred keys by touch is pushing the limits of my own abilities. And when I cut on FCP, I never tried to use all of the modifier key options that I had available to me. The problem with my keyboard mapping theory is that getting used to a completely customized keyboard layout means that you MUST carry your customized settings with you at all times if you want to be able to edit on someone else's Avid system. I've had to sit down at Avid systems with "default" keyboard layouts, and I have to admit that I have absolutely no clue how to edit on them. And if you help train people, you definitely don't want to use one of those massively customized keyboards.

The second theory is to edit with all of the default keys in their correct positions and then customize only the function keys and possibly any unmapped keys. This allows you to jump onto "default" systems relatively easily. It also means that if you have a keyboard with the default labeling on the keys themselves, you have a nice visual reference for keyboard functions that you haven't yet memorized.

The third theory is to split the difference. Use as many of the default keys as possible and customize the rest of the keys relying on mnemonic placement of keys. For example, putting the Matchframe function on the M key or the Title function on the T key.

Before I get too deep into specific mapping strategies, let me explain how to map your keys and what's possible, including mapping menu commands.

Mapping the Keyboard

The main element in customizing your keyboard is the Command Palette (see Figure 2.23). This is a good Command/Control function to memorize: Command-3 (Mac), Control-3 (PC). The Command Palette has the full complement of mappable buttons and tools.

While the Cohen/Pappas book is not available any more, Cohen has since written a truly great Avid editing book called *Avid Agility: Working Faster and More Intuitively with Avid Media Composer* see Figure 2.22 (ISBN-10: 1477654356, ISBN-13: 978-1477654354). The book is in its third edition as of the writing of this book and is available on Amazon. It is full of sage, hard-won advice from an experienced LA editor, and if you are a visual learner, you'll love the heavy reliance on lots and lots of lovingly annotated screenshots.

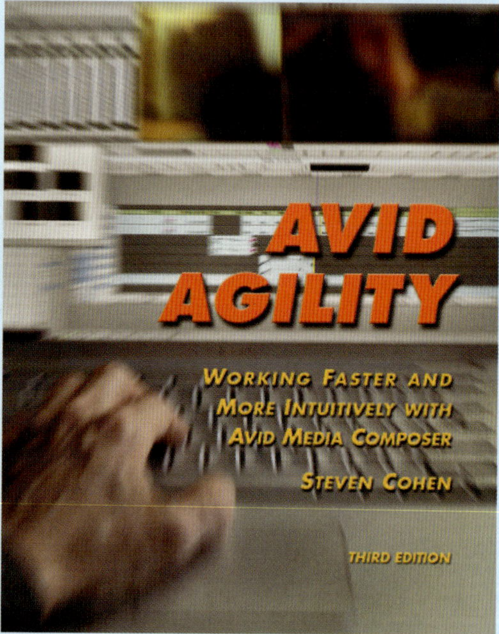

Fig. 2.22 Avid Agility by Steven Cohen.

Steven Cohen was one of the first to use Avid on TV movies (HBO's *Teamster Boss*) and then cut the first studio feature to use Avid's Film Composer (*Lost in Yonkers*). He was the first chairman of the editing department at the American Film Institute.

The tabs of buttons in the Command Palette are Move, Play, Edit, Trim, FX, 3D, CC, Tracks, Smart Tools, Other, More, and Workspaces. The functions organized into those tabs are not necessarily intuitive or obvious. Even after two decades, I basically have to guess where stuff is. My recommendation is to spend some time going through all the tabs and learning what all the buttons do. Many of the buttons are things I've

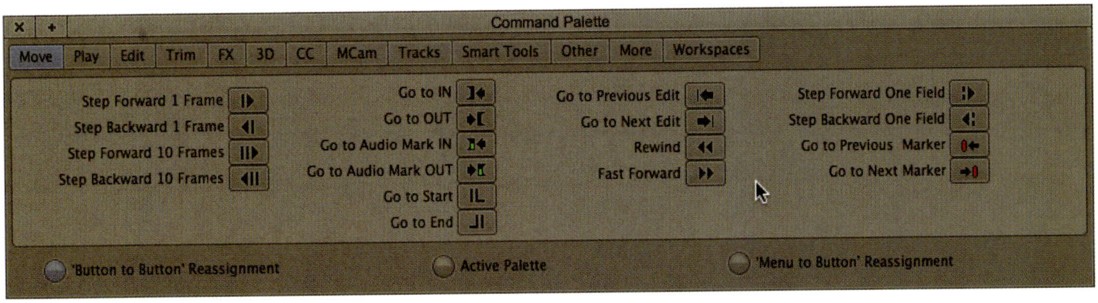

Fig. 2.23 Command Palette.

never chosen to map, but every editor is different and you may find great uses for buttons that I never use.

To map the functions from the Command Palette to your keyboard, call up the Command Palette with Command-3 (Mac) or Control-3 (PC). Then from the Settings tab of the Project window, double-click on Keyboard Settings. At the bottom of the Command Palette are three radio buttons: Button to Button Reassignment, Active Palette, and Menu to Button Reassignment. With the Button to Button Reassignment choice active, it's as easy as drag-and-drop to move any function in the Command Palette to any key in your Keyboard Settings. To move a function to a shift key, just hold down the shift key as you drag the function to Keyboard Settings.

While we have the Command Palette open, you can also map those functions to any of the spots on the UI. Specifically, Command Palette functions can be mapped to the one or two rows below the Source and Record monitors or to the row at the top of the Timeline. They can even be mapped to multi-function mice keys using the Mouse setting.

The other very cool and highly useful thing that you can do with the Command Palette is to map menu functions to keys. There are several great examples of this that I use on a regular basis. There is no Command-key combo that allows you to call up the Audio Mixer or to Render In to Out. These are two important functions that I need to have access to at the touch of a button, and the ability to map menu commands to the keyboard allow me to do this.

For some menu commands, before you can map the menu function, you have to create the proper conditions. For example, you can only map the Render In to Out button if you have an un-rendered effect in the Timeline between an in and an out mark. So if you want to map Render In/Out to a key, first you have to create this condition in the Timeline.

To map menu commands, the radio button at the bottom of the Command Palette must be active to use Menu to Button Reassignment. Simply click on any key in your Keyboard setting or UI and your cursor will turn to an odd-shaped little "menu" icon. With the cursor in "menu icon"

Video Tutorial

I created a video tutorial that walks through the tabs and functions available in the Command Palette. Check it out if you're trying to decide which functions you may want to map to your keyboard, www.focalpress.com/cw/hullfish.

mode, choose any menu command to map it. To map the Audio Mixer to your A key, for example, click on the A key in your Keyboard setting, then mouse up (don't drag) to the Tools menu, open the Tools menu as you normally would, and click on the Audio Mixer selection. You'll notice that the A key now has a small icon of a mixer on it. Most menu commands are indicated on the mapped button by its initials, like RI for Render In to Out. This ability to map menu commands can save a lot of time and mousing. Mapping Render In to Out is a very good one for FCP editors who are used to Option-R for rendering all effects. Though you can't map Render in to Out to the Option-R key, you can map it to R or Shift-R.

It is possible to map some Fast Menu commands to buttons. Later in this chapter, I describe a special keyboard setting that I created for FCP users. On that keyboard, I mapped the Show Entire Sequence functionality from the Timeline Fast Menu to the Shift-Z key on the keyboard, which is the same functionality that the Shift-Z key has in FCP. The methodology for mapping this is similar to the example above, except that, using the Menu to Button Reassignment option in the Command Palette, you first hold down the shift key and then select the Z key in the Keyboard setting. From there, move the strange-looking "menu" cursor down to the Fast Menu in the Timeline and select Show Entire Sequence. When this has been done correctly, the shifted Z key in the Keyboard setting will have the letters SE (Show Entire) on the key.

My suggestion for determining which functions you should map is to "journal." Basically, keep a pen and paper handy and mark everything that you mouse to. Maybe you use JKL for movement but always mouse up to hit the edit buttons, or maybe you regularly choose Render In to Out from the Clip menu. Keep track of these things. If you mouse to any button or menu more than five or six times a day, I would map that function. Keep track of the things you do the most and keep them closest to the "home" position of your nondominant hand (because your dominant hand is one the mouse, trackball or pen, usually).

Download

I've created a blank version of the keyboard (see Figure 2.24) that you can download, or you can make one pretty simply by mapping the "blank" button (on the far right of the Other tab) to every single key, then do a screengrab of the blank keyboard. Use this to pencil in various options for mapping.

Another element in determining how you will map your keyboard is whether you are left-handed or right-handed as well as whether you will be planning to use a mouse, tablet, trackball, or other manual user interfaces on a regular basis. The big question is, where will each of your hands regularly be? This makes a difference when it comes to mapping the most critical keys. Do you want them on the left side or the right side of the keyboard? In some ways, I wish that when I originally customized my keyboard, more than twenty years ago, I had mapped my critical keys to the left side of the keyboard instead of the right side. My reason for this is that I am right-handed, so I use the mouse on the right side. By having the most critical editing keys on the right side, I either feel a little "cramped"

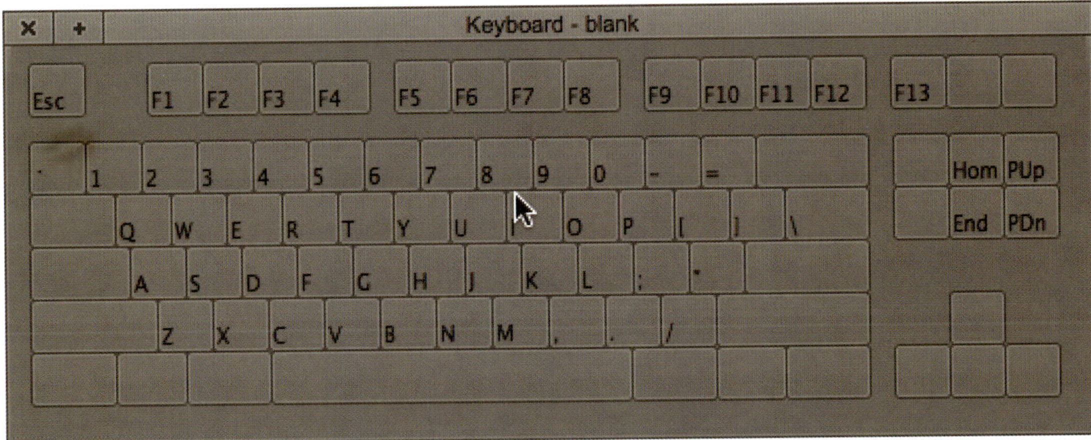

Fig. 2.24 Blank keyboard.

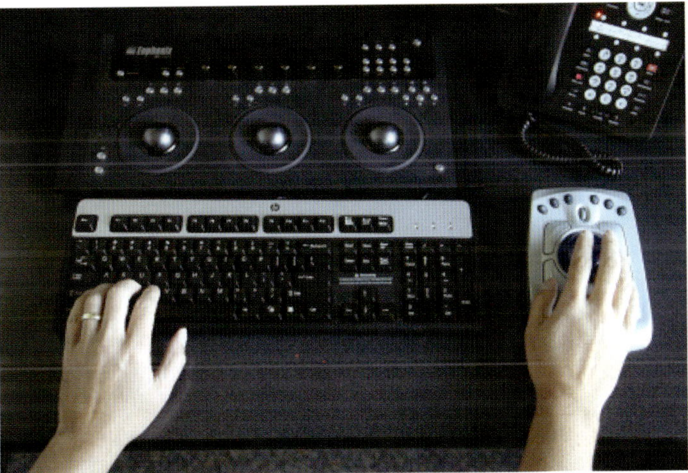

Fig. 2.25 This image shows the benefits of mapping your keyboard for improved ergonomics.

because my left hand is forced across my body to the right side of the keyboard, or it means I have to slide my keyboard to the left, so that I can edit comfortably. But then if I want to use the keyboard for regular typing, it means both of my hands have to shift over to the left or I have to move my keyboard back to the right. Obviously the goal is to get your body and your hands into a comfortable ergonomic position and then map the keys you use the most wherever your nondominant hand comfortably rests (see Figure 2.25). This could mean that the well-used JKL keys that are familiar to so many editors need to be moved over to ASD, perhaps.

In my heavily customized keyboard, I also moved the JKL keys down a row to <>?. I was very intentional in moving them to this position because I didn't want to have to stretch my hand any further up the keyboard than I had to. Nowadays, many keyboards are very flat, but in the 1990s, the typical keyboard was quite raised, and each row was higher than the next. The relief of using the lower keys is obvious. I also placed them there so I could find those keys by touch, based on where they were positioned in relation to the spacebar and the shift key. This helped me keep my eyes on the screen without having to glance at my hands to get them repositioned. Considering this, if you're right-handed, you might try replacing JKL functionality to ZXC.

The only consideration for keeping the JKL keys in the center row of the keyboard is that it allows you to "touch type" more easily because you can more easily move your fingers down a row or up a row. By placing the JKL functionality at the bottom row, reaching more keys from your "home" position means moving up a row or two.

Play vs. Play Forward Functionality

Proper JKL functionality is dependent on the difference between the Play key and the Play Forward key (see Figure 2.26). They both look like your basic triangular Play icon unless you examine the Play Forward key a little more closely. Then you can see that it has a tiny slit on the left side. There is also a difference between the Play Reverse and the Rewind key. I usually refer to the Play Reverse key as the Rewind key in generic discussions, but this is wrong. The Play Reverse key has the same slit that the Play Forward key has. These two "slitted" key icons are the basis for the typical JKL functionality, plus the Pause key, of course. These keys allow you to forward-scrub using KL or backwards-scrub using JK. There is also a regular Play key, which, in the Avid default mapping, sits on the space bar. The regular Play key (without the slit) toggles back and forth between Play and Stop. The Play Forward key does not toggle with multiple button pushes but instead increases the speed of playback with each key press, effectively acting as a shuttle command.

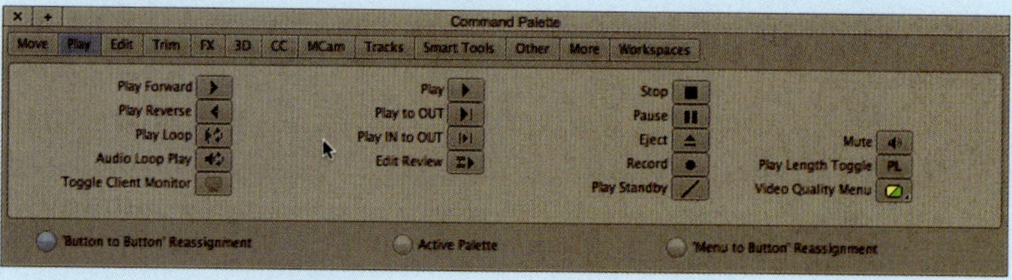

Fig. 2.26 Play Tab Command Palette.

My Keyboard Settings

My keyboard settings (see Figure 2.27) center around eleven critical functions that are mapped to what I call the home position of my non-dominant hand. My dominant hand rests on my trackball. The eleven critical functions include JKL, which are rewind, pause, and play. On Avid, these keys also allow you to scrub backwards and forwards at quarter speed (by simultaneously holding the pause and play key or the pause and rewind key) as well as to shuttle forward and backward at high speeds (by double-, triple-, or quadruple-tapping the play or rewind key). The next two critical keys are Mark In and Mark Out. The default mapping for this is on the IO keys. Since I moved my JKL keys down a row to <>?, I also moved my IO keys down a row to L:. The next most important keys to me are the Lift, Extract, Overwrite, and Splice buttons. The default positions for these keys are at the ZXVB keys

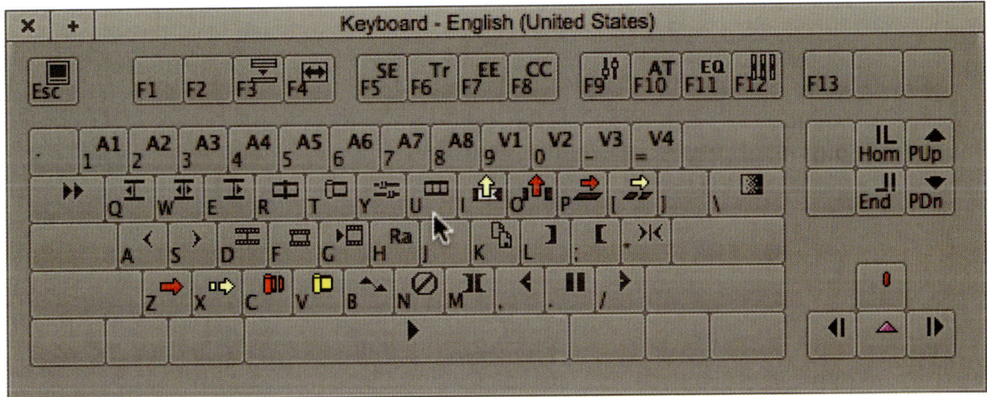

Fig. 2.27 This represents the remapped positions of keys on my highly customized Avid keyboard.

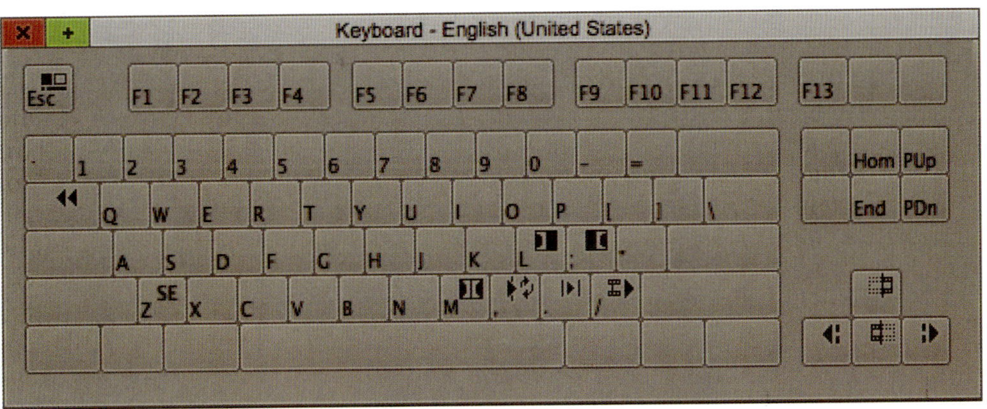

Fig. 2.28 These are my customized shifted keys. I have very few. Also, note that I mapped the "blank" key onto all of the default shifted key so that I wouldn't accidentally edit with them.

(leaving the C key in the middle for the mnemonic Copy to Clipboard). This makes sense if you have both hands on the keyboard, but I wanted them where I could reach them with a single hand, so I mapped them above my Mark In and Mark Out keys (L:) at IOP{. I keep the "red" editing keys in the middle and the "yellow" editing keys—that "ripple" the Timeline—to the outside.

Those are the primary nine keys I use. Additionally, I mapped the Mark Clip function to the mnemonic M key. I also use the Extend key a lot, so I mapped that to the " key. What completes this "home position" mapping are the functions mapped to the SHIFT keys under the main functions (see Figure 2.28). At the shifted position under the Mark In and Mark Out keys, I map the Clear In and Clear Out buttons. I also map the Clear Marks functionality to the shifted Mark Clip key (M and Shift-M for me). Avid has functionality built in for the Mark In and Mark Out keys, which, when used holding down the option key on the Mac or the alt key on the PC, turns them into Go To Mark In and Go To Mark Out. This option/alt functionality works for wherever the Mark In and Mark Out keys are mapped. This means that using the shift and option modifiers with just the L: keys on my keyboard, I can access six different functions: Mark In, Mark Out, Clear Mark In, Clear Mark Out, Go To Mark In, and Go To Mark Out.

My Specific Keyboard Mapping

Step Forward and Step Backward One Frame = Left Arrow and Right Arrow (Avid default)
Step Forward and Step Backward One Field = Shift Left Arrow and Shift Right Arrow
Toggle Audio tracks 1-8 = 12345678 (numeric keys above QWERTY)
Toggle/Enable/Disable Video tracks 1–4 = 90-= (numeric keys above QWERTY.
Delete Effect = N (mnemonic for No)
Add Edit = D (mnemonic for AdD EDit)
Match Frame = F (mnemonic for Find Frame)
Fast Forward (which I use to jump to the next Locator or next Head frame) = Tab
Rewind (which I use to jump to the previous Locator or Previous Head frame) = Shift-Tab.
Lift Segment Mode (SmartTool) Toggle = Z
Extract Segment Mode (SmartTool) Toggle = X
Overwrite Trim (SmartTool) Toggle = C

Ripple Trim (SmartTool) Toggle = V
Keyframe (SmartTool) Toggle = B (keeps SmartTools together, and mnemonic for RuBBerband)
Trim One Frame Backward = A
Trim One Frame Forward = S
Trim A Side = Q
Trim AB Sides = W
Trim B Sides = E
Source/Record Workspace (Mode) = R (mnemonic of Source/Record)
Trim Mode = T (mnemonic for Trim)
Effects Workspace (Mode) = Y
Color Correction Workspace (Mode) = U
Render at Position = G
Render In to Out (menu mapped) = H
Play/Stop toggle = Spacebar (Avid default)
Step Out = Page Up key (next to the Home key)
Step In = Page Down key (next to the End key)
Add Keyframe = Down Arrow key
Add (red) Locator = Up Arrow key
Top = Shift-Up Arrow key
Tail = Shift-Down Arrow key
Audio Mixer (menu mapped) = F9
Audio Tool (menu mapped) = F10
Audio EQ (menu mapped) = F11 (In truth I rarely use this, but all audio functions are together.)
AudioSuite (menu mapped) = F12 (In truth I rarely use this, but all audio functions are together.)
Edit Review = shift/ (For me, this is under my Play Forward key in default JKL functionality.)
Play Loop = shift, (For me, this is under Play Reverse in default JKL functionality.)
Play In to Out = shift. (For me, this is under Pause in default JKL functionality. I rarely use this.)
Fit to Fill = F5 (I rarely use this, but it's nice to have.)
Collapse = F6 (I rarely use this, but it's nice to have.)
Copy to Clipboard and Add to Source Monitor = K using the mnemonic "Kopy" (see sidebar)
Toggle Source/Record = Escape (Avid default)
Toggle Source/Record in Timeline = Shift-Escape
Go To Start = Home (Avid default)
Go to End = End (Avid default)

> **Copy to Clipboard**
>
> Whenever I use the Copy to Clipboard Command, I always want to edit with it immediately, so I want it in my Source monitor without having to mouse to the Clipboard Contents option of the Source monitor pulldown menu. That functionality exists if you press the option button (on a Mac) or the Alt button (PC) simultaneously with the Copy to Clipboard command. To minimize the keystrokes, you can make this "option" functionality the default by "co-mapping" the Add Option functionality from the Other tab in the Command Palette. To do this with the Copy Clipboard function, drag Copy to Clipboard from the Edit tab of the Command Palette to whatever key you want it to go on. Then drag Add Option from the Other tab on top of the same key. A small "." will appear on top of the icon to indicate that it now operates with the option functionality instead of the normal functionality. The press of a single button copies to the clipboard and then automatically adds it to the Source monitor, ready for editing.

JKL—Play, pause and play reverse = <>?
IO—Mark In and Out = L:

The J key is mapped with the Blank key because it's the one I most regularly "fat-finger" so by mapping the blank key, it ensures that I don't accidentally trigger an unintended function or operation while working quickly without looking down at the keyboard.

Some other critical functions are part of the default Avid modifier key functionality, and even though they also have Command Palette functions that can be mapped, I got used to the default before these were available in the Command Palette. Specifically, the ability to zoom in and out on the Timeline or in the Source or Record monitor is Command-[] (Mac) or Control-[] (PC). For the Final Cut Pro editors out there, this is identical to Command-+ or - and is close in proximity on the keyboard. Then, instead of Shift-Z on a Mac to zoom back out to the full view, on Avid it's Command-/ (Mac) or Control-/ (PC). You could also use Command+Opt-M (Mac) or Control-Alt-M (PC), which gives a little bracket icon that can be dragged across a section of the Timeline to indicate the exact section you want to zoom in to.

Things I Don't Have Mapped

Step Forward or Back 10 Frames: (see Figure 2.29) I either want the granular control of One Frame or I play through or scrub to find spots.

Go to In and Go to Out: I use the Option (Mac) or Alt (PC) modifier on the Mark In or Mark Out buttons to achieve this, so I don't have specifically mapped buttons. I do use the functionality, but rarely.

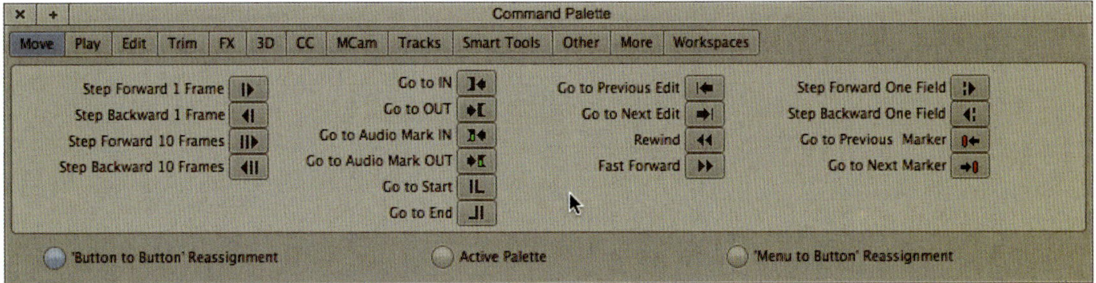

Fig. 2.29 Command Palette Move tab.

Audio Mark In and Out and Go to Audio Mark In and Out: I'm sure some people love this feature. It got added long after I started on Avid, so I never got used to using it. I'm also far more likely to do a regular edit and then trim out an audio "L" cut than to think it out ahead of time.

Go to Previous Edit and Go to Next Edit: These keys automatically send you into Trim mode on the transition you jump to. Often, this is not the behavior I want, so I use the Fast Forward and Rewind keys to serve this purpose (The correct behavior has to be set up in the FF/REW tab of the Composer setting.).

Go to Previous Marker and Go to Next Marker: Similar to Go to Previous Edit, the Fast Forward and Rewind keys do the same thing as long as the setting in the FF/REW tab of the Composer setting is set up to do this. I am usually using the FF/REW keys to jump to the head frame of the next clip. When I'm jumping to Locators, I usually do this with multiple Locators, one after another. I could map these, maybe to the UI, but I'll often just jump to the Composer setting and set the behavior to jump to Locators.

Replace Edit: (see Figure 2.30) I tend to just use Mark Clip and Overwrite to accomplish basically the same thing, but I'm sure there are many Replace Edit fans out there. Replace Edit allows you to mark an in and out on either the Source or Record side and then park on a specific frame on both the Source side and Record side that you want as a common point. When you hit Replace Edit, it goes into the Timeline in sync at that matching cursor point. For example, syncing a cymbal hit on the Sequence side with a visual impact in a scene from the Source side.

This is handy in the example of matching a beat or in replacing stock footage without matching code, for example.

Select Left, Select Right, and Select In to Out: (see Figure 2.37) These are recent additions to Avid. I'm sure that if you're coming from FCP, these are buttons you'll want to map. For me, instead of selecting and moving all of the clips before or after a certain spot, I use ripple trims. There are certainly times when Select Right would be useful, though. For things that you don't use all the time, consider placing them in the UI.

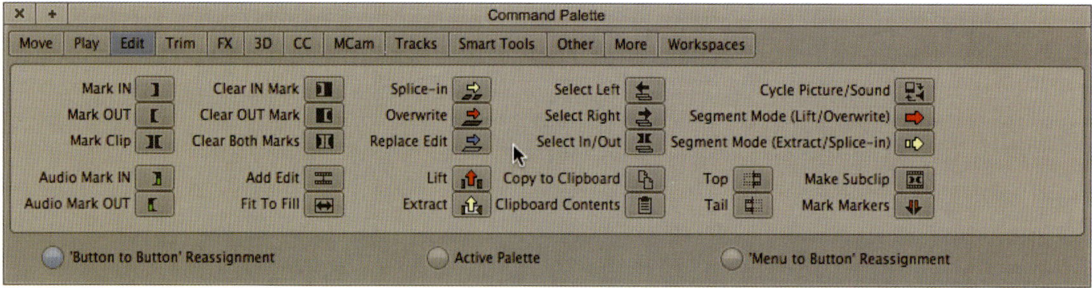

Fig. 2.30 Command Palette Edit tab.

Select Left and Select Right are Timeline operations, so it would make the most sense to place them somewhere in the buttons at the top of the Timeline window. They are definitely useful keys, but I would probably not use them every day, and that's my criteria for placing things on the keyboard. Starting in MC7, opt (Mac) or alt (PC) + select right or left selects everything EXCEPT the filler in enabled tracks. This is super handy for condensing footage down from upper tracks. Remember you can add the option/alt functionality to a key by adding the little "." button from the Other tab in the Command Palette.

Segment Mode arrows: I used to have these mapped, but these are legacy from the Avid systems before SmartTool was added. I would suggest that the SmartTool versions of these keys are preferred now.

Make Subclip: There are times when I sit down and create dozens or even hundreds of subclips. In those instances, I'll temporarily map that button to my keyboard or even duplicate my main keyboard and replace the button on the duplicate keyboard. Typically, I am marking in and out points and then making the subclip, so it makes sense to place this to the right of the In and Out keys. For me, this means replacing my Extend key. Otherwise, for occasional creation of subclips, you can do this by simply marking an in and an out on a clip and then dragging the clip from the Source monitor back to a bin or by dragging the Make Subclip icon at the top left of the Source monitor into a bin.

Mark Markers: This is another great function, though I tend to use it infrequently. If you like the idea of making multiple in and out points on a single clip like FCP-X can do, this is a way to accomplish the same basic concept. You can place multiple Markers on a clip at the points you want to have Mark Ins and Mark Outs. Then to use those Markers as Mark Ins and Mark Outs, you simply leave the Timeline locator between the Markers and hit Mark Markers.

Because I use this infrequently, it's a great candidate to go in the row of buttons under the Source monitor, which is where I would usually look for it. You could also use it in the Timeline, but I have never used it there.

Slip Left and Right: (see Figure 2.31) I tend to do these now as Timeline Trims with SmartTool functions.

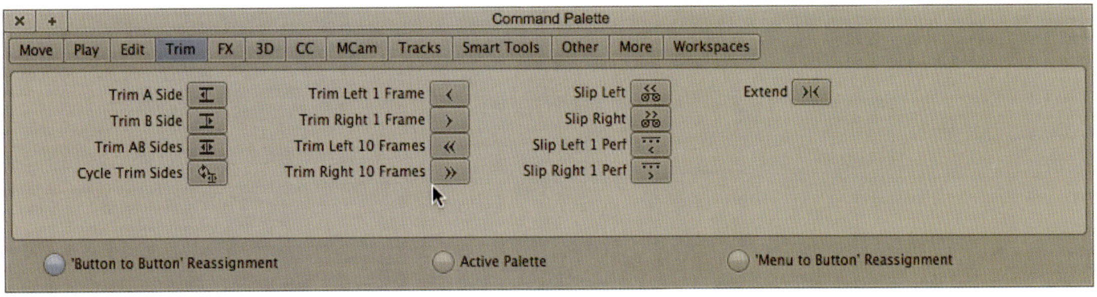

Fig. 2.31 Command Palette Trim tab.

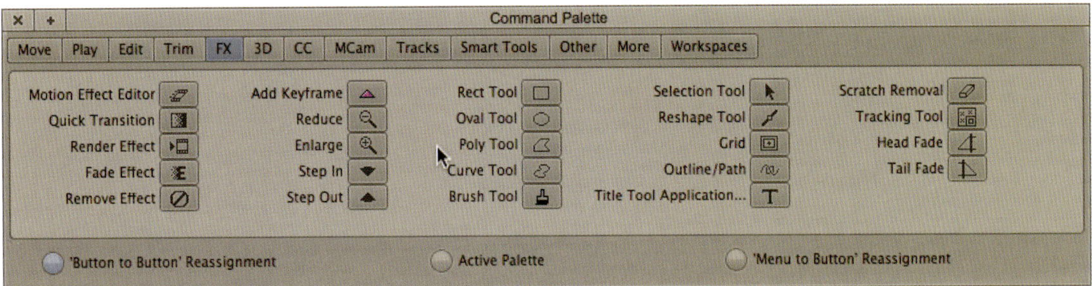

Fig. 2.32 Command Palette Effects tab.

Slipping Left and Right 1 Perf: This is a film editor's tool to fix slight audio sync issues. Unless you're editing features, this won't be too handy. This tool will be usable, however, only if you select the check box for Film when you create a project.

Almost all keys in the FX Tab or 3D tab: (see Figure 2.32) Other than the few that I have mapped, all of these keys fall into the category of not being used enough on a daily basis to warrant a spot on the main keyboard mapping. If you do a lot of effects work, this may be an option. You also may consider the possibility of creating a special Keyboard setting that gets enabled when you go into the Effects Workspace. Then you could map all of these keys to the keyboard, perhaps mnemonically, and they'd only be "under your fingers" when you go into the Effects Workspace. This just involves memorizing multiple positions of keys. For me, doing Effects involves so much mouse movement anyway that using the mouse to click a few buttons isn't really that big of a deal.

All Keys in the CC Tab: (see Figure 2.33) I do a lot of color correction, but I tend to do it all at once at the end of a project, so I have a specific color correction keyboard that I call up just when I go to color correction mode. Specifically, I map the color correction buckets and the Save Correction button, as well as the Next Uncorrected button and Previous Uncorrected button.

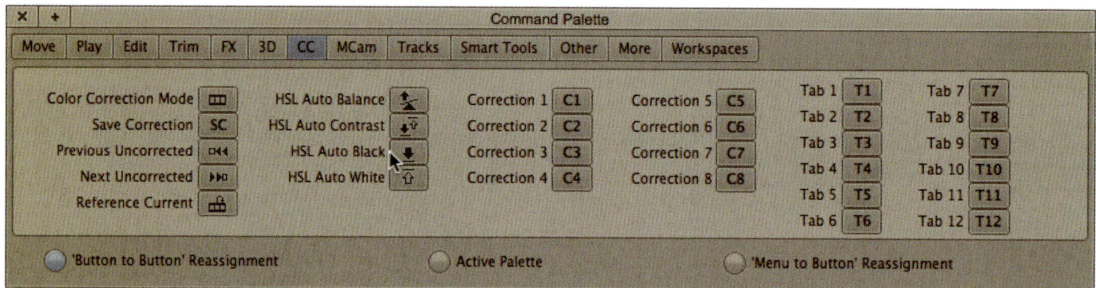

Fig. 2.33 Command Palette CC tab.

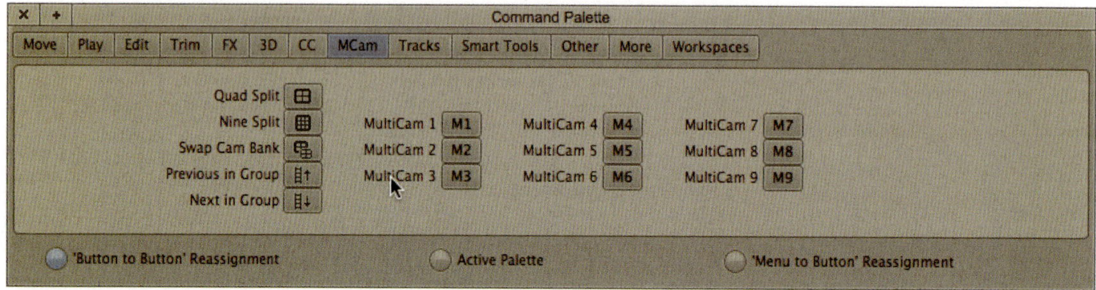

Fig. 2.34 Command Palette Multicam tab.

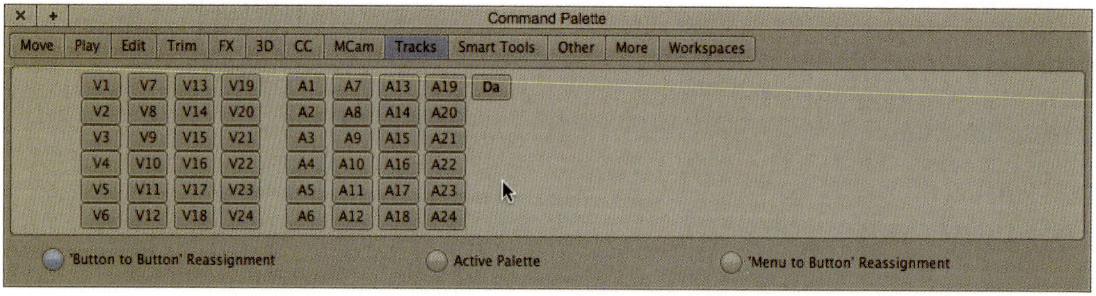

Fig. 2.35 Command Palette Tracks tab.

All Keys in Multicam tab: (see Figure 2.34) I do about one Multicam job a year. When I do, I map the Multicam 1–9 keys along the bottom of my keyboard on a specialty keyboard setting.

All Keys in Tracks Tab: (see Figure 2.35) Much of my editing involves only eight tracks of audio and four tracks of video, so having those tracks mapped is enough for me on many jobs. On jobs when I exceed that number, I just tend to click the mouse to toggle these tracks.

Smart Tool Toggle: (see Figure 2.36) I tend to use the individual buttons instead of the global toggle.

Link Selection Toggle: (see Figure 2.36) This is a key that many FCP editors will want to have mapped. I tend to use the mouse to select

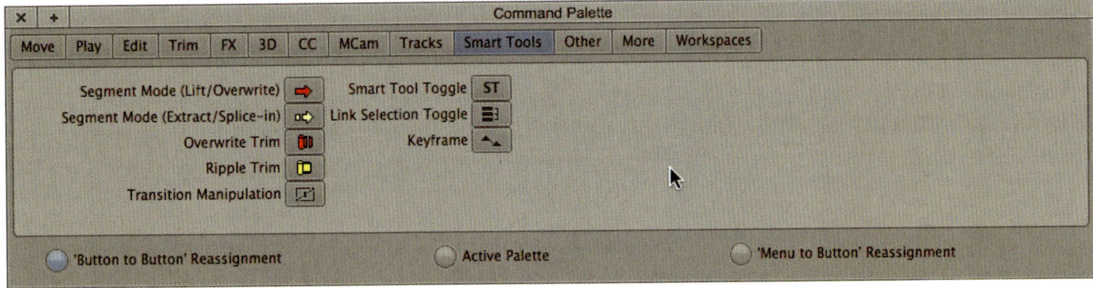

Fig. 2.36 Command Palette Smart Tools tab.

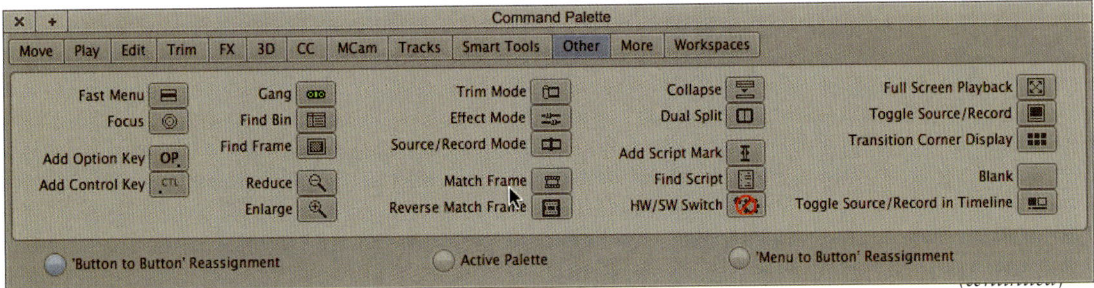

Fig. 2.37 Command Palette Other tab.

the clips that I want to move. However, for FCP editors, you may be used to having any video and audio that is recorded together move together, unless otherwise directed. So if you move a video track, the linked audio tracks will automatically move too. Use this button to either make the behavior mimic FCP (enabled) or to switch it to the "old" Avid behavior; in other words, if you want to drag clips in the Timeline, you need to select the exact clips on the exact tracks you want to drag.

Transition Manipulation: This is a very cool key for customizing the length of a transition using the mouse in the Timeline. Again, it's a fairly late addition to Avid, so I haven't adopted it, but this will feel very familiar to many FCP editors. If you can find a spot for it, definitely add it. However, since you already have to use the mouse to use this tool, you could easily just map this button to your Timeline.

Fast Menu: I tend to mouse to the Fast Menus I need, since I don't do this regularly. Also, the Fast Menu button is based on the active window, which is not always obvious, so I prefer to just click on the Fast Menu I want.

Focus Button: This is actually a pretty good button that zooms in and centers, or "focuses," the Timeline on the Timeline locator. I tend to do this more specifically with the Command-[] keys and the Command-/ key, but the Focus button toggles this back and forth with a single button click. I might have to consider adding this to my Keyboard settings.

Add Option and Add Control keys: These are not actually meant to be mapped by themselves, but in conjunction with other keys in instances when you want the option or control functionalities of those keys to be the default without having to hold down the option or control keys at the same time. As I explained earlier in this chapter, this is what I did with the Copy to Clipboard command. I always want the functionality of Option-Copy to Clipboard, so I mapped the Option key on top of the Copy to Clipboard key, so now I don't have to hold the option key to get the functionality I want.

Gang: This let you gang monitors together so that as you play or scrub in one monitor, the other monitor follows by the same amount. This is very useful sometimes, but I do it so rarely that I tend to just add it when I need it.

Find Bin: Another very useful key for finding and opening the Bin that contains a specific clip, but since I don't do this regularly, it's not mapped.

Find Frame: This is a fairly antiquated key. Don't confuse this with Matchframe. Find Frame allows you to trace a frame in a clip back to the same frame on the *Physical tape* from which it was captured. Basically, this is good to use if you have a tape hit on a clip and you want to see if the hit is actually on the tape or whether it just happened during playback for some reason. Since almost everyone has moved to file-based shooting and since tape hits are fairly rare, this key is basically pointless now.

Dual Split: I use this only in Color Correction Mode and when I'm color correcting, I do have this mapped.

Add Script Mark: This is for use only with ScriptSync. If you don't have ScriptSync, you don't need it. Even if you do have ScriptSync, you probably use this button only at the beginning stages of the edit, so you don't need it mapped all the time. If you're a big ScriptSync user, though, you may want this mapped to a permanent spot on your keyboard.

Find Script: Again, this is only for people who have purchased the ScriptSync option.

HW/SW Switch: My machines tend to stay in their hardware or software configurations, so I don't need this key mapped. But for anyone running multiple applications with a third-party card, this is handy. Avid will not "release" a card to another app when you switch apps. If you bounce back to FCP or Premiere, for example, the card won't be grabbed unless you release it from Avid first.

FullScreen Playback: I tend to work with a large client monitor, but I don't really mind editing with smaller playback, so I don't have this mapped. If you like to watch your edit back on the computer monitor at full size, then this would be a great key for you.

Transition Corner Display: This is an awesome function that lets you see the beginning, middle, and end frames of incoming and outgoing video through a transition. In a dissolve, it's the best way to be able to spot problems. I edit a lot of trailers, which, of course, use pre-edited movies. If I cut in a shot that's too close to an edit, then add a dissolve to that, the cut in the movie can make the dissolve in my transition seem like a jump cut. This can be hard to see or diagnose with regular playback of the transition. If all three frames don't match in Transition Corner Display, then you'll have a "jump cut."

Blank: This key has no real function. It's good to map to some keyboard layouts because the goal would be to have only keys mapped that you specifically assign. If you start with the Avid default mapping, you may want to place "blanks" over the keys with functions you won't be using. This helps keep "fat finger" mistakes from happening. If keys are mapped with "blanks" they won't do anything.

More Tab Keys: (see Figure 2.38) These are different-colored locators and Timeline views . For many people, these could be very useful. When I do on-line editing sessions, I often use multiple colored Markers to indicate problems or issues I have to deal with before the show can go to air. In those instances, I usually make a custom-mapped keyboard and put the different-colored markers on the function keys. The Timeline View keys could also be useful. I tend to simply mouse down to the bottom of the Timeline and choose these views by name. I probably switch Timeline views only a couple of times a day, if ever, so mousing to a specifically named view makes more sense than trying to memorize which view name is mapped to which key.

It is useful to map Workspaces (see Figure 2.39), but you can also set your settings so that when you go in to the various modes—Source/Record, Trim, Effects, Color Correction—the Workspace associated with that mode will be called up. Mapping buttons in the nonmodal Workspaces, like audio or Capture, would be an excellent idea.

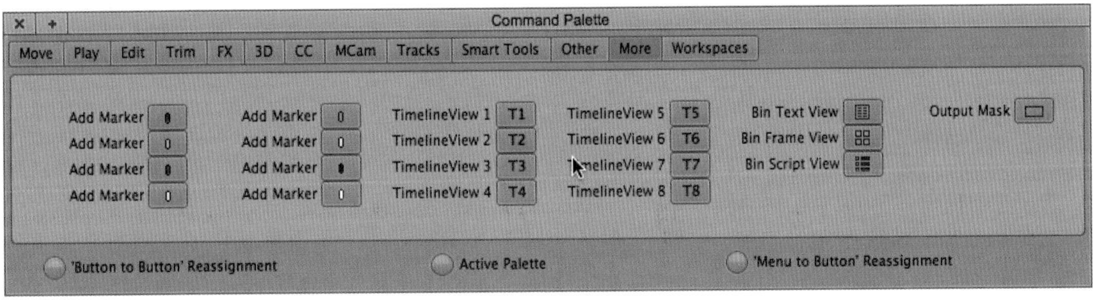

Fig. 2.38 Command Palette More tab.

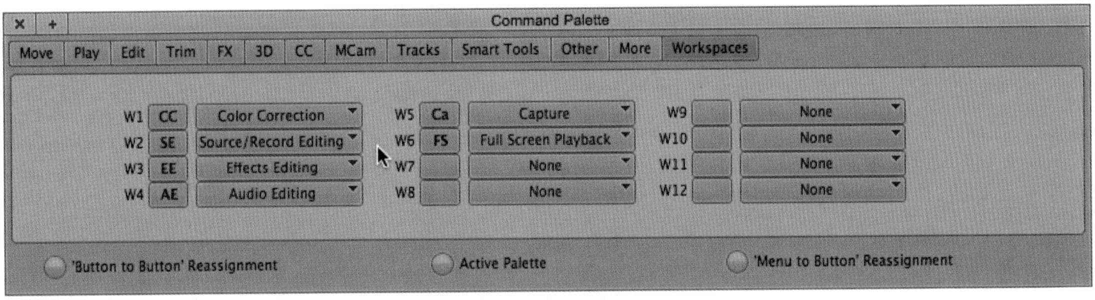

Fig. 2.39 Command Palette Workspaces tab.

FCP Keyboard

Some people would advocate that when you're editing in Avid, you should not try to force it to work like FCP. But many editors find that the muscle memory of pressing familiar keystrokes from FCP is something that's hard to leave behind. So, to help out, I went through the main keystrokes on the standard nonmodified FCP keyboard and mapped as many of them as possible to an Avid keyboard (see Figure 2.40).

To best explain what I did and why I did it, I created a table to accompany the FCP–Avid keyboard setting. The list starts at the bottom of the keyboard (space bar) and moves up from there and left to right.

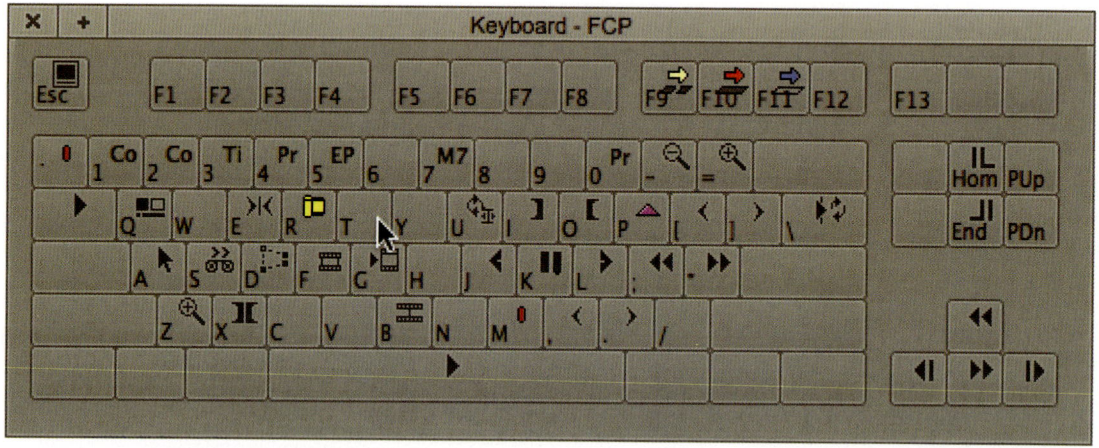

Fig. 2.40 FCP keyboard mapped on to an Avid keyboard setting.

Key	FCP	Avid (All Keys Are Mac Specific)
Space bar	Play	Default is the same
z	Zoom Tool	Zoom IN = z, Zoom OUT = Shift-Z)*
x	Mark Clip	Mark Clip = M, Remove Marks = shift x
C	Crop Tool	Can't map—In Effects Palette—3D Effect or PIP
V	Select Edit	No need—In Effects Mode click on edit transition
B	Blade Tool	Add Edit—Works similarly, but not quite the same
N	Toggle Snapping	No Need—Command-drag enables snapping
M	Add Marker	Marker = M
<	Trim Back	Default is the same
>	Trim Forward	Default is the same

Key	FCP	Avid (All Keys Are Mac Specific)
/	NOTHING	NOTHING
Left arrow	One frame back	Default is the same
Right arrow	One frame ahead	Default is the same
Down arrow	Next Edit	Go to next edit—Must set Composer setting FFWD
Up arrow	Previous Edit	Go to Last edit—Must set Composer setting REW
A	Selection Tool	No such concept in Avid, but Select for Effect = A
S	Slip and Slide	Slip 1 Frame FWD = S, Slip 1 Frame Back = Shift-S
D	Distort Tool	Corner Pin = D, Effect Palette > 3D Effect > CornerPin
F	Match Frame	Match Frame = F
G	Edit Selection	No need for this. Lasso transition enables Trimming
H	Hand Tool	No such concept in Avid, drag in Timeline Ruler
J	Play Reverse	Default is the same
K	Pause	Default is the same
L	Play Forward	Default is the same
;:	Prev Edit	Go to Last edit—Must set Composer setting REW
'"	Next Edit	Go to Next edit—Must set Composer setting FWD
Q	Toggle Windows	Toggle Source/Record = Q, not quite the same
W	Toggle Wireframe	No such concept in Avid
E	Extend Edit	Extend Edit = E, use with Mark In and Out
R	Ripple/Roll	Ripple = R, Roll = Shift-R
T	Track tool	No equivalent. Lassoing can select all in a track.
Y	Not mapped	Not mapped
U	Toggle Edit	Toggle Trim Sides = U
I	Mark In	Default is the same
O	Mark Out	Default is the same
P	Pen Tool	No such concept, Mapped Add Keyframe = P
[	Trim Backwards	Trim Forwards 1 Frame = [
]	Trim Forwards	Trim Backwards 1 Frame =]
\	PlayAroundCrrrnt	Play Loop = \
~ (tilde)	Marker	Marker = ~ (tilde)
Command - 1	Viewer	Composer Window = 1 view/canvas same in Avid
Command - 2	Canvas	Composer Window = 2 view/canvas same in Avid
Command - 3	Timeline	Timeline Window = 3
Command - 4	Browser	Project Window = 4, Project similar to Browser
Command - 5	Effects	Effects Palette = 5, Effects Editor = Shift-5

(*continued*)

(*continued*)

Key	FCP	Avid (All Keys Are Mac Specific)
Command - 6	Favorites	No such concept in Avid
Command - 7	Trim Edit	Go To Next Edit—7, Go To Previous Edit—Shift-7
Command - 8	Log and Capture	Go To Capture Mode—8, (Avid default is Cmd-B)
Command - 9	Item Properties	Control-click on Source window—Get Info
Command - 0	Sequence Settings	No real equivalent, Project Window > Format = 0
-	Zoom In	Zoom in = -, but really Command-L to zoom in Avid
+	Zoom Out	Zoom out = -, but Command-K to zoom out in Avid
F1	NOTHING	NOTHING is the default because F1 = HELP
F2	Log Clip	Logging and Capturing uses same tool = Capture
F3	Find Next	Find = F3, Find Again = shift F3, or just Cmd-F
F4	Lock Video	Can't map—Control-Click track > Lock Track
F5	Lock Audio	Can't map—Control-Click track > Lock Track
F6	Target Video	VideoTrack 1—F6. In Avid each track has patch #**
F7	Target A1	AudioTrack 1—F7. In Avid each track has patch #**
F8	Target A2	AudioTrack 2—F8. In Avid each track has patch #**
F9	Insert	Insert = F9
F10	Overwrite	Overwrite = F10
F11	Replace	Replace = F11
F12	Superimpose	Mapped option—Mark Clip here, then patch & edit***
Home	Jump to Start	Default is the same
End	Jump to End	Default is the same

* This can be up to the individual user. It is possible to map either zoom in to z and zoom out to Shift-Z, or it could be mapped more similarly to FCP, with Shift-Z for zooming out to show the entire Timeline. This is the way I prefer it. Since there are zooms elsewhere on this FCP keyboard, I would use zoom in on the z key and zoom out to full sequence on Shift-Z. This is described earlier in this chapter in the Mapping the Keyboard section.
** For F6, F7, and F8, Avid can't patch using a keyboard choice. The closest thing to this is to map the individual Enable Track buttons for all of the tracks that you may want to patch from the Command Palette (Command-3) in the Tracks tab. These keys allow you to select or enable tracks using the keys to which they're mapped. In order to actually patch to the tracks from the keyboard, you need to have Auto-Patching enabled in the Timeline Setting > Edit tab. Otherwise, in Avid this is something that simply has to be dragged in the Track Selector Panel, which is in the Timeline immediately to the right of the SmartTools.
*** The FCP Superimpose button selects the in and out point of the track below where you want to superimpose a title and then cuts it in directly above to the same duration. Avid has nothing exactly like superimpose, but you can use Option-Mark Clip to mark an in and out based on the duration of the clip below. On this button I have mapped Mark Clip and then added the Option modifier to it, so the only thing that you have to do is hit F12, patch the title or graphic that you want to superimpose to V2 (or higher), and hit F10 to Overwrite. And if you have Auto-Patch enabled, you just have to select the track you want to cut the superimposition on to, hit F12, and then F10. You could also put the Overwrite button on Shift-F12; then you could just hit F12 followed by Shift-F12 to get the same functionality.

Timeline

> **TIP**
>
> It is possible to add tracks and choose the numbering, similar to Target Video and Target Audio. Option-Command-Y (Mac) adds a specific video track that you can numerically assign. Option-Command-U (Mac) adds a specific audio track that you can numerically assign.

Many editors spend a lot of time in the Timeline. Having the correct settings and views for the Timeline can, therefore, be a huge help in making the most efficient use of your editing time. Earlier, I bragged that all of the options for Avid are in the Avid Settings tab of the Project window instead of spread throughout various menus, like they are with FCP, but the Timeline settings are the exception to the rule. The options for the Timeline are in three basic places: the Timeline Setting of the Settings tab, the Contextual (right-click or Option-click) Menu for the Timeline, and the Fast Menu in the Timeline.

All of these settings affect how the Timeline operates and how it displays various information. The Timeline can become quite "busy," so using the Timeline settings to keep the visual clutter to a minimum while still showing you the information you need is the key to these settings. Because the tasks you accomplish in the Timeline change and evolve as a project progresses, you also need to change what the Timeline is showing you throughout that progression. Having these Timeline views handy make it more likely that you will change these settings as you need to.

There are two tabs of options in the Settings list for Timeline, and there are even more options in the Timeline Fast Menu.

In the Display tab of the Timeline Settings (see Figure 2.41), the options are the following:

Show Toolbar: If you like to mouse-click on the UI to enable various functions, check this option. If screen real estate is at a premium, then uncheck it. The default is checked. Basically this provides you with a row of buttons in the UI directly above the Timeline. The more you use your keyboard, the less you need this option, especially if you have one or more toolbars under the Source/Record monitors in the Composer window.

Show Marked Region: This shows a colored region indicating the Mark In to Mark Out on the selected tracks. The default is checked. Having this on can be confusing when using various Clip Color options because the Show Marked Region makes everything blue between marks on selected tracks, obliterating your ability to see the other colors.

Show Marked Waveforms: This allows you to see audio waveforms in the Timeline between the Mark In and Mark Out. The default is unchecked.

Highlight Suggested Render Areas after Playback: This gives colored indicators letting you know what should probably be rendered. The default is checked.

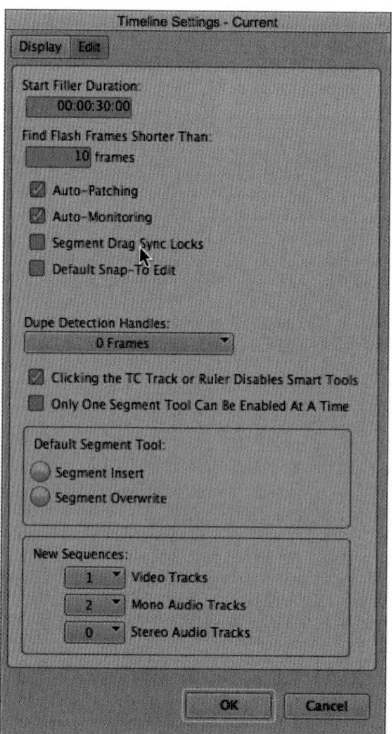

Fig. 2.41 Timeline Settings Display tab.

Show Position Bar: This adds a display at the very bottom right of the Timeline representing the entire length of the sequence and the position of the play head within that. This is helpful when you are zoomed in on the Timeline. To me, it's distracting. To an FCP editor, this will feel comfortable and desirable. The default is unchecked.

Show Effect Contents: This provides you with more information about the effects placed in the Timeline. I keep this turned off because I don't feel I need to see that in the Timeline and it increases redraw time and visual clutter. The default is unchecked.

Double-Click to Show Nesting: When you double-click on any effect or nest, the nest opens up, showing the nested/collapsed tracks. Some editors love this. I often open this by accident when I have it enabled, so I uncheck it. The default is checked.

Show Four-Frame Display: This is a relatively recent addition that makes drag-and-drop Timeline work more easily. It changes the Composer window to a four-frame window that displays the head and tail of the clip you're dragging plus the tail from of the "outgoing" clip before it and the head frame of the "incoming" clip that will come after it. As you drag through the Timeline, the exterior two frames will update depending on your position in the Timeline.

Use Fast Scrub: I prefer to have this turned on. Without it, scrubbing feels very slow. With it on, it's a little more choppy, but acceptable. The default is checked.

Movement during Play: The options are Page, Scroll, and None. The default is Page. This has an effect only when you are zoomed in on the Timeline. As the Timeline reaches the far right, it jumps (or "pages") to the next segment of Timeline with the same amount of zoom. With Scroll, the Timeline stays in place in the middle of the Timeline area and the Timeline scrolls underneath it. This used to be very processor intensive (it probably still is, but the computers are better now), so if you like this behavior, then give it a shot. I have become accustomed to Paging. Choosing None allows the play head to play right off the screen and doesn't update until the Timeline reaches the end or until playback is paused or stopped. You can still see the position of the play `head in the Source/Record Timeline.

In the Edit tab (see Figure 2.42), you will find the following options:

Start Filler Duration: The default is thirty seconds. I rarely add filler at the start of a sequence. If you do, how much do you usually add? In the old days, if you added filler at the beginning, it would add only one second at a time. I might set this at 5:00.

Find Flash Frames Shorter Than: This setting helps catch unintended flash frames. How you set this depends on the type of editing you do. If you cut promos or sizzle reels or music videos, you might want to set this at one or two frames because anything longer than that could easily be an intentional edit. If you cut longer form, or you edit at a more leisurely pace, you can set this at almost a second because any edit under one second is probably a mistake. Having at least one frame is useful to catch the occasional mistake. When this is set, any edit shorter than what is set will be flagged in the sequence. The default is ten frames. I'll occasionally edit five–eight frame edit, so the default is too long for me.

Auto-Patching: This allows automatic patching between the enabled tracks between Source and Record in the Track Selector panel. Any

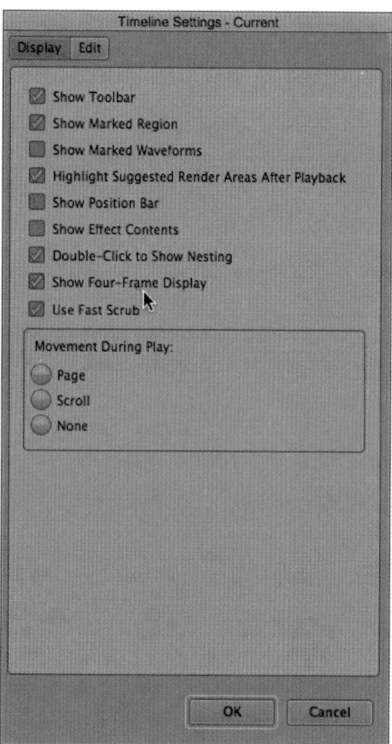

Fig. 2.42 Timeline Settings Edit tab.

enabled tracks on the Source side are automatically patched to any enabled tracks on the Record side. Generally, this is a good and efficient thing to turn on.

Auto-Monitoring: This setting automatically switches the track monitoring to the enabled tracks. Sometimes, when dealing with nested tracks, this can confuse things. If you notice that your monitoring is often on a track where you don't want it, switch this setting. With the release of MC7, I find that I sometimes monitor a track lower than the one that I want. If you don't see effects that you're expecting to see in the Timeline, make sure that you're monitoring the track that you expect—in most cases, the highest track.

Segment Drag Sync Locks: The default is unchecked. I leave it that way. This was a big feature request for several years, and Avid added the ability to lock sync when segment dragging. If you are worried about knocking things out of sync when segment dragging, turn this on.

Default Snap-To Edit: This is a very FCP-like setting. If you liked having magnetic Timeline behavior in FCP, then you'll want this turned on.

I prefer to turn "snapping" on and off by Command-dragging to enable snapping (magnetic Timeline) or regular dragging to not snap. The default is unchecked.

Dupe Detection Handles: This is set to zero as a default. The Timeline can flag shots that are already in your Timeline. In the "old days" of film editing, it was very helpful to get a warning if you were using frames that were even close to frames that were already in the sequence since it would be necessary to actually cut the negative. This allowed you to determine how much "handles" the dupe detection warns about. I don't cut things that have film negative, so I don't care as long as I'm not using the same shot twice in one sequence.

Clicking the TC Track or Ruler Disables Smart Tools: This is usually unchecked, but it is really a personal preference. This is actually the way Trimming works. If you are in Trim mode and you click on the Timecode track, it bumps you back to Source/Record. I tried editing with this on for a while and didn't like it, but I can see how some people would prefer to have it on.

Only One Segment Tool Can Be Enabled at a Time: This seems restrictive to me because you already have control over the number of segment tools that are enabled in the SmartTools area, and since I have the toggles of the two segment modes mapped to my keyboard, I can determine this at the touch of a key. Many FCP editors won't want this enabled.

Default Segment Tool: This can be set to Insert or Overwrite. For me, Insert has the potential to knock things out of sync, so I keep this on Overwrite.

New Sequences: This allows you to set up how many audio or video tracks are automatically created when you create a new sequence. The default is one video track and two mono audio tracks. As a default, I set the audio to four mono audio tracks and one video track. This depends on whether you work on a variety of projects that use different track options or always edit on a single type of project with distinct track requirements. Even though most of my projects end up with probably three video tracks and eight audio tracks, I'd rather start with less and add them as I go.

Timeline Contextual Menu

By Control-clicking (Mac) in the Timeline or right-clicking (Mac or PC) in the Timeline, you can call up the Timeline contextual menu (see

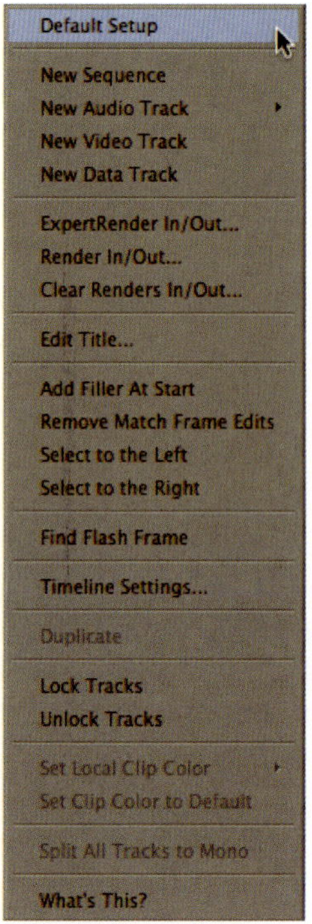

Fig. 2.43 Timeline contextual menu.

Figure 2.43). This menu allows certain behavior modifications and appearance modifications to the Timeline. The options here are the following:

Default Setup: This gets you back to Avid's factory default for the Timeline.

New Sequence: This is available in the Clip Menu as well. I've never accessed it from here.

New Audio Track: While I usually use the standard Command-u (Mac) or Control-u (PC) to add new audio tracks, this contextual pulldown allows you to add mono, stereo, or even surround tracks. The surround tracks are the only ones that you can't create from a keyboard Command. Shift-Command-u (Mac) or Shift-Control-u (PC) will create new stereo audio tracks in the Timeline.

New Video Track: Again, I usually add these with Command-y (Mac) or Control-y (PC). But if you're a mouser, here's another place to add video tracks.

New Data Track: Adds a data track to the Timeline.

Render Options: There are three rendering options available from the contextual menu. This is relatively convenient, but I prefer to map these and trigger them from keyboard commands.

Edit Title: A convenient option, Edit Title allows you to edit titles that are in the Timeline without going to Effects mode first.

Add Filler at Start: This option is available from the menus, but this is a convenient place to find it.

Remove Match Frame Edits: I rarely use this option, but if you need to remove any match frames that don't separate sections of audio or video that have had changes applied on either side, this is the place to access the capability. A good use of this is to clean up sequences from other editors, especially when doing color correction where you might apply a sequence correction and have it miss a piece of the shot due to an add edit.

Select to the Left and Select to the Right: Both are accessible from here but are also available as keyboard mappable commands as well as from the default Timeline toolbar. This location is the least convenient.

Find Flash Frame: Available elsewhere, this is a relatively rare thing to need to execute.

Timeline Settings: It's nice that these settings can be accessed from here, but this is the same settings list available from the Settings tab of the Project window.

Duplicate: You can select Duplicate to duplicate the current sequence, but again, this is doable from the keyboard with a simple Command-d (Mac) or Control-d (PC).

Lock and Unlock Tracks: I rarely lock tracks, but for many people it's an important capability to be able to access.

Timeline Fast Menu

This is where most of the real Timeline customization happens. Access to these options is through the Fast Menu (hamburger menu) at the bottom left corner of the Timeline window (see Figure 2.44). These are the options:

Default Setup: This allows you to return a heavily customized Timeline to the factory default at the click of a button. Excellent for troubleshooting Timeline issues.

View Type: The choices in the submenu are Time, Heads, and Heads Tails. This actually completely replaces your Timeline with thumbnails representing your clips. If you choose Heads, each clip is represented by a single head frame thumbnail. If you choose Heads Tails, each clip is represented by head and tail thumbnails (see Figure 2.45). This is really useful only in the early stages of a rough cut or selects reel. In this mode, you can very easily drag

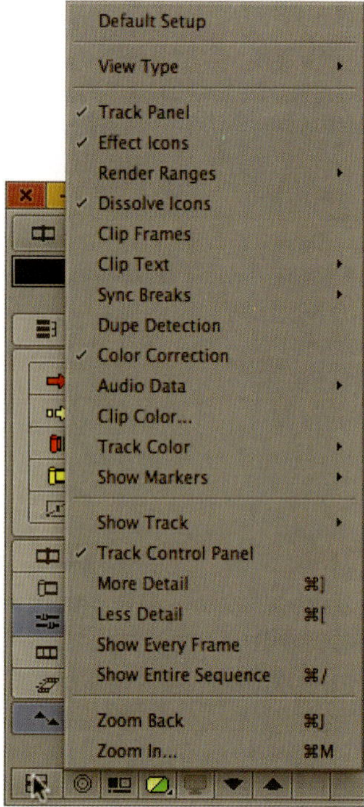

Fig. 2.44 Timeline Fast Menu.

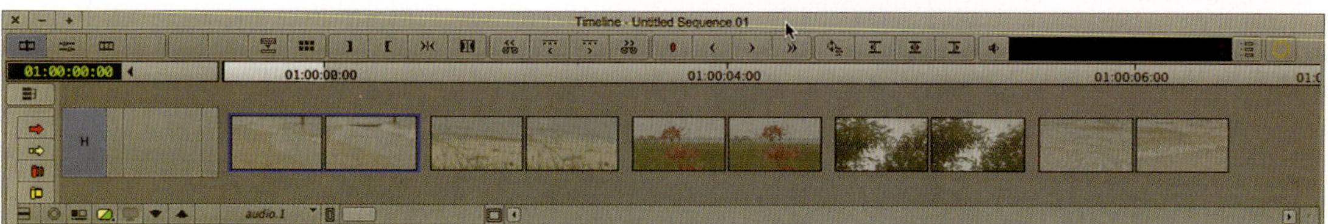

Fig. 2.45 This shows the timeline with the view type switched from the default of time to Heads Tails.

clips in the Timeline to rearrange their order. The Time option is the default. It's not the greatest name for it, but Time basically refers to the standard Timeline display.

Track Panel: Displays or hides the Track Panel, which is where you patch and control the tracks in the Timeline. For many editing functions, this is a critical thing to be missing, but there are certainly phases of an edit when the Track Panel is not necessary, and removing it gives you a lot more space in your Timeline for the sequence.

Effects Icons: Many people like the fact that they can see in the Timeline which clips have effects on them. But if you like the FCP Timeline, or you just want a cleaner Timeline with less clutter, you can eliminate the effects icons from the Timeline. The effects are not removed, just the icons.

Render Ranges: Use this to display colored indicators showing clips and sections of the Timeline that need to be rendered. The options are None, Partial Only, or All. None keeps the Timeline clean but gives you no information about needed renders. Partial Only shows you clips that have been partially rendered. All shows anything that needs to be rendered.

Dissolve Icons: Disabling effects icons will still show dissolves, which are much more common. If you want to hide dissolve icons too, this menu choice gives you the option.

Clip Frames: Allows you to see thumbnails of each segment in the Timeline (see Figure 2.46). To me they're distracting and unhelpful. And you really have to zoom in to your Timeline and expand the video tracks to make it useful.

Clip Text: The default is to display clip names in the Timeline. For a very clean Timeline, you can also choose None. Other options include showing Clip Tracks, Clip Duration, Clip Full Duration, Comments, Source Names, Media File Names, and Clip Resolutions (see Figure 2.47).

Sync Breaks: The default is Both, which refers to the fact that any discrepancies in sync between the audio and video tracks are shown both in audio and in video. Sync breaks are shown as small white numbers indicating the number of frames that separate audio and video. You can also choose to show only the sync error indication in the video track, the audio track, or none. I'd leave the default.

Dupe Detection: Colored indicators in the Timeline show display places where the same section of a clip has been used more than once. This can be a useful thing unless you're editing something that intentionally reuses footage multiple times.

Color Correction: Displays colored indicators in the Timeline showing which clips have had color correction applied. I don't care to see these indicators, but they may help you in certain circumstances. This is available with Symphony source side color correction only.

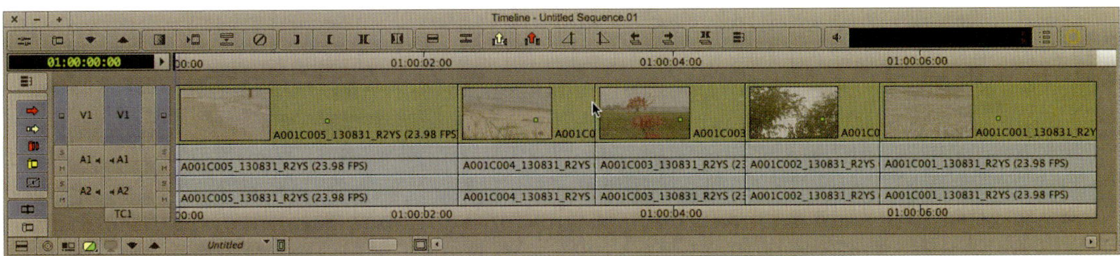

Fig. 2.46 This shows the Timeline with Clip Frames enabled.

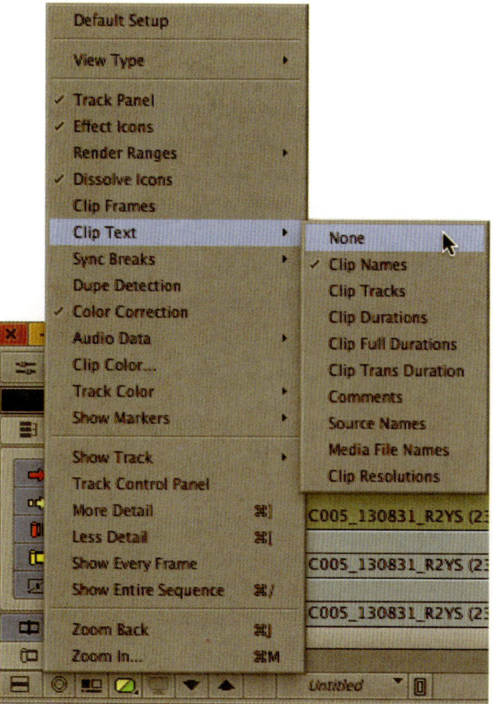

Fig. 2.47 Clip text in Timeline Fast Menu.

And it is the only way you would know there was a color correction on the clip with a source side correction.

Audio Data: (see Figure 2.48) This is one of the settings I visit most often. The default is None and Allow Per Track Settings. None shows no audio data. Allow Per Track Settings defers the Audio Data choices to a per-track decision that you can make in the Track Selector panel. Waveform displays audio waveforms in all of the audio tracks in the Timeline. Clip Gain allows for a small "fader" icon in each track segment. Clicking on the icon reveals a fader for accurate setting of levels. If you want to do rubberbanding of audio, choose the Volume option.

Clip Color: (see Figure 2.49) Allows for monitoring a number of states using colored clips in the Timeline. The options are to see colors for Offline media, Proxy footage, AMA footage, Mixed Frame Rates, SD/HD, Timeline Local colors, and Source colors. For each of these options, you can pick a specific color to indicate that state or use the default. Most of these are fairly self-evident. The two that I use the most are Timeline Local and Source. Timeline Local lets you pick specific clips in the Timeline and assign a color to them. This is

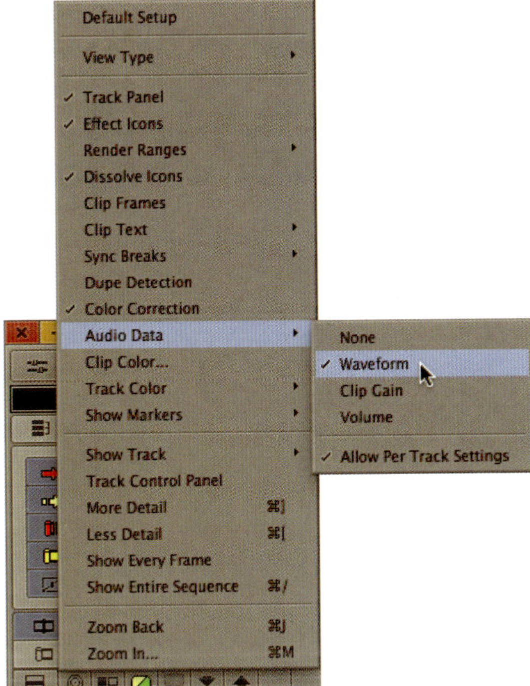

Fig. 2.48 Audio Data in Timeline Fast Menu.

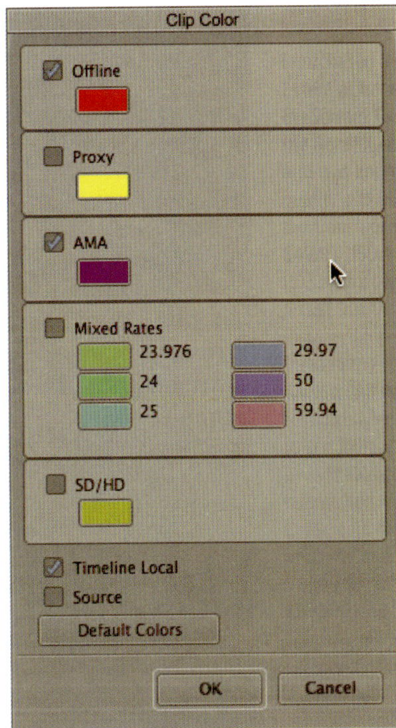

Fig. 2.49 Clip Color window.

useful to indicate things like the location of "bumpers" or commercial breaks. I also sometimes use this when creating selects reels to categorize certain subject matter by color. Then I can drag all the similarly color-coded clips into groups in the Timeline. Source allows you to color-code clips in the bin and see where those clips are in the Timeline. I like to color-code interview subjects and be able to see where each interview subject is in the Timeline.

Track Color: This is something I usually leave as the default. But if you want to indicate each track with a specific color, this is where you assign the colors. To me, the darker and less saturated the better. There's a big palette of colors to choose from beyond the spin-down arrow. You can also hold down Option (Mac) or Alt (PC) when selecting the preset color to open the system color picker dialog.

Show Markers: Allows you to display the markers and spanned markers that you've assigned. The default is to display them.

Show Track: It's possible to "hide" tracks. This has gotten me in trouble in the past, but the spin-down menu gives you the opportunity to hide specific tracks, selected tracks, show all (the default), and to hide or reveal various timecode tracks. The option for Film is similar to the Head Tail or Clip Frame options. It shows you a representation of your Timeline with visual thumbnails. However, for complex or lengthy Timelines, this option doesn't show every image because there's not enough space. This is also the place to eliminate a track, such as EC, if you don't want it there. This simply hides the tracks. If you actually want to delete them, it's as simple as selecting the track and hitting the delete button and then confirming that you want to delete that track(s).

Track Control Panel: Track Control Panel toggles between the extended Track Control Panel and the simplified panel. (The simplified panel was from the original Avid UI.) This control is a little pointless because the easiest way to toggle this on and off is using the spin-down arrow next to the small timecode window in the upper left corner of the Timeline. If you don't need to see all of the additional controls, this is a good thing to toggle off.

TIP

It is possible to map the Show Entire Timeline functionality that is available in the Timeline Fast Menu to Shift-z, just as it is in FCP. Follow the instructions from Mapping the Keyboard section on page 132 of this chapter.

More Detail, Less Detail, Show Every Frame, Show Entire Sequence, Zoom Back, and Zoom In all have keyboard shortcuts that make a lot more sense that executing these options from here.

Workspaces

Workspaces (see Figure 2.50) help you quickly access the proper tools when doing a specific task. While some editing hops between various tasks with a good deal of frequency, many editors work on a specific task—like effects, color correction, or audio—for an extended time. Workspaces makes the most of both of these working methods by allowing the editor to quickly move between tasks without having to set up special windows or call up special tools over and over again.

A good example of a workspace is the Audio Workspace. At the touch of a button it recalls all of the audio tools and positions them in the exact spot the editor wants them. It can also create linked settings that have a specific, common name. This would make it possible to link, for example, a timeline

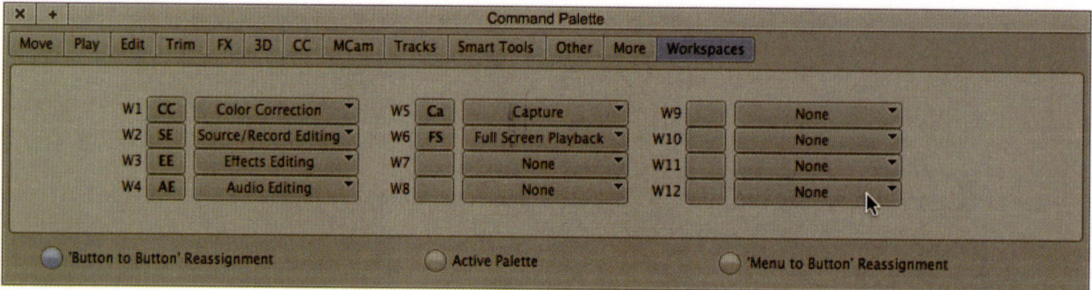

Fig. 2.50 Command Palette Workspaces tab.

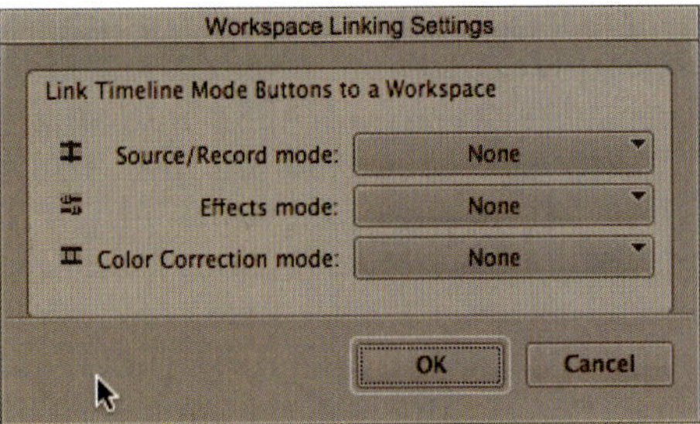

Fig. 2.51 Workspace Linking Settings.

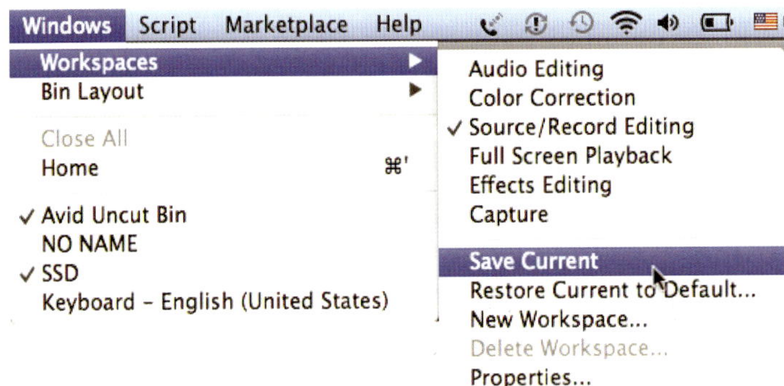

Fig. 2.52 Workspaces Menu.

with enlarged audio tracks and skinny video tracks, a keyboard setting with specialized buttons for doing audio stuff, and a Bin View with specialized audio columns with special audio information. More on that in a moment.

In some ways, it actually makes sense to replace the use of the mode keys (Source/Record, Trim, Effects, Color Correction) with workspaces that invoke those same modes because you get the same functionality as the mode with the added benefit of a workspace that helps you work in a contextual way. Proof of this is that Media Composer "from the factory" comes with workspaces that are matches for all of the modes except Trim, and that's probably because the Trim workspace would be very similar to the Source/Record workspace. Or you can use the Workspace Linking setting to automatically launch specific workspaces from the Mode buttons themselves (see Figure 2.51).

To begin exploring the use and set-up of workspaces, let's start with those "factory" defaults. Activating a workspace from the menu bar (Windows > Workspaces) is pretty easy (see Figure 2.52), but the real value comes when you map the Workspace buttons so you can jump between workspaces at the press of a button.

As of the writing of this book, the default workspaces are Audio, Color Correction, Source/Record Editing, Full Screen Playback, Effects Editing, and Capture. Seriously, I could understand if Avid dropped the Capture workspace sometime in the near future, though there's really no reason to other than to declutter the menu a bit. Meanwhile, people are still capturing from tape and laying off to tape, so, like any other workspace, its purpose is to simplify the task of tape capture.

As I briefly mentioned earlier in this chapter, I have four of these workspaces mapped to my keyboard: Source/Record, Trim, Effects, and Color Correction. I rarely use Full-screen Playback or Capture, and I tend to use my Audio workspace only occasionally, so I'm fine with calling that up from the menu.

Create a Custom Workspace

Earlier in this chapter I described how to map buttons from the Command Palette and from the menus to your keyboard. In the Command Palette, there is an entire tab devoted to Workspace buttons. Notice that many of the options are blank. If you would like to create custom workspaces, beyond those indicated in the Workspace tab, you first have to create them using the Workspaces menu options (Windows > Workspaces > New Workspace). When you map the workspaces to your keyboard, you will be much more likely to use them regularly.

There is no trim workspace, so let's make one. Later on, we'll link several other settings to this workspace to make it even more useful and efficient.

The basics of the Trim workspace will be fairly similar to the Source/Record editing workspace, so before we start to customize the Trim workspace, start by recalling the Source/Record editing workspace.

Most of the elements of the Trim workspace will be done by linking other settings to this workspace, for example, a customized keyboard and Timeline. But we can also add some features to the UI that will be helpful, especially if you tend to alternate between keyboard and mouse operation.

The big danger with starting with an existing workspace is that you could accidentally modify the workspace that you start with. To prevent this, we'll begin by naming and saving our new trim workspace before we actually do any editing or modification of the workspace. That will protect the Source/Record workspace that we started with. To do this, go to Window > Workspaces > New Workspace (see Figure 2.53). This will open a dialog window and you can name your new workspace. I called mine Trim. Now the check mark indicating which workspace is active will be on your new workspace instead of next to Source/Record (see Figure 2.54). Now any changes you make to the workspace will be saved to the Trim Workspace.

One advantage of a Trim workspace is that when you're trimming, you don't need to see anything other than the Source/Record windows and the Timeline, so if you're editing on a single monitor your Trim workspace can fill the monitor with just the S/R monitors and Timeline (see Figure 2.55).

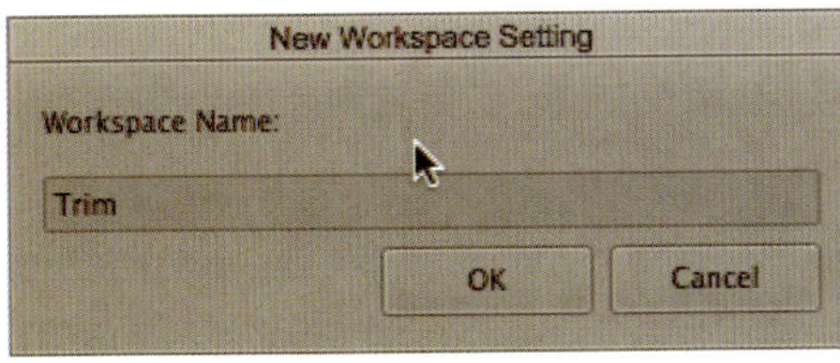

Fig. 2.53 New Workspace Setting.

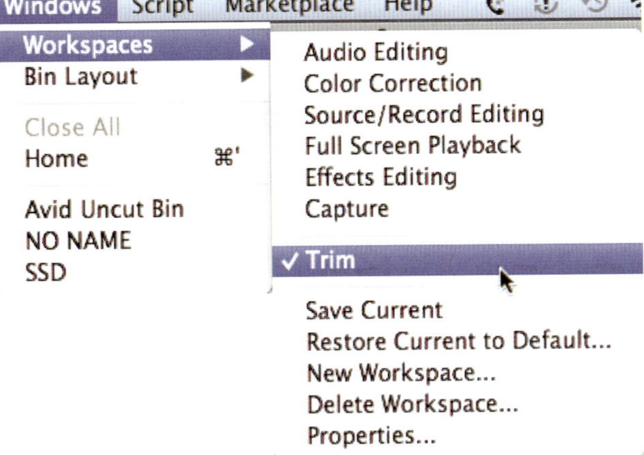

Fig. 2.54 Trim workspace in windows menu.

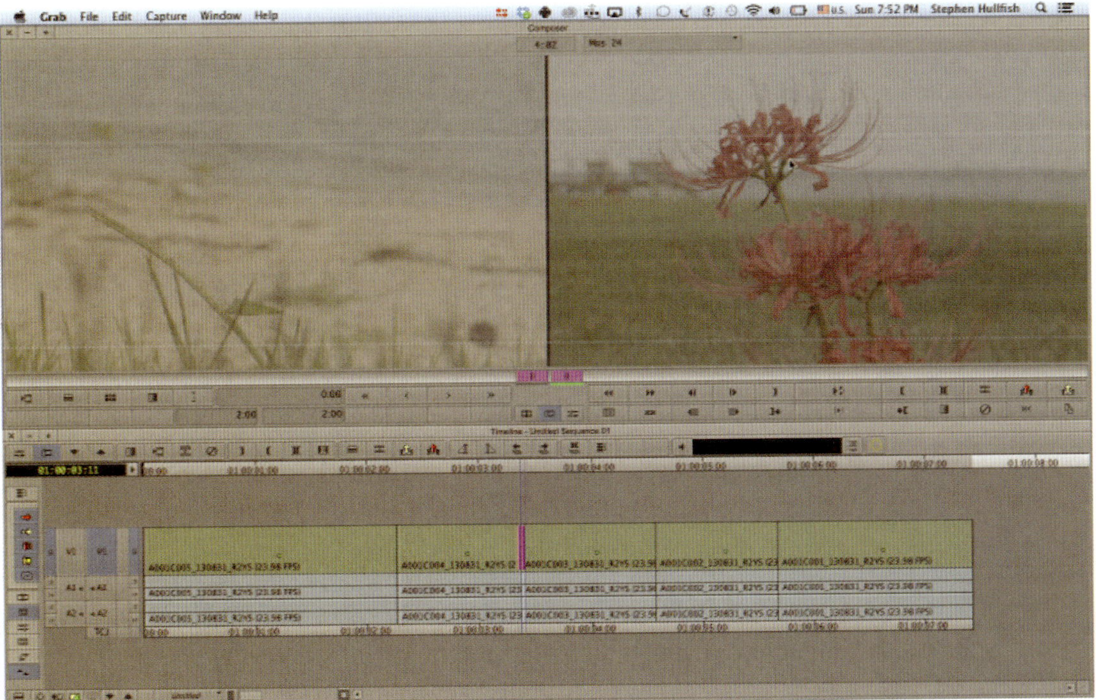

Fig. 2.55 Trim workspace.

Most of the other stuff you'll want to do with the Trim workspace will be based on Linked Settings, which will be explained later in this chapter.

Reconfiguring and Saving an Existing Workspace

The factory defaults are pretty well designed and thought out, but because we all have different expectations, workflows, and comfort zones, making a workspace your own will also increase the odds that you'll use it and that it will provide the most efficiency to you.

Let's customize the Audio workspace to add some extra capabilities and tools to those provided in the factory default Audio workspace.

The first thing to do is to simply call up the existing Audio workspace from the menu bar: Windows > Workspaces > Audio editing.

The factory default for this workspace is fairly simple. It launches the Audio Mixer and the Audio Tool. You can resize and reposition these tools and arrange them on your screen or screens.

Then, from the menu bar choose Tools. In the Tools menu, there is a series of audio tools near the top of the menu: Audio Mixer, Audio EQ, AudioSuite, RTAS, Audio Tool, Audio Punch-In. You've already got the Audio Mixer and Audio Tool. Audio Punch-In is more of a recording tool than an audio tool, so let's open the others: Audio EQ, AudioSuite, and RTAS (see Figure 2.56).

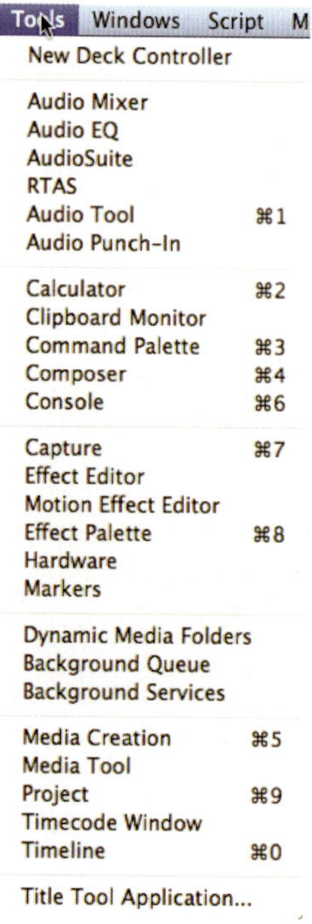

Fig. 2.56 Audio Tools.

Organize, size, and position the various audio tools on your screen (see Figure 2.57) and then, from the menu bar, choose Windows > Workspaces > Save Current. This saves the layout. Give it a try by switching to a different workspace and coming back to the Audio workspace.

Linked Workspace Settings

The way to make these workspaces really useful is to link other settings to the workspace, so that every time the workspace is recalled, other settings, such as Timeline, Bin Views, and Keyboard, are also enabled.

I want to create a Bin View and a Timeline View that automatically switch when we go into the Audio workspace. Also, when you're in the Audio workspace, you may not need to actually look at video at all on the monitors, or, at least, you could shrink the video monitors down substantially to make room for a bigger Timeline.

If you have a video card, Nitris hardware, or an external video box, such as AJA IO XT or Xpress, it's possible to completely eliminate the Composer video monitors because you can monitor video on your "client monitor" which automatically switches between the Source monitor and the Sequence monitor. To completely eliminate the video monitors but still keep the rest of the Composer UI, click at the bottom of the Composer window to get a cursor that looks like a double-headed arrow. Drag straight up and the two video windows will disappear, leaving all of the tracking info, Timelines, and buttons below the Timelines. If you

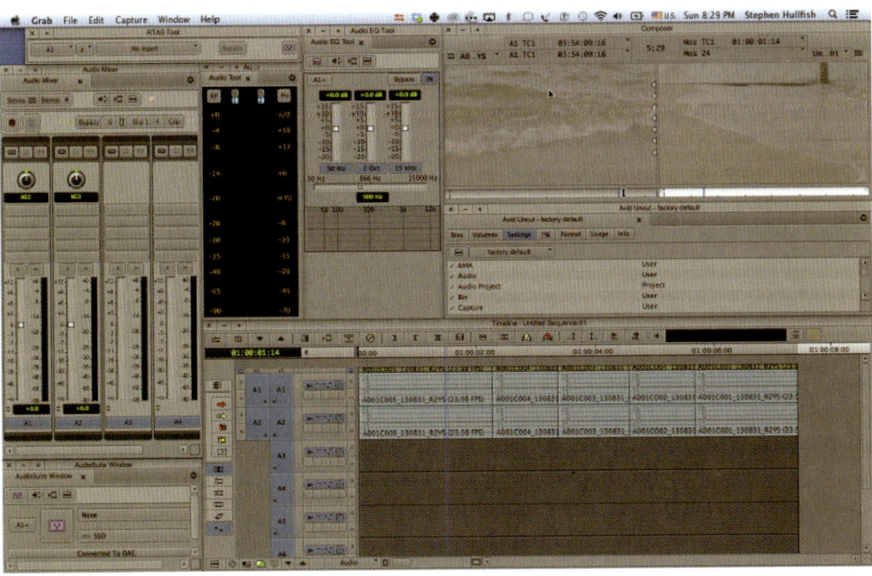

Fig. 2.57 Audio workspace.

don't have external video display, you can make the Composer window as small as possible by clicking on the bottom right corner and dragging diagonally up and to the left until the window can't get any smaller.

With that done, drag your Timeline out the same way so that it fills up more of the screen, making room for nice, fat audio tracks. Then, let's use Command-L (Mac) or Control-L (PC) on enabled audio tracks to make the tracks nice and thick. We also want to enable audio waveforms (Figure 2.58). Since we won't be doing effects in this Timeline, we can turn off Effects Icons in the Timeline Fast Menu. Turn on all of the Audio data choices in the Timeline Fast Menu in the Audio Data options. With the Timeline customized the way you like it for doing audio work, go to the Timeline view menu, which is at the bottom left of the Timeline (it may say "Untitled" if you've never saved a Timeline view), and select the Save As option and name the Timeline view "audio" or something similar.

We can create any other settings that we want to have launch with our audio workspace. Some ideas are a Keyboard setting with lots of audio options or a Bin view with lots of audio-related columns.

To create an audio keyboard, go to the Settings tab in the Project window and type K to jump to the Keyboard setting. Select the Keyboard setting and hit Command-d (Mac) or Control-d (PC) to duplicate the current keyboard. On the duplicated keyboard, click to the right of the word *keyboard*, probably where "US English" or something like that describes the name of the keyboard. Type the same exact word that you named the Timeline view. Open the audio keyboard and the Command Palette and map as many audio tools as you can find to the keyboard (see Figure 2.59).

Then with a bin selected, choose Choose Columns from the Bin menu and clear all of the columns using the All/None until all of the columns are deselected. Then add back every audio-related column you can find.

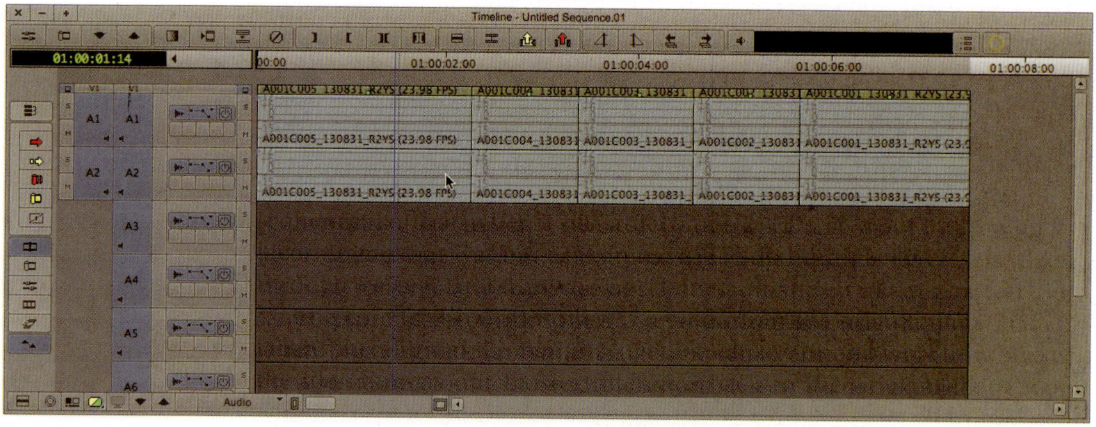

Fig. 2.58 Audio Timeline.

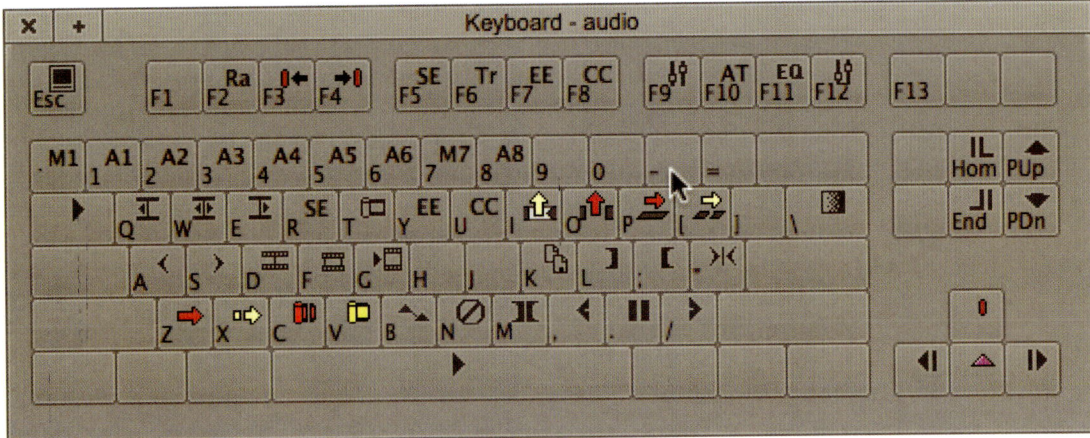

Fig. 2.59 Audio Keyboard.

* NO NAME

Name	Audio Format	Audio SR	Tracks	Audio Bit Depth	Soundroll	Sound TC
00045	PCM (MXF)	48000	V1 A1-2	16		
00044	PCM (MXF)	48000	V1 A1-2	16		
00043	PCM (MXF)	48000	V1 A1-2	16		
00042	PCM (MXF)	48000	V1 A1-2	16		
00041	PCM (MXF)	48000	V1 A1-2	16		
00040	PCM (MXF)	48000	V1 A1-2	16		
00039	PCM (MXF)	48000	V1 A1-2	16		
00038	PCM (MXF)	48000	V1 A1-2	16		
00037	PCM (MXF)	48000	V1 A1-2	16		
00036	PCM (MXF)	48000	V1 A1-2	16		
00035	PCM (MXF)	48000	V1 A1-2	16		

Untitled

Fig. 2.60 Audio Bin view.

When you're done, choose OK in the Bin Column Selection window. But that doesn't save your bin view. You still have to go to Bin View menu pulldown at the bottom right of the bin (see Figure 2.60). The pulldown could say "Untitled" if you haven't saved any bin views before. Choose Save As from the pulldown and give the bin view the exact same name as both the Keyboard and the Timeline view.

With those three settings named the same, go back to the Settings tab of the Project window and choose the Workspace view for the Audio Editing workspace (see Figure 2.61). Click on the Link to Named Setting button and type the name that you gave to the three settings (probably "audio") in the text window below. Then click OK.

With those settings done, try switching back and forth between your Audio Workspace and another workspace. When you go to Audio Workspace, notice that the Bin view, Timeline view and Keyboard all switch along with the workspace.

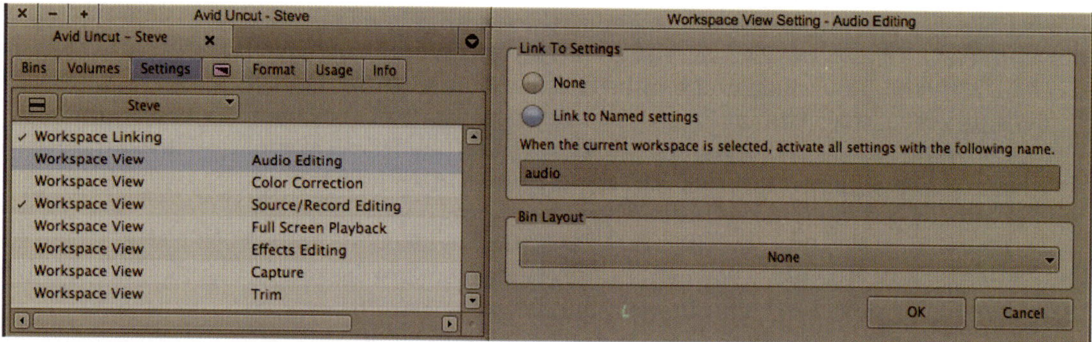

Fig. 2.61 Audio Workspace Linking.

Sharing Settings between Users or Projects and Site Settings

The ability to share settings between users and machines is a very useful thing. I don't do it a lot, but when I do, I'm glad for the time it saves me.

There are two ways to share user settings. One way is to export the entire user setting from one machine. Save it to a CD, DVD, USB drive, or network location and then import it into the target Avid. To do this, go to the Settings tab of the Project window. Click on the User Name pulldown menu and choose Export User or User Profile (see Figure 2.62). Choose the file path to the storage location and hit OK. Move the media to the target Avid. In the Settings tab of the Project folder, go to the User Setting pull-down and choose Import User or User Profile and navigate to the media where the exported user setting is, select the User folder, and choose Open.

The other way to do this more selectively is to create a New Settings file. Select the Settings tab in the Project file. Then, in the menu bar File menu, choose New Setting. This opens a small window that appears empty. You can drag specific settings from the Settings list into this window. Try dragging the Keyboard setting, a Bin View, and an Export Setting into the New Settings file window (see Figure 2.63). When you're done, click on the close button (x) in the top left corner. This calls up a Save dialog. Type a name for your saved settings and use the pulldown menu next to Where and navigate to a location, such as the Desktop, where you want to save the settings (see Figure 2.64). It creates an XML file. Save this XML file to a USB drive or something to get it to the target Avid.

On the target Avid system, in the Setting tab of the Project window, from the menu bar, choose File > Open Settings file and navigate to the XML file on the USB drive and choose Open. This opens a window with your short list of settings. You can drag these files from the window into the Settings list in the Project window. This will call up a dialog asking to Add or Replace. Replace is always a dangerous choice, so I'd use Add, but the choice is up to you.

TIP

Whenever you start using a new version of Media Composer, the recommended procedure is to start off by completely rebuilding your user settings from scratch. If you are a freelance editor that often moves between versions of Media Composer at different facilities, you should carry a copy of your user settings from each version. The risk of not doing this is that user settings can become corrupted. It's rather hard to tell when your settings are corrupted unless you're an experienced Avid editor. The reason for this is that corrupted user settings create strange behaviors in Media Composer that you might mistake for bugs in the software. Back up your user settings so that you can recall them from your safe location if they become corrupted.

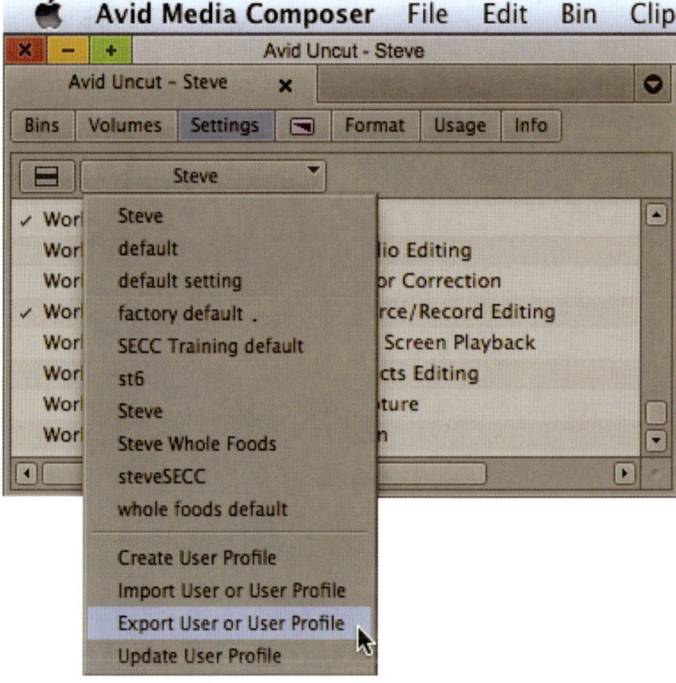

Fig. 2.62 Export User setting.

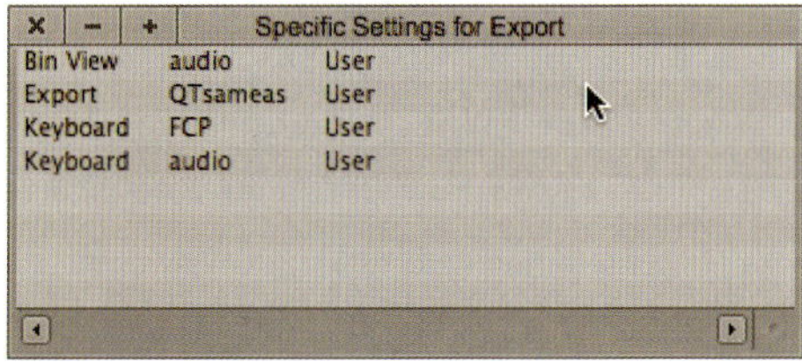

Fig. 2.63 New Settings file.

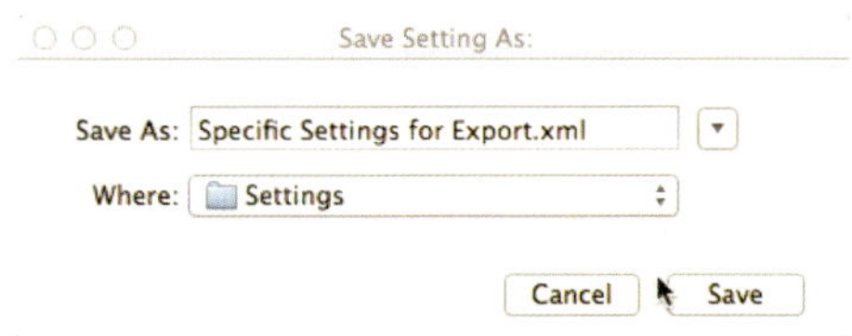

Fig. 2.64 Specific User setting.

Import Settings

I see some pretty horrendous video on broadcast TV that I attribute directly to poor Import settings, especially for video.

One of the things that FCP definitely does better than Avid is to guess the appropriate input settings. With Avid, you have to know what the frame size, frame rate, color space, and field order are before you import a file. Editors are often forced to bring assets into their projects for which they don't know any of these details. If you get a video file or audio asset from someone you can contact, ask the person to clarify all of the options in the Import dialog. I just brought in files from a feature film mixer and sent a quick note to confirm that the audio was 23.98, not 24fps. If you work with an outside 3D animator or After Effects artist, ask for the exact render specifications for the file. This will insure that the file is imported properly and will look its best.

The Options for Import and the Effects of Common Mistakes in Import Settings

Check out the Import Settings options. To do this, select a bin. Then from the menu bar, choose File > Import. This opens the Import dialog. Most of this dialog window is simply the navigation to the file. At the bottom, there are a few options, including an Enable pulldown. The Enable pulldown has options to choose the type of file you plan to import. The most obvious choice is to leave the default of "any." The other choices include an option for which media drive to store the imported media on. That option also allows you to click on the strange little icon of the color bars and speaker. That icon allows you to store the audio and video from the imported file on separate drives. There is also a pulldown menu from which to select the resolution for saving the imported file. There are several options for saving as DNxHD, ProRes, and XDCAM among others.

But the crucial section is actually under the Options pulldown. This launches the Import Settings dialog window (see Figure 2.65), which has five tabs: Image, OMFI/AAF, Shot Log, Audio and XDCAM. Of these, the Image tab is the one with the most variables that can go wrong.

In the Image tab, there are four sections: Image Size Adjustment, File Pixel to Video Mapping, Frame Import Duration, and Alpha Channel.

In Image Size Adjustments, quite a number of things can go wrong. There are four choices: Image sized for current format, Crop/Pad for DV scan line difference, Do not resize smaller images, and Resize image to fit format raster. These all have to do basically with the shape, size, and aspect ratio differences between the source file and the imported image.

Image sized for current format means that you must be certain that the aspect ratio and size of the file to be imported must be the same as the project. For many graphics files and photos, this will be a bad choice because the shape of the file will be squished or stretched to match the current project.

Crop/Pad for DV scan line difference is really necessary only if importing DV footage or if bringing footage into a DV project, both of which are becoming rarer and rarer. The reason for this is that DV footage actually has a slightly different number of vertical lines of resolution from standard broadcast standard def video. Making a mistake here delivers an odd-looking artifacting that is often mistaken for a field-ordering issue.

Do not resize smaller images is a good choice for smaller graphic or photographic files because otherwise the files will be up-scaled to the size of the project. Failure to use this with smaller files will cause the imported files to look very blocky and aliased because they will scale up beyond the resolution they can support.

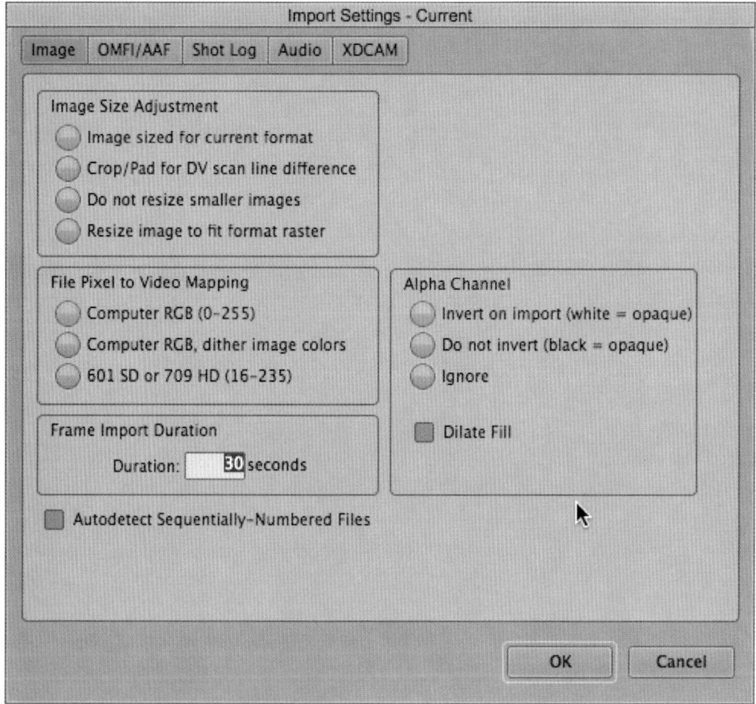

Fig. 2.65 Import Settings dialog.

Finally, Resize image to fit format raster is the best solution for any graphic that you know is not the same aspect as the current project or if you're unsure of the aspect of the file. A good example of this is if you want to import a photograph that was taken in "portrait" orientation. Using this option will import it so that the vertical size of the photo is matched to the vertical size of the current project, and the horizontal size stays in the proper ratio of the original file.

Often, the best way to determine the proper import setting is a little trial and error. But the rule of thumb is that if you know you're importing 16:9 footage, go with the first choice and if you know that it's not 16:9 or if you're not sure, choose the last option.

File Pixel to Video Mapping is all about color space. The choices here are Computer RGB (0–255), Computer RGB, dither image colors, and 601 SD, or 709 HD (16–235). The point of these choices is to correctly map the color of pixels between images that have been captured or created using 8-bit color space of 0–255 and true REC709 or CCIR601 color space (meaning "real video color space"). The problem here is that you have to know ahead of time if the file you are importing is actually using 0–255 color space. Almost all graphics and photos will be in this color space. The problem comes from imported video footage, which could have been saved as REC709 or as pure RGB, using the full 0–255 range. If you're importing footage from someone you know, like a vendor, ask whether the image is 0–255 or 16–235. A lot of stock footage, including ArtBeats, is RGB 0–255 (though ArtBeats does have RED R3D files available, and those could be imported as REC709). Pond5 is also RGB. Footage imported directly from a Canon 5D MII is also RGB. Footage from one of my favorite After Effects artists always comes to me as RGB. And footage from my favorite 3D vendor comes to me from Maya as RGB. Certainly, it's possible to export from After Effects or Maya as REC709, but you need to ask your vendors how they're exporting footage. When in doubt, do a test. Import it both ways. Using RGB on a REC709 file will always look slightly more washed out. But using REC709 on an RGB file will result in illegal colors. If you know you'll be doing color correction on it, you can import as REC709 and use your color correction tools to fix the illegal levels.

Check out the following chart for the errors that can occur:

	Imported as RGB	Imported as REC709
Source is RGB	Perfect colors	Illegal—blacks too low, whites too high
Source is REC709	Washed out—blacks too high, whites too low	Perfect colors

The other choice includes the ability to "dither" the incoming file. This is a mathematical function that helps smooth the conversion of colors between the source and imported file. It is supposed to help smooth gradients. I imported an After Effects file that had a gradient and some very fine detailed images with RGB, and RGB dithered. I didn't notice any difference in the smoothness of the gradient or anywhere else in the image. The imports took about the same time, so I suppose using the dithered option makes the most sense.

The default in Frame Import Duration, is 15 seconds. If you want to import a graphic and use it as a bug that sits in the bottom corner of your sequence the whole time, you do not want to import it as a 15-second clip that you have to edit into the sequence 120 times. The crazy thing is that no matter how many seconds this setting is set to, the size of the media file on the hard drive doesn't change. I imported a still at 15 seconds, 2 minutes, and 3 hours 20 minutes (12,000 seconds) in length, and all of the files took up 1,025KB on the drive.

The default in Alpha Channels is to import without alpha. I think that's a good default. It's actually a bit of a pain if you import a graphic that you want to have come in as a standard clip, and it imports as a keyable clip (with the little purple key icon). If you want a clip instead of an "effect," choose Ignore. If you have a graphic that imports and keys using an alpha channel perfectly on FCP, then choose the Invert on Alpha option. FCP and Avid see the alpha channel and react to it in different ways. FCP sees white as opaque in the alpha channel and Avid sees black as opaque. Do not complain about this. Do not worry about how it is illogical. This is the way it is, and for Avid to be able to have backwards compatibility with previously created sequences, it needs to stay this way. Avid deals with transparency, which it converts to alpha, perfectly though, which is why I try to create .PNG files for import into Avid. Stick with the Invert option when importing files with alpha channels unless your clip comes in with the wrong stuff keyed out. Do not invert is for importing alpha-channeled graphics that are specifically made for Avid. Most graphics artists, however, will deliver files with alpha that will need to be inverted.

Checking the box in front of Autodetect Sequentially-Numbered Files will allow you to import any sequence of stills as a continuous video file by simply selecting the first file in the sequence. Oftentimes, with my 3D animation vendor, if the 3D is going to be keyed over something, I'll ask him to deliver a TIFF sequence instead of a QT file. He will export every frame of the animation as a TIFF with alpha, and I will import the animation by turning on this option and selecting the first file in the folder. This works with many different still image file types. I have used TIFFs, PNGs, and JPEGs. I know that many other file types work as well. The files must be numbered in the file name and sequential.

If you ever import a still that you are expecting to be fifteen seconds long, but it comes in as a single frame, it's probably because this option was chosen, and the Avid system thinks you wanted to import a sequence of files that is only one frame long. Also, make sure you only select the first file of the sequence in the folder. If you do this in Avid you will end up with a new sequence for every frame selected starting with that selected frame.

OMFI/AAF Import: The next tab controls how AAF and legacy OMFI is imported (see Figure 2.66). You can choose to import the footage using the source's file resolution or with the current import resolution that is set in the Import dialog. You can also have the Avid system ask you what to do with every file that is not already at the current import resolution. I've never used this option.

Shot Log: This is used more for big productions with footage coming from some kind of dailies producer. It allows you to import a shot log and have all of the metadata from the shot log merge with master clips that have already been imported into your system (see Figure 2.67). It's a method of getting a lot more metadata into the master clips than is possible through any other means. You can either ignore this option because you aren't playing at this level, or you can have your postproduction supervisor tell you what option to choose.

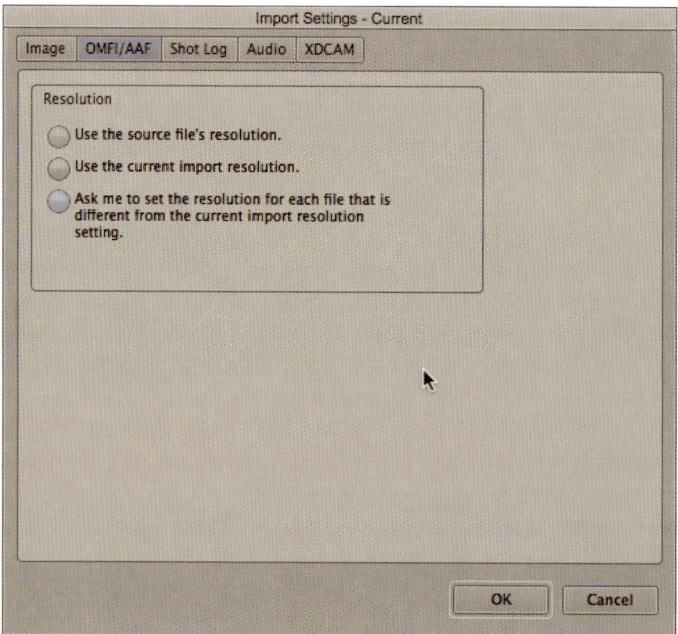

Fig. 2.66 OMFI tab.

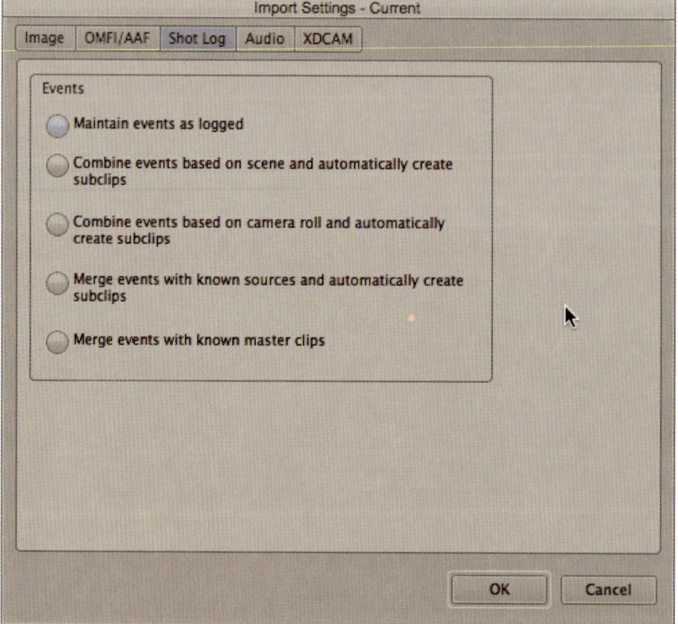

Fig. 2.67 Shot Log tab.

Audio: This gives options for multichannel clips to import so that they go to specific mono and stereo tracks in the source (see Figure 2.68). You can ignore this unless you're getting your audio from an on-set audio mixer recording to a specialized recording device or camera that allows multiple audio tracks. You also have options to convert the sample rate from the source sample rate to the current project sample rate. This is a good idea to do. If you know that your footage is already at the correct sample rate, then you can ignore this. You can also convert to the project bit depth—16-bit or 24-bit. This is also a good idea. I tend to stick with 16-bit, but higher-end productions may go with 24-bit. There's also an option to automatically attenuate (lower) the audio levels on import. This is ideal for many mixed audio files coming from CD or from stock audio houses. These files tend to max out the headroom of the file, recording as hot as possible without clipping, but you will probably want these files brought in at -12 dB. This is where you can make that choice. It helps to keep from blowing out your speakers and also saves you time in setting levels.

XDCAM: I have worked on an annual concert video in Chicago that uses XDCAM isos (isolated cameras) from all of the cameras. This section helps bring in those XDCAM disks properly (see Figure 2.69). When I

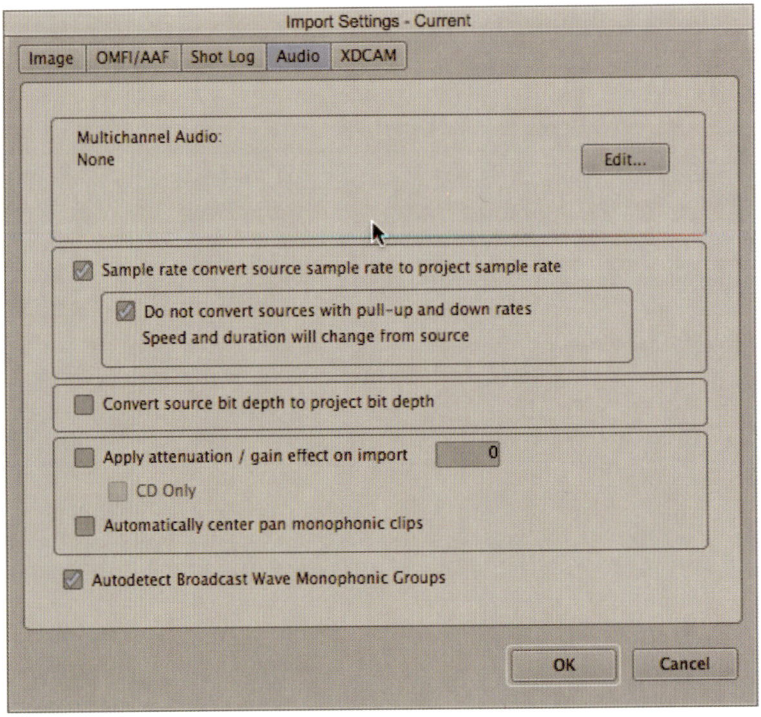

Fig. 2.68 Audio tab.

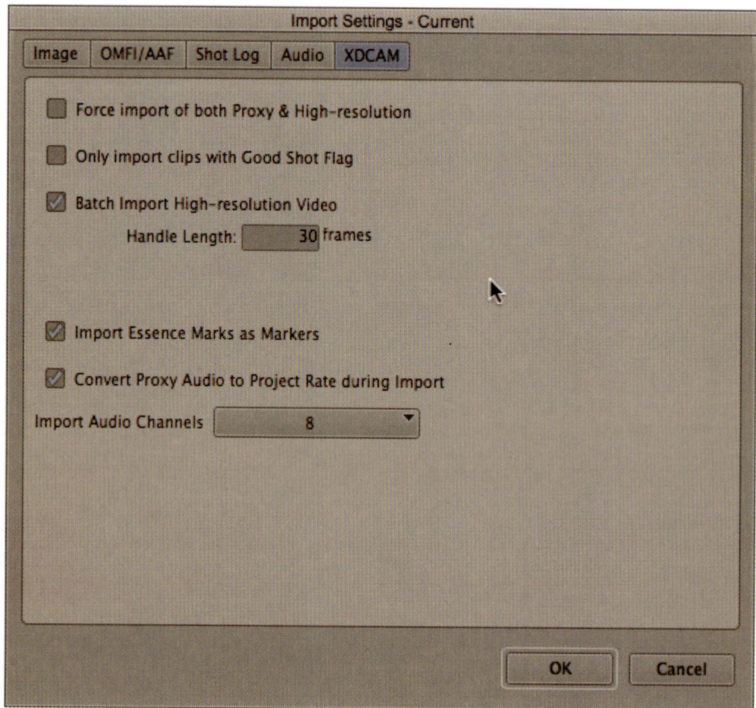

Fig. 2.69 XDCAM tab.

first import, I do not want the high resolution images because they take a long time to import and I just want to get started editing, so I don't force the import of both proxy and high resolution. If you have the time and the space, though, go for it. In my XDCAM workflow, we edit with the proxies and up-rezzing at the end of the project. If someone, like a camera operator, has been logging clips with a Good Shot flag, you can elect to import only the good takes. This seems dangerous to me. My XDCAM footage usually comes from a truck and no one has flagged anything, so I don't use this option. I do use Automatically Import Proxies when the disk is inserted. I don't use Import Essence Markers as Locators, but if someone took the time to place Essence Markers in the original footage, then this would be a great way of bringing in valuable metadata. These are done from the camera, so in my instance of footage from a truck, I don't have these. You can also import the proxy audio at the project rate, which I would recommend. XDCAM proxy audio is notoriously bad. But at least it's a reference. You can also determine how many tracks of audio you want to import. In my case, I chose just two because I knew that the live mix was on those two tracks and all of the others were blank or had erroneously recorded audio on them.

Export Settings

I don't know if you ever had to make video tape dubs, but a perfectly edited project and an ecstatic and appreciative client at the end of a day can turn very nasty if the end result is dubbed wrong. I personally once put a 5:30 sequence on a 5-minute BetaSP tape and shipped it to a TV station. Not a happy thing. The modern day version of dubbing is exporting. Making sure that the beautiful project you and your client saw in the edit suite is delivered to the myriad video and computer screens for the end audience, still looking pristine, is largely dependent on your Export settings.

Each project is so different and it changes so fast that there's no hard and fast rule, but the main point here is that by creating and using custom-made Export settings, you have the ability to deliver consistently. And if there are problems, you have the original setting to help troubleshoot whatever the error is. I have developed a specific Export for clients for whom I've developed a successful workflow and delivery mechanism,. There's no sense reinventing the wheel and trying to find a successful combination of export variables every time you send something off to a particular client.

For 90 percent of what I do, I make "same as source" QuickTime files because they're very quick and they create an actual file that can be saved, moved and transferred on a network. Many people prefer QuickTime reference files because they are very small, but they work only when played on a computer attached to the original media. If that works for you, that's great. For me, the "same as source" QuickTime is often archived and saved and provided to the client. QT Reference files can't be used for those purposes.

Use the directions from the Sharing Settings Between Users section to share your export and import settings between members of your team so that everyone is using the same proven settings.

A quick glance at the Export Settings dialog window shows that some choices still need to be made, even if you only ever use Same as Source.

At the top, you have an Export As pulldown menu with choices to export AAF, AFE, QuickTime Reference, DV Stream, QuickTime movie, HDV, Audio, Graphic, Avid Log Exchange, and Tab Delimited (see Figure 2.70). If you don't see these options, it's probable that you had a bin selected. Certain export options are only available in certain circumstances. You can't export a QuickTime of a bin, so the only options for exporting .ALEs are available with a bin selected.

AAF is great for exporting files to other editing systems or color correction systems. It is basically a souped-up EDL, providing the information on the sequence. You can choose to include the video and audio or send the AAF without it.

AFE delivers metadata but isn't as widely used, at least in my travels. Bins can be exported in this manner.

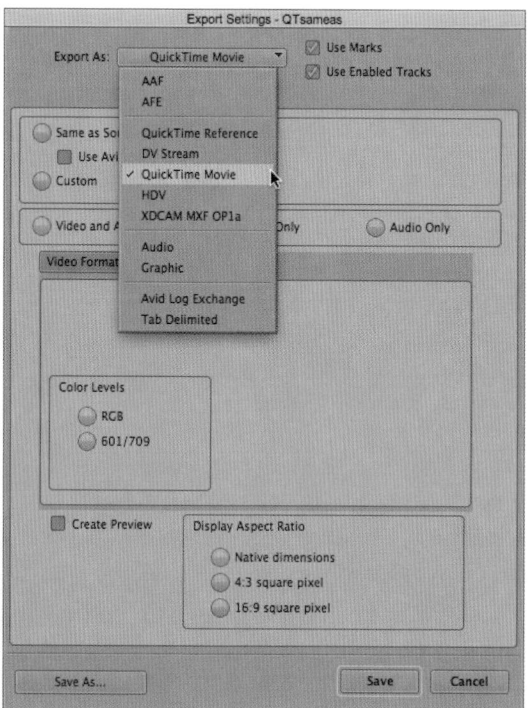

Fig. 2.70 Export Settings with the Export As option pulled down to select QuickTime Movie

QuickTime Reference makes a very small, quick export that does not actually create a useable, viewable file that can be exported and used anywhere. It looks to the system like a regular QT file, but it contains only pointers back to the original source material instead of creating a self-contained file. This can be very useful and beneficial, but it can also screw you over, because if you don't have access to the original media, the QTRef file is useless.

QuickTime file is probably the most used Avid motion export setting. If you export using the Same as Source option, the export will be very fast. Same as Source creates the resulting QT file in the Avid codec, so it's only viewable by machines with the Avid codec, but the file is self-contained and can be shared and transcoded from. One of the things that is different about FCP and Avid is that Avid allows you to selectively export specific tracks and specific sections of the sequence in the Export setting. You do this with the Use Mark Clips and Use Enabled Tracks check boxes. Another useful setting is buried in the Audio tab: you can export the audio tracks directly or as mono, stereo, or surround. I sometimes export to RGB instead of 709 if I know the clip is only destined for the web or computer viewing. And I almost always export as 16:9 square pixel. When you have set all of your settings for a specific export, choose the Save As button to save the setting itself. Give it a descriptive name because you'll end up with dozens of export settings and it will be useful to distinguish between them at a glance. Once the setting is saved, then use the Save button to actually start the export of the file itself. If you want to export as something other than "same as source" check out the Custom button. When you choose that, you get a Format Options menu button with hundreds of choices for encoding video and audio and applying filters.

The DV stream and HDV options allow you to export transport or elementary streams of that exact flavor of video. DV is becoming pretty useless, but the HDV transport and elementary streams are valuable export resources.

Audio exports are something I do on a regular basis, whether for client approvals or various deliverables or even to send stuff off to transcription. The options are pretty self-explanatory and simple. Remember to save your Export settings for things that you think you'll end up using as deliverables time and time again. For example, I have one that sends to mono audio with selected tracks and enabled marks that I use to send to my transcriptionist. This file then gets sent through Squeeze to create MP3 files for the transcriptionist.

Graphic export is very handy. You can create a wide assortment of still files from either source clips or sequences. Use some caution with the Use Marks here because otherwise the export will happen at the Timeline indicator. Often that's what people want to do, park on a frame and send it to someone. If that's what you want to do, make sure that Use Marks and probably Use Enabled tracks are disabled. This is a big difference for FCP users. In Avid, you can Export just parts of the sequence—based on the setting a mark in and mark out—or just export selected tracks by making sure they are enabled in the Timeline. There are lots of good uses for enabling both of these options, but you certainly need to be aware that if they are enabled, then you have to be cognizant of what tracks are enabled and where you have mark ins and mark outs set. Choose one of the many image types from the Graphic Format pulldown menu. There are a ton of options. I probably use JPEG the most often. You can even choose to export Sequential Files based on a sequence. These are usually good for sending to effects houses or animators. As for MC7, DPX is not an option, unfortunately.

Avid Log Exchange is good for exporting bin data. If you want to share a database of your shots in a sequence, including a lot of metadata about your shots, this is a good option. There are many workflows that use ALE to get metadata from shoots into and out of Avid. It is essentially a database file. There are ways to "hack" or "jailbreak" an ALE file to be able to add or extract material from it. However, it's really just a database at heart.

Tab Delimited is another way to export the same information that's in an ALE file, but it isless proprietary. Basically The Tab Delimited option creates a text file in which each of the columns from a bin is stored in a tab-delimited style. This allows it to be opened by most database or even text applications.

CHAPTER 3

Off-Line

Off-line is almost a quaint and archaic term at this point in the evolution of nonlinear editing. *Off-line* is the concept that one editor cuts the story together artistically and aesthetically, while another editor, in another process, uses the edit decisions by the first editor and reassembles the story with an eye to the technical and "finishing" details in the on-line edit. There are still obviously many workflows that use this concept, but for many, there is no off-line editing and on-line editing—there is simply editing. Depending on the workflow there are many gray areas that divide the two types of editing. For the purposes of this book, though, *off-line* refers to the basic editing process—no color correction, no technical quality control, no output. It is simply the act of cutting a project together from the raw material, telling the story, finessing the pacing and preparing it for on-line. I have also included a discussion of audio in this section, even though for many high-end workflows, audio would be done by a separate specialist, often at a separate facility.

This is an advanced book, but this is also a book for advanced editors who may primarily have experience with Final Cut Pro (FCP) or some other NLE (nonlinear editor), and for editors who may have had prior experience on Avid but are unfamiliar with some of the most recent tools and workflows enabled by the latest advancements and additions to Avid Media Composer. Also, even the most experienced Avid editors have odd holes in their knowledge of certain things and will benefit from an introduction to even the most basic concepts. A few years ago I found the original binder for the Media Composer 101 class that I took back in 1992. I flipped through it casually, looking at the notes I had taken in the margins. And there, in my own handwriting, was a tip that I would swear that I'd never heard before. Yet, I'd obviously thought that it was significant enough at the time that I had written it down myself. So, it's also possible for experienced editors to simply forget some very important and basic concepts.

Avid is the predominant, overwhelming choice for editing feature films and primetime TV drama and comedy shows. In the last four years, three shows nominated for the Oscar for Best Editing were edited in Final Cut Pro and two on Lightworks. The other fifteen were edited on Avid. Over the last decade, forty-nine of fifty Emmy nominations for Best Comedy Editing were edited on Avid, except *Scrubs*, which started on Avid and later moved to FCP. Of the primetime drama Emmy nominations over the last ten years, forty-two of fifty nominees were definitely edited on Avid. I couldn't find conclusive evidence on the other eight, but they were probably edited on Avid.

Moving from FCP to Avid

The most prevalent nonlinear editor that people may be coming from is Final Cut Pro. In many ways, the two systems are very similar. So, building on those similarities and explaining the differences is an easy way to get a Final Cut Pro editor comfortable with the Avid editing platform.

I have great respect for Final Cut Pro. I co-edited the Sony/Tristar theatrical feature film, *Courageous* (Figure 3.1), using Final Cut Pro alongside the director/editor Alex Kendrick (Figure 3.2) and my good friend, and fellow editor, Bill Ebel.

Most of the skills and many of the methodologies used in Final Cut will transfer seamlessly to Avid because the functions are identical. With

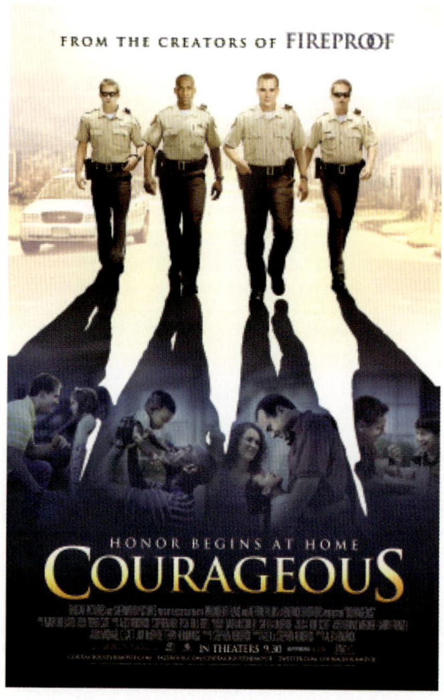

Fig. 3.1 Image used courtesy of Provident Pictures.

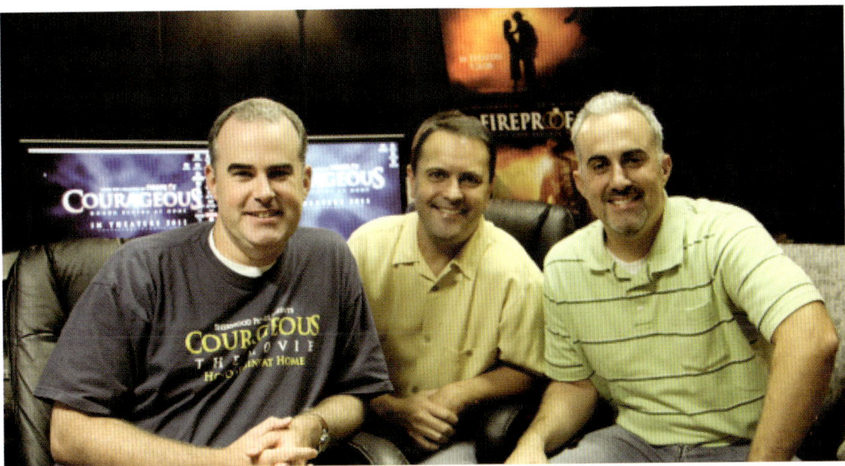

Fig. 3.2 Left to right: Alex Kendrick (writer/director/editor), Stephen Hullfish (editor), and Stephen Kendrick (writer/producer) in the edit suite of the 2011 Sony/TriStar feature, *Courageous*.

some features, you may be disappointed in the way Avid works. With many things, I will show you superior ways in which Avid is able to work.

When I was learning Final Cut, I received some very good advice: don't try to force Final Cut to work the way you're used to working on Avid. The same advice goes in reverse for Final Cut Pro editors. It is be possible to map your Avid keyboard in ways that are similar to Final Cut Pro. For some things, this will make the transition more comfortable and will allow you to maintain some of the muscle memory your fingers have acquired from years of editing. For other things you will be far better off to unlearn Final Cut methods and learn the ways that they are best done in Avid. Keep an open mind. It's pointless to have a negative attitude about the things that are different or "worse." The most productive thing is to understand the way Avid works and to take advantage of its many world-class features. But let's start with features that are identical between the two systems. This should give FCP editors a feeling that they can feel "at home" with Avid.

Identical Features

On the keyboard, JKL and IO work very similarly between FCP and Avid.

Trim one frame forward and back are the same.

Cutting, copying, and pasting segments work very similarly between FCP7 and Avid. If you're on a PC-based Avid, you'll just need to remember that instead of Command-C, V, and X, you need to use Control-C, V, and X for copy, paste, and cut.

If you're used to FCP7's magnetic timeline, the equivalent in Avid is to Command-drag (Mac) or Control-drag (PC). Dragging with the Command (Mac) or Control (PC) key held down will "snap" to transitions.

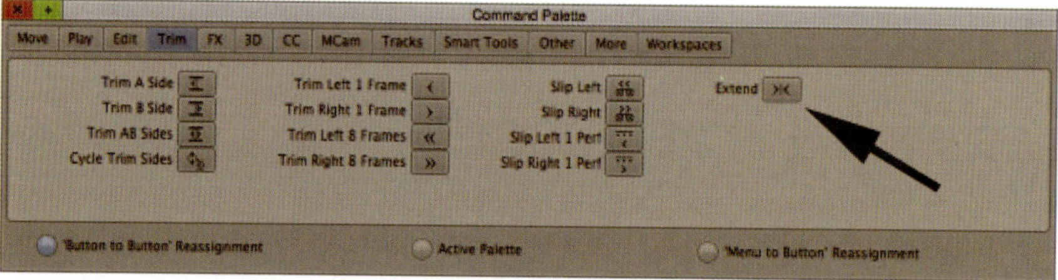

Fig. 3.3 The Extend key can be mapped to anywhere on the keyboard or user interface. This shows the Extend key in the Avid Command Palette's Trim tab.

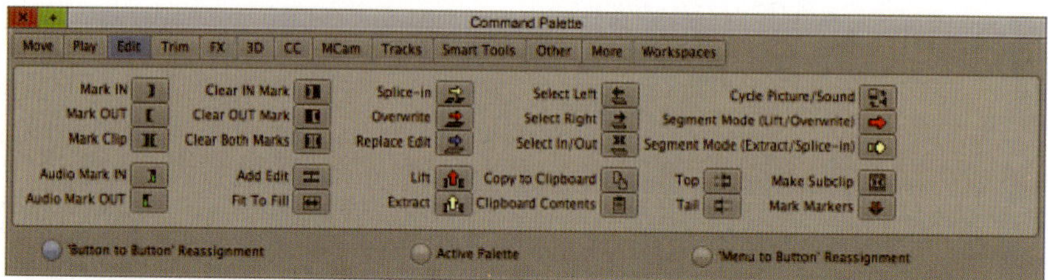

Fig. 3.4 The Overwrite, Splice-in, and Replace keys are the same. Here they're pictured in the Edit tab of the Command Palette. But the most prominent spot for the Splice-in and Overwrite buttons is centered at the bottom of the Composer window, between the Source/Record monitors. They're also on the V and B keys on the default keyboard.

Avid and FCP both have Extend keys (see Figure 3.3). If you have never used this key, check it out. It's very handy. It basically lets you "extend" an edit point either forward or backward, by putting an in point *before* the cut or an out point *after* the cut and hitting Extend to move the cut to the mark. In FCP the Extend key is mapped, mnemonically, to the E key. I'll touch on this more later in the chapter.

Other trimming tools from FCP, such as Slip and Slide, are also on Avid and work essentially the same way, except that they're accessed in the Timeline in Avid instead of using the keyboard.

The Overwrite, Splice-in, and Replace keys are the same, though mapped to different spots (see Figure 3.4). As is the Mark Clip button. If you don't like Avid's placement of these buttons, take another look at chapter 2, where I describe mapping the keyboard. Or look at the Project files for the book on-line for an importable FCP keyboard that you can use to edit FCP-like in Avid.

Things You'll Love about Avid

This is really difficult for me, because there are so many things that I take for granted with Avid. A huge plus in simple functionality is that video and audio tracks can be "soloed" or muted without having all of the effects on the track becoming unrendered. Soloing a video track can be done by Command-clicking on the track monitor (Mac) or Control-clicking (PC). Or for audio or video muting, use the mute and solo buttons in the Timeline track selector (see Figure 3.5). If you don't see a little M or S to the far right of the track selector panel, turn them on with the little On/Off toggle that looks like a power switch on each track. (Some of these things I already mentioned in chapter 2.)

In the Timeline, I also love the ability to rename my tracks, so they don't just have to be V1, V2, A1, A2, A3, A4, and so on. (see Figure 3.6). Sometimes there's no need to complicate the Timeline with names, but often, especially when working with other editors, it's helpful to name individual tracks so that all editors know which audio tracks are for music

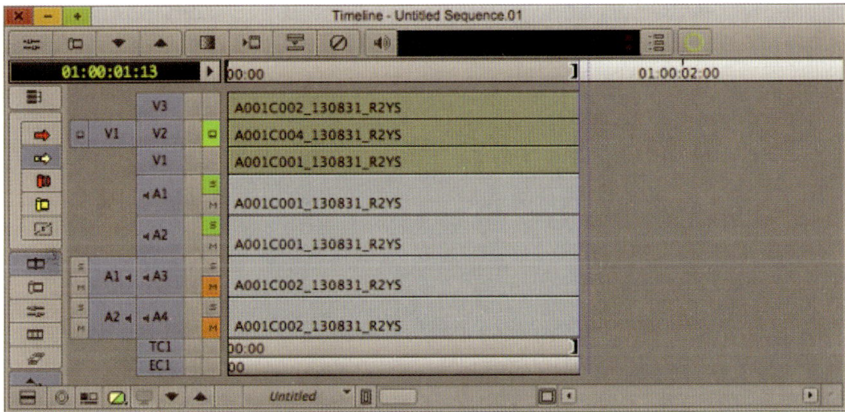

Fig. 3.5 Tracks can be soloed or muted without losing renders. Note the bright green monitor icon to the left of video track 2 and the green solo icons and orange mute icons in the audio tracks.

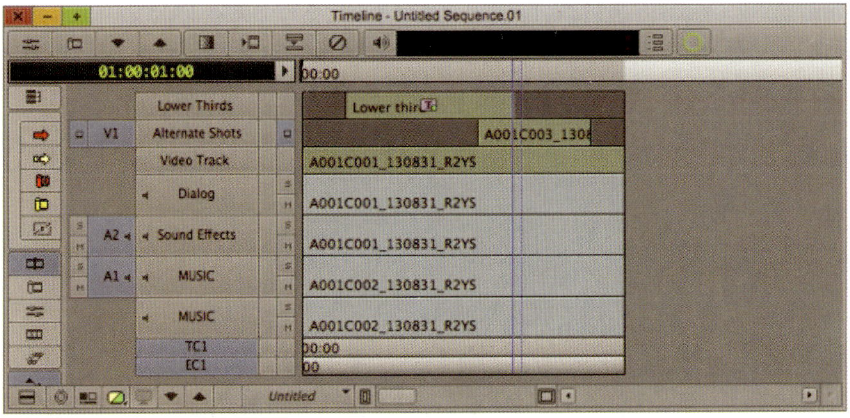

Fig. 3.6 You can rename tracks in the Timeline to assist with organizing and sharing of projects.

or dialog, or which video track has the Lower Thirds on it. When creating edits that eventually need specific mixdowns, such as for dialog, M&E, music minus, sound effects, and so on, it is helpful to name these tracks to maintain order and remind the editing staff where certain types of music need to be placed to make the mixdown process easier at the end. To rename a track, simply Control-click (Mac) or right-click (PC) on the track and a contextual drop-down menu appears with Rename as a choice. Select Rename and type the track name in the dialog box that appears.

Another cool thing is the ability to see the original source track for an audio clip in the Timeline. For example, if you patch source track 1 to track 14 in the Timeline and then later forget which track it came from, you can select Clip Text > Clip Tracks from the Timeline Fast Menu and the number of the source track will appear on each segment in the Timeline (see Figure 3.7). This can be handy in troubleshooting audio problems or in reminding you which audio track you used the last time you cut audio in from a specific clip.

Also very handy is Avid's ability to actually e-mail you when a render, export, or consolidate/transcode operation is complete, so if you have to render and it's time for a nap (in the old days, renders took a lot longer and this would have been very handy), you can set up the e-mail settings to e-mail your smartphone to alert you at the end of the render. I have a tendency to start working with AMA-linked media and then consolidate it when I leave for the night. If it's critical that the media gets consolidated or transcoded overnight, you could have the system e-mail you so that you'd know if it failed. Then you would know to either call your assistant to go in and fix whatever's wrong or do it yourself. It's better than coming in the next morning to find that your hard drive is full or some other little thing happened to hold up the entire process.

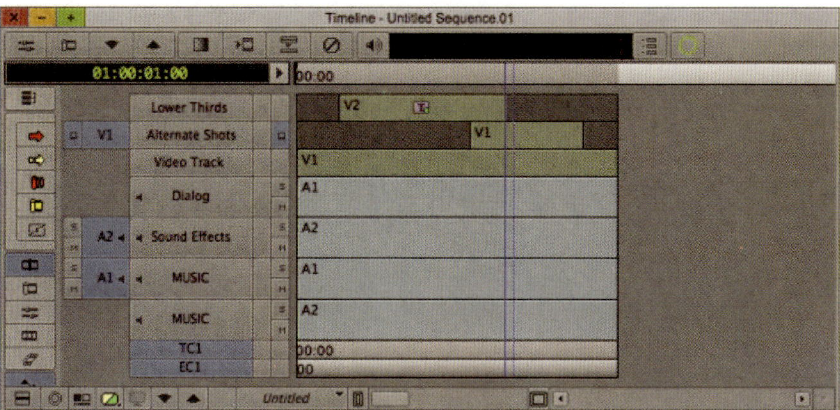

Fig. 3.7 In the Timeline Fast Menu, you can change the Clip Text > Clip Tracks to display the original source tracks in the timeline.

Comparable Tools and Methods

Luckily for editors familiar with Final Cut Pro, Avid has implemented several features that make the transition to Avid much easier. Primary among these is the collection of tools that Avid groups under the label Smart Tool. These tools are designed to make the Timeline editing functions—moving and trimming using the mouse in the Timeline—very similar to Final Cut Pro.

Smart Tool Functions

Avid editors have certainly used the Timeline for decades, but Avid has always been much more conducive to keyboard editing. Final Cut Pro editors—and I realize that I'm generalizing—are often much more comfortable moving and trimming their sequences using the arrow tool and the Timeline trimming functions to drag segments and transitions quickly and efficiently using the mouse (or trackball or tablet).

As more and more editors switched from Final Cut, Avid added tools to make them feel comfortable and to offer the benefits of Final Cut's Timeline-based editing features.

Smart Tool is available in the UI (user interface) in the Timeline window to the far left (see Figure 3.8). The main icons are a red arrow, a yellow arrow, a red trim icon, and a yellow trim icon. There is also an icon

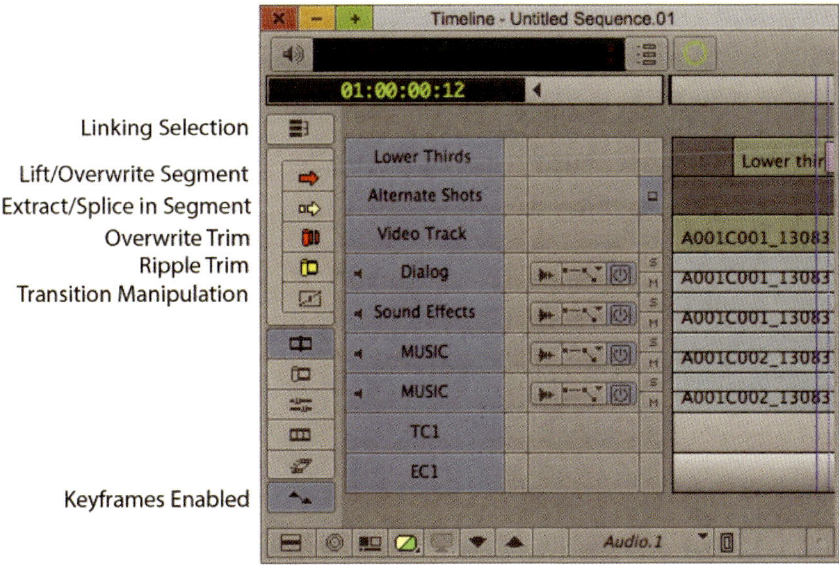

Fig. 3.8 Avid's Smart Tool is an easy way for Final Cut Pro editors to feel more comfortable in the Avid Timeline. Smart Tool adds even more flexibility to the way clips and transitions can be manipulated in the Timeline.

above those tools that allows you to link audio and video pairs (similar to the Link button in FCP), and icons below that allow you to drag the length of a transition and to "rubberband" audio levels.

Let's discuss these tools one at a time. First is the Overwrite Segment tool: this is indicated by a red arrow, and it functions similarly to the red segment mode in Final Cut. When it is selected, it allows the editor to drag segments in the Timeline in overwrite mode. This means that dragging a segment leaves behind a hole in the spot where it was, and it overwrites any other clips in the spot where the drag is completed. Next is the Splice Segment tool, indicated by a yellow arrow. When it is selected, it allows the editor to drag segments in the Timeline in splice mode. This means that dragging a segment causes a "ripple" in the sequence, not leaving a hole where the clip was moved, but closing up the gap with the next clip in the sequence. Then, when the drag is complete, the segment pushes segments further down the Timeline. The Splice Segment tool is valuable for switching the order of segments.

The red Overwrite Trim icon allows you to do trims that do not ripple or change the length of the sequence. The yellow trim icon allows you to do "ripple" trims that extend or reduce the length of the sequence.

If you want a Timeline editing experience that is similar to Final Cut, turn all of the Smart Tool buttons on and leave them on. This allows you to switch back and forth between all of the tools, depending on whether you click near the top of the segment in the Timeline to activate the Lift/Overwrite Segment Tool or near the bottom for the Extract/Splice Segment Tool. Then by clicking close to a transition, you can access the trim tools—red Overwrite Trim near the top of the transition and yellow Ripple Trim near the bottom.

One of my biggest complaints about Final Cut was that the Timeline was very "fiddly." The function of the cursor was highly dependent on the exact position of the mouse in relation to various clips and transitions. When editing in the Timeline in Avid, I feel as if I can edit much faster because I don't have to be so careful about where I position my cursor when I click or drag.

At first I didn't like Smart Tool at all because it brought one of the worst features of Final Cut (in my opinion) to Avid, but then I mapped the Smart Tool toggle buttons to my keyboard and discovered that I could have the best of both worlds. By keeping one hand on the buttons that I assigned to the Smart Tool functions and the other hand on my mouse, I could switch tools instantly with the keyboard without having the "fine-target mousing" required by Final Cut. Even if you are not a fan of Timeline-based editing, this is something you may want to experiment with. You may find that Timeline editing is very efficient and intuitive.

Video Tutorial

To understand Smart Tool, you really need to see the functions in motion, so check out the Smart Tool video tutorial on the companion website, www.focalpress.com/cw/hullfish.

Some Things That You'll Need to Get Used To

One of the big things for Final Cut Pro editors to get used to is Timeline zooming. And I have to say that after editing in FCP for a while, I really miss the shift-z jump to full Timeline view. In Avid, it's Command-/, which is just as easy, but for me, shift-z feels right, even after decades on Avid. This is especially true because I jump back and forth between Mac-based and PC-based Avid systems, so the finger positioning to go from Command-/ to Control-/ screws up my muscle memory. Avid does granular Timeline zooms with Command-[] instead of FCP's use of Command–/+. It is possible to map this shift-z keyboard from FCP to Avid.

Additionally, there's the very cool ability to zoom in on an exact section of the Timeline with Command-M (Mac) or Control-M (PC), which allows you to lasso the exact part of the Timeline you want to fill the screen. Hit Command-J (Mac) or Control-J (PC) to return to the previous zoom.

In FCP, instead of right–clicking or Control-clicking on a transition to get the contextual menu for Add Transition "Cross Dissolve," Avid uses the default key of \ (above the Return key) to add a dissolve to a transition. I prefer this personally. This key calls up the Quick Transition window (see Figure 3.9). This window allows you options to set the length of the dissolve, the tracks that the dissolve will apply to, where the dissolve starts, and a few other transitions and allows you to add the dissolve or

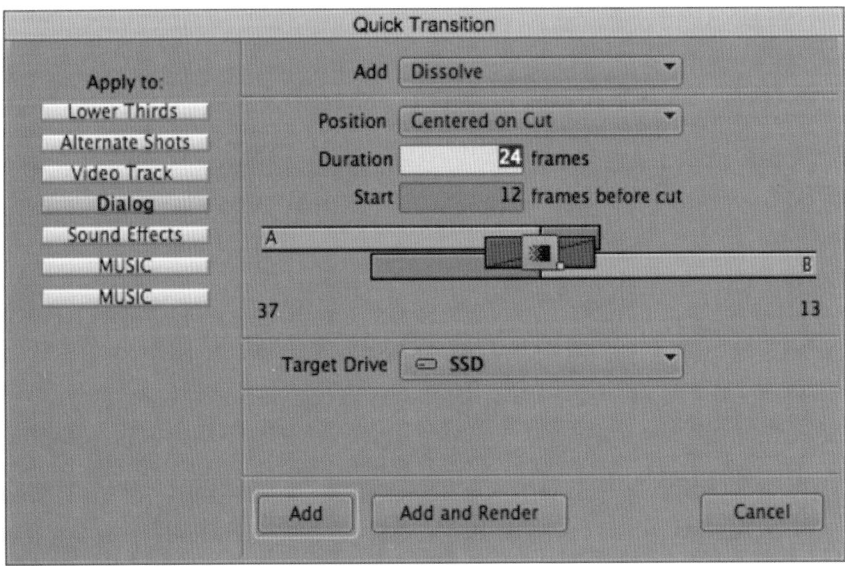

Fig. 3.9 Quick Transition dialog is called up with the \ key.

add and render the dissolve. It's very nice in the latest versions of Media Composer that you can select tracks in the window. Prior to this, it was exclusively based on the selected tracks, so if you didn't have the right tracks selected, you'd have to quit out of the window, deselect the wrong tracks and go back into the window. But if you just want to execute a simple, quick dissolve, as long as the options are correct, all you have to do is hit the \ key and Return (Mac) or Enter (PC).

Avid doesn't have many of the "tools" that Final Cut has simply because it works differently.

Avid doesn't have a "pen" tool. To adjust and add keyframes to the Timeline, the track or tracks that you want to add keyframes to must be selected, and the Volume Audio Data option has to be chosen in the Timeline Fast Menu or in the Track Control Panel area (see Figure 3.10). Then press the Keyframe button or key. Dragging a keyframe up and down—in volume, for example—just involves dragging the keyframe vertically. To move the position of a keyframe in the Timeline to change the timing of a rubberbanded transition requires you to Option-drag (Mac) or Alt-drag (PC). You can also select keyframes and move them via the trim buttons, one or ten frames forward or backward. Avid's MC version 7 brings even greater mixing abilities to the Timeline, allowing much more accurate levels via a minislider.

Another cool thing with moving or adjusting keyframes is that they respond only on active tracks and they are also affected by in and out points. If you mark an in and out point at each end of your sequence and enable all of your tracks, all keyframes will move when you move just one. Sometimes this is a great asset, and sometimes you'll do this by accident, trying to move only a single keyframe between marks. (Fair warning!) Also, when "ducking" audio between sound bites, marking in and out points around a smaller section of keyframes let's you just

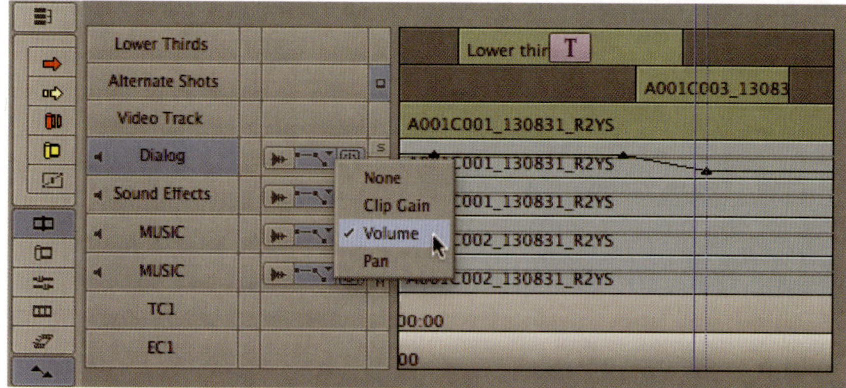

Fig. 3.10 With Volume enabled on audio tracks in the Track Control Panel, adding keyframes offers capabilities similar to FCP's Pen Tool.

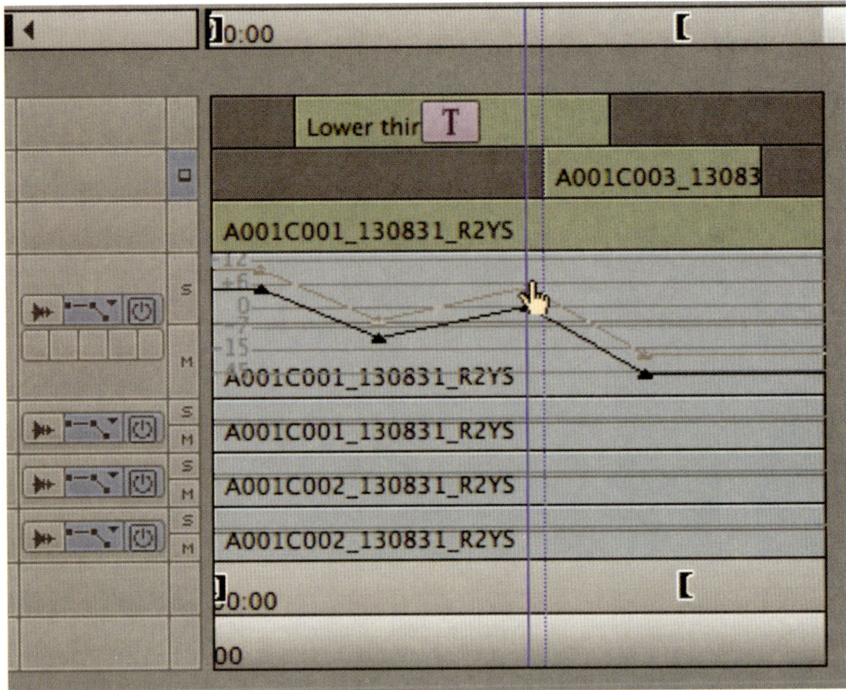

Fig. 3.11 Between a Mark in and Mark out on enabled tracks, moving a single keyframe moves all keyframes.

isolate a small section of multiple keyframes and raise them equally (see Figure 3.11). This is one of the things I prefer about FCP. On Avid, with multiple keyframes, you can't just grab the level itself. You have to move both the keyframes on either end of a section where you want to adjust the level. You can also slide numerous keyframes of rubberbanded audio forward or backward in time the same way.

Another thing I miss when I move from FCP to Avid is that on FCP, hitting the delete button will "lift delete" a selected segment or segments (leaving a hole in the Timeline) and using shift-delete will "ripple delete" the segment or segments, shortening the length of the Timeline. Using the buttons in Smart Tool, you have this exact same functionality, but you simply need to get used to using the keyboard equivalents of the Lift and Extract in Smart Tool to select the segments before using the delete key. It's the same number of button presses, just different muscle memory.

If you have rubberbanding on audio tracks or effects on video tracks and you segment select and then hit the delete key, the first tap of the delete key removes the rubberbanding or effects. If you want to actually delete the segment itself, you'll need to hit delete again.

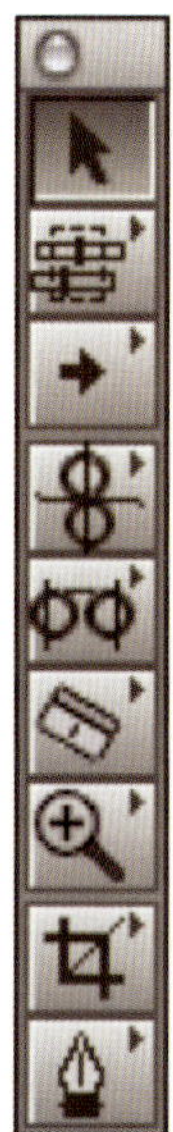

Fig. 3.12 Final Cut Pro has a palette of editing tools. Avid has equivalents of most of these.

In FCP there is a palette (see Figure 3.12) that includes the following tools. I've included their Avid equivalents:

Selection Tool—Arrow: This is basically the same as Avid's main editing mode. Edit Selection Tool: This is similar to Avid's Trim Mode.

Select Track Forward Tool: This is similar to Avid's Select Right tool, which allows you to select all of the clips to the right (downstream) of the Timeline location on enabled tracks. Avid also has a tool called Select Left that allows you to select everything to the left of the Timeline locator.

Roll Tool: This is similar to Avid's Trim Tool or the red Smart Tool trim.

Slip Tool: This is similar to Avid's Slip tool in Trim Mode.

Razor Blade Tool: This tool has the same functionality as Avid's Add Edit tool (see Figure 3.13), though the actual operation is different. The Razor Blade tool is a mode in which, anywhere you click in the Timeline sequence clips—on any track—it cuts the segment at that point. Avid's Add Edit is not a mode, so you don't have to go into and out of the tool. You have to enable the track that you want to cut and park the Timeline locator at a spot, and when you press the Add Edit button, it cuts the clip into two pieces at that point. This can also be done on-the-fly if you have the Add Edit button mapped to a keyboard key. When you tap the Add Edit key, the sequence will be cut wherever the Timeline locator is on whichever tracks are enabled. Also, in Avid, you can use Add Edit to place Add Edits in filler, which is not possible with FCP.

ZOOM Tool: There are numerous ways in Avid to Zoom.

Crop Tool: In the Timeline, this basically works like Smart Tool for selecting segments. For effects, this is done with an effect, like a Picture-in-Picture, and then edited in the Effects Editor.

Pen Tool: As explained earlier in the chapter, in editing, the Pen tool is used mainly for adding keyframes, which Avid does with the Keyframe button. In Effects, it can be used for drawing and cutting masks. In Avid, this is only really possible in the Animatte Effect and needs to be edited using the Effects Editor once the Animatte effect is added.

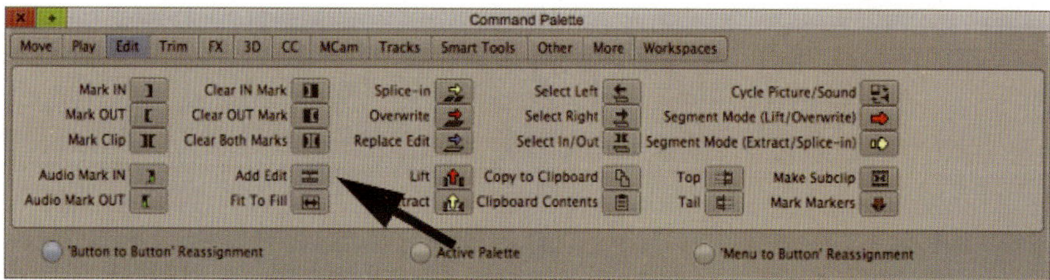

Fig. 3.13 Avid's Add Edit offers functionality similar to the Razor Blade tool in FCP.

Effects Differences for FCP Users

One of the biggest things to wrap your head around as an FCP user moving to Avid is that effects are done completely differently. This is one of those times where you just have to give up your FCP concept of effects and attempt to embrace the way Avid does it. I will forewarn you that Avid is not as good as FCP at effects, especially without third-party help. The company has made some attempt at fixing this by incorporating Boris RED into Avid as "AvidFX."

So, to understand effects in Avid, let's start with the concept of a single layered effect applied to one clip on one track of the Timeline.

Let's say you want to add a little blow-up to a shot in the Timeline. In FCP, effects are generally added from the menu bar and manipulated in the Source Window. Instead, effects are found in the Effect Palette (see Figure 3.14), which can be found or called up in a number of ways. Command-8 (Mac) or control-8 (PC) calls up the Effect Palette. It's also accessible from the Effect tab of the Project window or from the Tools Menu > Effect Palette. The Effect Palette is simply a list of the available effects. On the left side of the Effect Palette are the categories of effects. On the right are the individual effects that you can apply from the selected effects category. To find the effect you want, you have to search through the categories. To be honest, not all of them are intuitively categorized.

For our example of doing a blow-up (resize) on a segment, there are at least three effects that will allow you to do this. One would be the

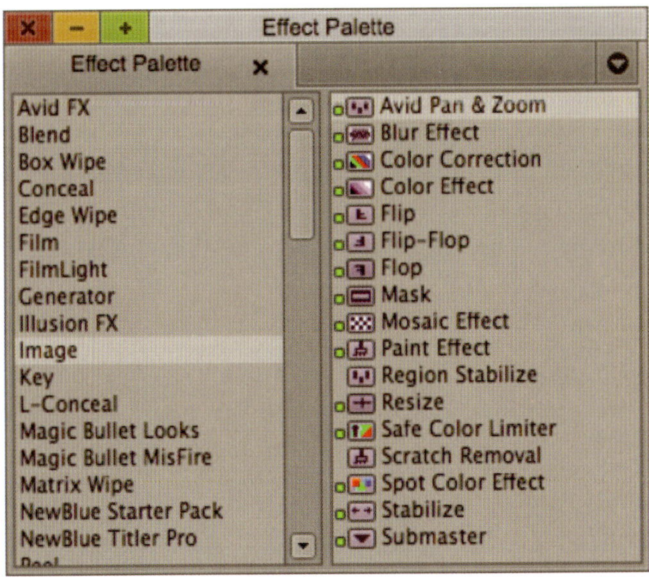

Fig. 3.14 This is the Avid Effect Palette. It is similar to FCP's Effects menu.

Resize effect in the Image category or the 3D Warp or Picture-in-Picture effect, both in the Blend category. In twenty plus years of using Avid, I still often have to hunt for the correct category of effect to find what I'm looking for, and sadly there is no "find" command for effects, as there is in Premiere.

The first step in doing the resize, or blow-up, is to drag and drop the effect from the Effect Palette (dragging the purple effect icon) to the segment in the Timeline where you want to add the effect. You can also select the segment in the Timeline with one of the Smart Tool segment tools and then double-click on the icon in the Effect Palette. Adding effects to transitions works the same way: either drag and drop, or select and double-click the effect. You can also add effects to multiple segments or transitions by having them selected and double-clicking to apply across the selected transitions or segments.

With the effect applied to the segment (see Figure 3.15), you need to call up the Effect Editor to be able to change the effect parameters. The Effect Editor does not have a keyboard command, so you may want to rectify that by adding it to a key on your keyboard. (Use the instructions from chapter 2 page 131.) The other options are to call it up from the Tools Menu > Effect Editor or from the icon (that looks like effects sliders) (see Figure 3.16) at the top or on the left side of the Timeline. You can also get to the Effect Editor by opening the Effect Editing Workspace from the Windows menu > Workspace > Effects Editor.

The look and functionality of the Effect Editor depends on what specific effect it is editing. With the Resize effect, it's quite simple. Spin-down the Scaling and Position parameters and either drag the sliders or use numeric entries to control the parameters. Explaining more about how

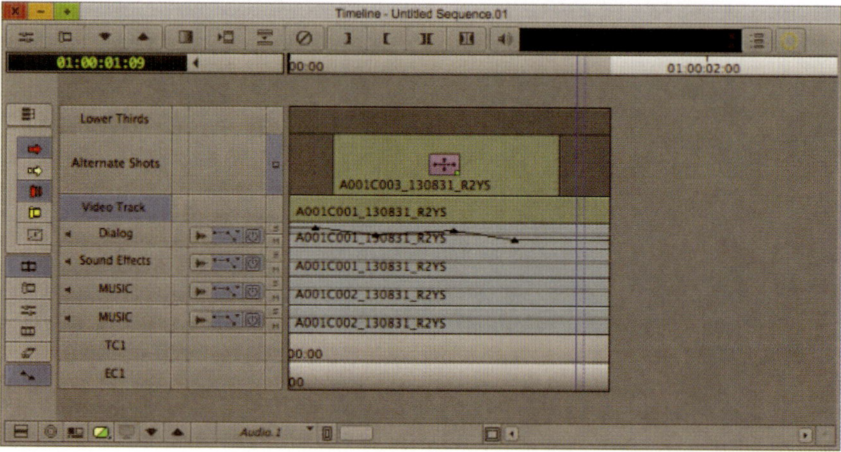

Fig. 3.15 Resize effect applied to a segment in the Timeline.

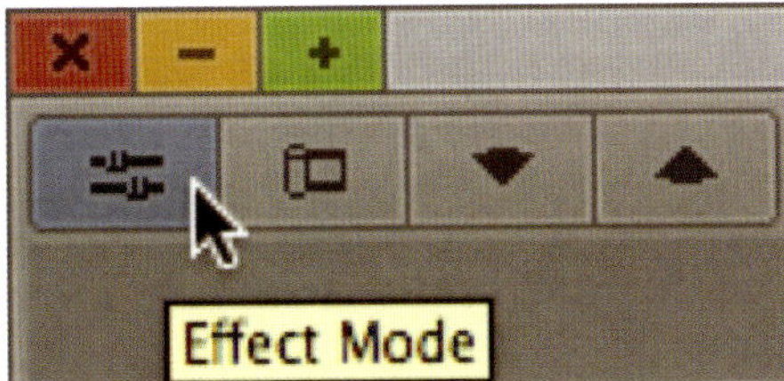

Fig. 3.16 The Effect Editor is called up by going into Effect mode which can be done from the little slider icon in the top left corner of the Timeline window.

effects work and about complicated keyframed effects are explained in the next chapter.

Trimming Concepts

Trimming is one of the strong points of the Avid system: the ability to refine your cut quickly and precisely. Trim mode gives you the most power, but using the trim functionality of Smart Tool gets you where you need to go most of the time quickly and effortlessly. Before Smart Tool, you had to go into Trim mode to trim. This was no big deal, getting into and out of Trim mode happened at the click of a button, but I have found myself barely using Trim mode now that I can adjust trims in Smart Tool. The big reason for Trim mode over Smart Tool trimming is that in Trim mode, you can work very intuitively while watching the trim happen in real time. In Smart Tool, you get a good graphical indication of how much you're trimming in the Timeline but not such a good idea of the trim in the context of when the cut happens in time and space.

I remember when I first moved to Avid from editing linear tape-based projects, and it was hard to wrap my head around some of the options. By this point, I think most people understand the basics of trimming. But there are some important and useful concepts that require some explanation.

With a simple Timeline, there's not much to understand about trimming. With a more complete, complicated Timeline, with multiple tracks and the edit points on most of the tracks not lined up, making ripple trims that change the length of the Timeline require some thought and an understanding of the way sync is maintained downstream of the trim.

I'm a big fan of turning on the sync locks on all of the tracks. If you're new to Avid, the sync locks are turned on with the column of controls just to the right of the track number. When sync locks are turned on, there is a small, bold slash mark in that column (see Figure 3.17).

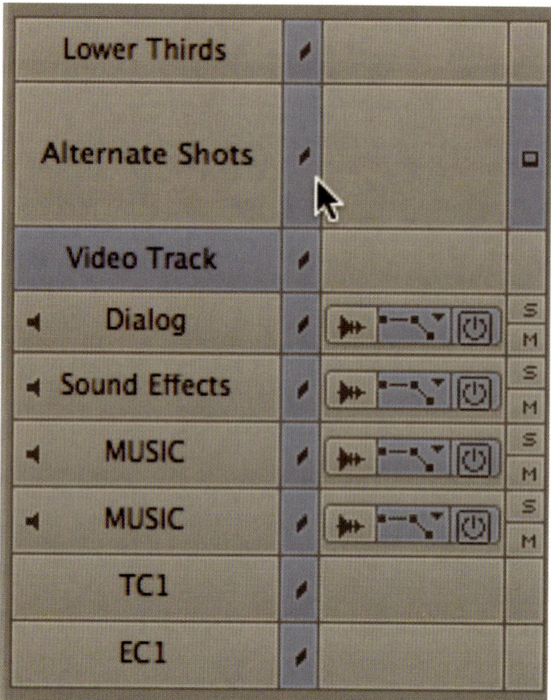

Fig. 3.17 With Sync Locks turned on all tracks, it is impossible to knock clips downstream of the trim out of sync. It does remove a slight amount of control over the trim. The Sync Locks are indicated with the blue-highlighted slash icons.

TIP

If you want to toggle sync locks on or off for all tracks, click Sync Lock in the timecode or EC1 track.

Video Tutorial

This is a difficult thing to explain with text, so check out the video tutorial Maintaining Sync During Trimming, especially with Option-Add Edit, www.focalpress.com/cw/hullfish.

Often, with complex Timelines, when doing ripple trims, it's easiest or most desirable to click on the trim points specifically in each track, but for maximum speed in selecting all transitions near a specific trim, you can use Command-A (Mac) or Control-A (PC) to add trim selections on all of the tracks. Doing this can sometimes accidentally trim a transition far from the main trim point on a track that doesn't have a lot of clips on it. In this case, it could take some time before the mistake is discovered, if ever. To be safe in these instances, the answer is to use Option-Add Edit (Mac) or Alt-Add Edit (PC) to add edits only to the empty tracks (filler) at the trim point. By doing this, the trims happen harmlessly in the filler, preventing accidental trims to distant trim points. Take a moment to give this a try. You'll need a Timeline with some filler on at least one track above or below a trim point.

The other trimming features I've seen editors fail to use are slipping and sliding. Slipping and sliding are trims that keep the segment length the same by simultaneously either adding frames to the head while

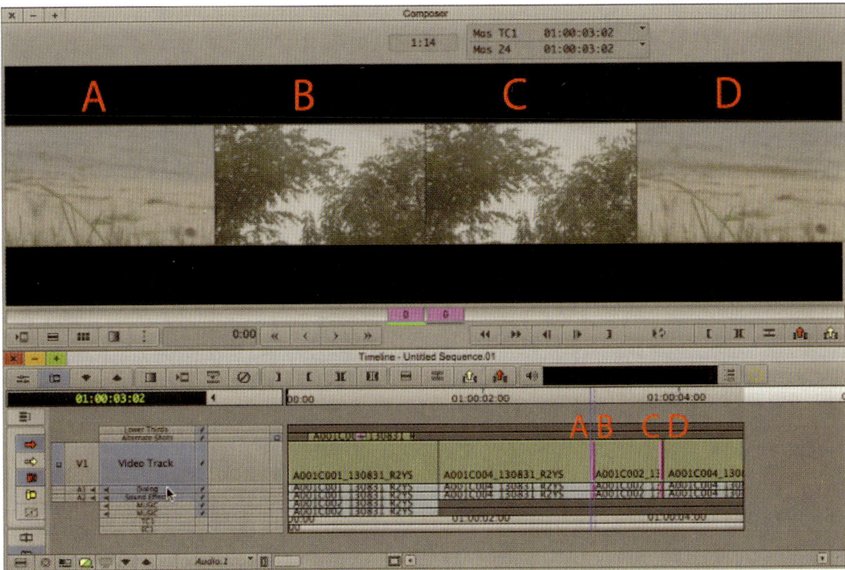

Fig. 3.18 Slipping and sliding is a great time-saving trimming method. The Slip trim being shown is inside the purple trim indicators in the timeline. The four frames of video displayed in the Composer window above it are (A) the outgoing frame of the segment before the slipped segment, (B) the incoming frame of the slipped segment, (C) the outgoing frame of the slipped segment, and (D) the incoming frame of the segment after the slipped segment.

subtracting frames from the tail, or subtracting frames from the head while adding frames to the tail. Slipping does this trim "inside" the clip, leaving the tail of the previous shot and the head of the next shot the same (see Figure 3.18). Sliding does this "outside" the clip, leaving the clip the same, but "sliding" it backward or forward in the Timeline. The easiest way to go into a slip trim is to click just above the topmost video track but below the timecode ruler, and drag a lasso from right to left on either side of the segment you want to slip. Two rollers should appear on the inside of the segment, one at the head and one at the tail. Using the trim back buttons, or negative trim numbers, will slip backward in the clip, revealing earlier footage. Using the trim forward, or positive trim numbers, will slip forward in the clip, revealing later footage.

Video Tutorial

Slipping and sliding are most easily explained with video, so check out the video tutorial, Slipping and Sliding, www.focalpress.com/cw/hullfish.

Using Extend Edit to Trim

As I touched on earlier in the chapter, one of my favorite ways to trim is not even considered a trim. At least you don't have to go in to trim mode to execute it. It's called Extend Edit. There is a virtually identical feature in FCP. Extend Edit allows you to place an in or out point on either side of any edit point and have the edit point, on selected tracks,

Video Tutorial

I love the Extend button, and explaining it is easier to do with video, so if you've never used Extend, please check out the video tutorial on the companion website, www.focalpress.com/cw/hullfish.

jump, or extend, to that in or out point. The trick with Extend Edit is that if you have both an in point *and* an out point, the Extend Edit assumes that you'd rather extend to the in point. Sometimes, you can have an in point marked accidentally and you think you're extending to the out point, but when you hit Extend Edit, it seems as if nothing happens. Unfortunately, if that in point is out of your field of view because you're zoomed in on your Timeline, you have just accidentally altered an edit point without knowing it. For this reason, I have become accustomed to hitting a series of keys in rapid succession when I use Extend Edit. If I'm extending to the out point, I'll mark the out point and then immediately hit Clear Mark In followed by Extend Edit. You could also do this by starting with Clear Marks, then going to your in or out point, and then to Extend Edit. On my customized keyboard, these keys are all together, so executing this series of keyboard commands is very nearly simultaneous. To me, after all these years, it almost feels like a single button push. Because this is often done while enabling and disabling tracks, it's also crucial to have the video and audio track enabling buttons mapped to the keyboard. Remember that when the Timeline is active, using Command-A (on a Mac) or Control-A (on a PC) enables all of the tracks, and using shift-Command-A (on a Mac) or shift-Control-A (on a PC) disables all of the tracks. That makes it faster to just enable the one or two tracks you want to extend when there are a lot of tracks that are active.

Trimming is Timing

While there are ways to pace a show or feature other than trimming individual shots—for example, cutting entire scenes—timing and pacing is critical to a well-edited show.

Allyson Johnson, an editor on the primetime drama, *Smash*, talks about the art of not cutting too much.

On *Smash*, the typical first cut of the show was 48–52 minutes. The finale was almost an hour. For Johnson, this is when the collaborative struggle begins: "The director comes in and the director doesn't want to cut it down too much because the showrunner* is going to want to see certain things. The showrunner tries to give the best show that may not be to time, then the network and the studio give their notes and then you can cut it to time because everyone has had a chance to give their input on what they like. So you know what can be cut out. You always had to worry about continuity of what scenes could be cut out. It was tough to get that show to time and still have enough dance numbers so people would think they'd seen enough musical numbers. We always shot the whole song or musical number and then cut it down in editing."

GLOSSARY

The showrunner is the person, usually the head writer, who is the ultimate "boss" on a show.

"We're always long because the scripts are always long," Johnson explains, "and the scripts are long because not everything is going to work, so you can cut the things that aren't working and you're still left with enough material. So the writers struggle with how much extra to give us. The editor's cut needs to keep everything in there, but you need to keep things tight. As tight as they want it, which is always tighter than you want."

Johnson, who has cut many feature films, including *Monsoon Wedding, Vanity Fair,* and a documentary for the twenty-fifth anniversary of The Who's *Tommy,* then drops this hard-won pearl of wisdom: "Because if you shorten everything, you're left with information, not entertainment."

The idea of pacing and tightening also struck a chord with film editor Wayne Wahrman. Wahrman has cut *I am Legend, All the Kings Men, Charlie's Angels I and II,* and *The Day the Earth Stood Still* and was one of the editors on *Last of the Mohicans.*

"There's a critical mass for a film and you have to whittle it down till you get to that point. Pacing really has to know its place inside the film," Wahrman explains. "It has to do with the scene itself. What's the scene trying to do? What's its purpose? Story comes first, then performance. Usually there's an internal pace or rhythm in the scene that you need to find and augment. As far as cutting goes, the less cuts the better. Every time you look at a scene you look at it differently. You have different ideas, a different attitude. Basically the first time through you're just trying to get all the scenes to play. A lot of times when I read a script, I can tell which scenes will not be in the final movie. The next pass through it you're tightening up, deciding what scenes need to be there and which don't. Sometimes, if a scene in a movie is dragging, it's not the problem with the scene itself, but the two or three scenes before it. So everything affects everything else."

Multicam

I can't tell you the distress I feel when I see "real editors" trying to do multicam on FCP or Premiere or Avid by putting each camera on a separate video track and then hacking away the cameras they don't want to see. I have witnessed this technique on numerous occasions by people who should know better.

Originally, I did multicam on beautifully engineered shoots where there was actually a line-cut of all of the cameras, directed from a truck or studio, and I also had all of the "iso" cameras at my disposal to tweak the director's cut and trim as was necessary. But now, with the prevalence of inexpensive DSLRs or other set-ups, like multiple RED cameras, editors need to edit multicamera footage that isn't synced together with timecode, or directed. Even with these more impromptu multicamera shoots, the multicamera functionality of the Avid system will really make cutting these projects a treat.

Video Tutorial

The process of editing multicam footage is best left to a video tutorial. Check out the companion website (www.focalpress.com/cw/hullfish) to see this great feature in action.

The basics of multicam for a real Hollywood production or a simple two- or three-camera DSLR multicam shoot are the same. Usually, the only difference is the methodology for syncing the cameras. For a more "produced" shoot, the cameras are all usually fed timecode, so the timecode syncs them together. For the DSLR shoots, they either have to be eye-matched or synced to some kind of front or back slate. On larger shoots, this is sometimes a camera flash. For simple two-camera DSLR interview shoots, I usually just ask the subject of the interview to clap in front of his or her face.

After the cameras have been imported into Avid Media Composer, assuming no shared timecode, a sync point has to be found and marked with an in point. On DSLRs, you want to be very careful NOT to match using audio. The internal sync of audio in DSLRs is notoriously bad, and using a loud noise or something on all of your cameras can mean that they are out of sync by several frames. This is also a danger of using something like PluralEyes to sync audio. PluralEyes does a fantastic job of matching the waveforms of a camera's audio track to another camera's audio track or to a digital audio recording device, like a Zoom, but the problem is that the audio waveforms of the camera might not actually match the video on the same camera. Test this on multiple clips from the same camera. Simply shoot yourself clapping your hands or using a slate. See if on multiple successive clips the sync between video and audio is consistent. Try this on longer clips with claps at the front and back of the clip. Try it when the camera is cool and when the camera has been in use for an extended time. It's possible that by the time you read this your new camera is better at this syncing than the cameras I'm working with now, but I see remarkable inconsistency in in-camera audio and video even between takes of the same camera, so you may want to even slip the sync of imported camera footage to get it to match. Slate every shot that you can.

With that caveat out of the way, use a visual reference on all cameras to set an in point. Ideally, this is a camera slate, but a camera flash or both cameras shooting someone clapping hands together will work. And, in a worst-case scenario, it's syncing an eye-blink, to a sound source, or when the drummer in a rock video hits a cymbal.

To sync the multiple angles visually, set an in point on each camera angle to the visual sync reference (like a camera slate closing). With the in points on each shot marked, select each of the angles in the bin by Shift-selecting or Command-clicking (Mac) or Control-clicking (PC). Then, select Bin > Group Clips or Shift-Command-G (Mac) or Shift-Control-G (PC) (Figure 3.19). This will call up the Group Clips dialog (see Figure 3.20). If you synced via timecode (the timecodes on all cameras match), use the Source Timecode option. If you are syncing with in points marked on a common visual marker, then use the In points option and click OK. This creates a grouped clip in the bin (see Figure 3.21).

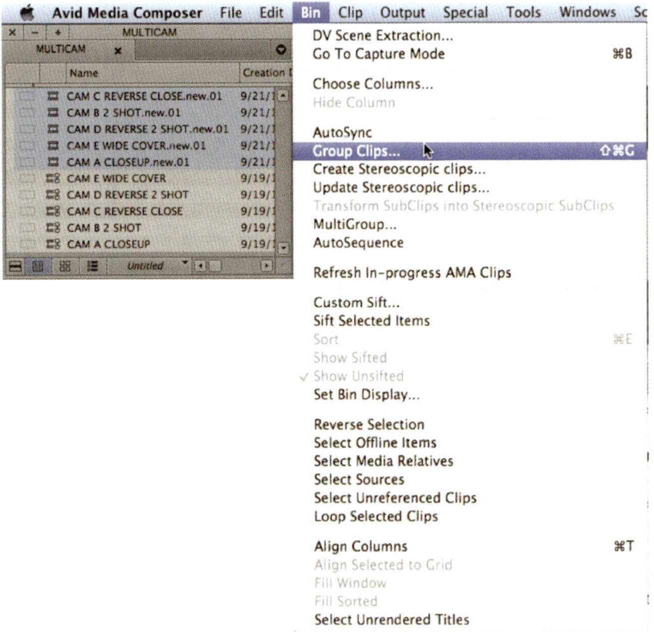

Fig. 3.19 The Multicam bin to the left of the pulldown menu shows the five clips selected to be grouped. The Bin menu is pulled down with the Group Clip option selected. This will create a grouped clip with the five angles.

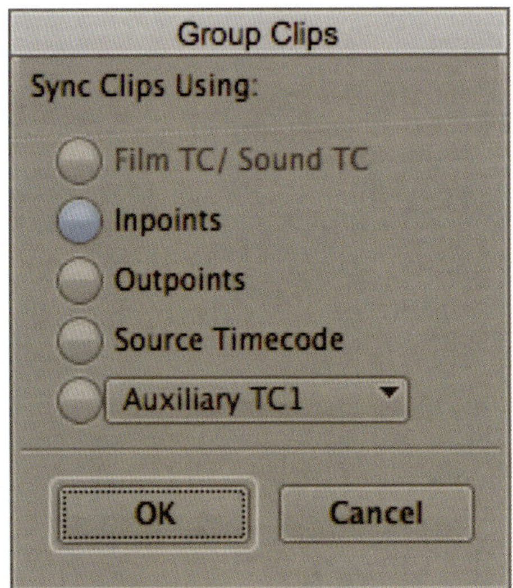

Fig. 3.20 The Group Clips window allows you to choose the syncing method.

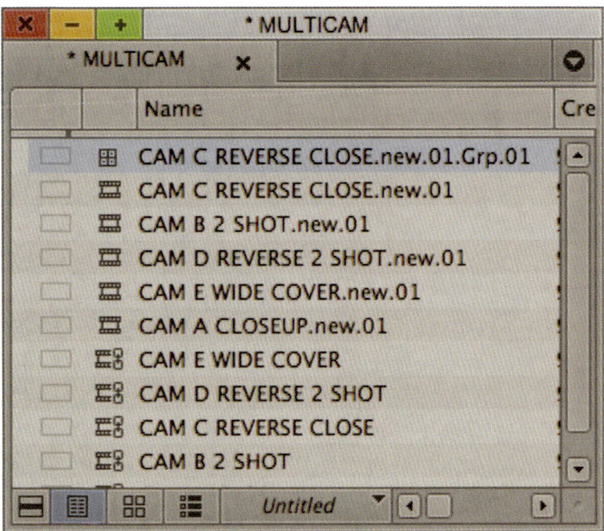

Fig. 3.21 At the top of the Multicam bin is the new grouped clip. Note the special icon indicating the grouped clip.

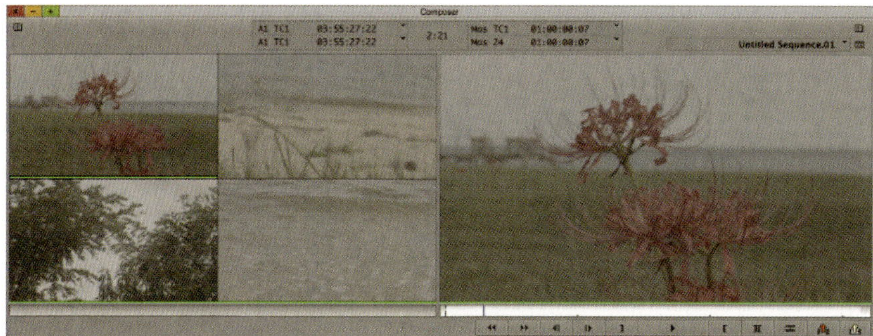

Fig. 3.22 The Composer window in Multicam Mode. Note that the "Source" side shows four different images in quadrants. The Green line under the top left quadrant shows which of the four cameras is currently selected.

A grouped clip has an icon that looks like four dots in four boxes and has the .GRP.XX suffix (XX is the number of grouped clips in the bin). Grouped clips can be loaded into the Source monitor and edited into the Timeline in the same way as regular clips.

To view the grouped clip in the Source monitor displaying all of the angles (or at least most of the angles), you need to go into Multicam mode. This can be done with a key command of Shift-Command-M (Mac) or Shift-Control-M (PC), or from the menu command Special > Multicamera Mode.

In Multicamera mode, you can see two to nine different angles in the Source monitor, depending on your settings (Figure 3.22). Choosing the number of angles is done from the Composer Setting/Multicam tab.

Another option is to choose an in and an out point in the grouped clip and cut the entire thing into the timeline just as you would any other clip. Choosing the angles and timing for the edits is done by being in Multicamera mode and playing through the sequence while using the multicamera buttons for each angle, much like a technical director would cut between cameras on a switcher in a live show. For a default user setting, the camera angles are mapped to the F9, F10, F11, and F12 buttons. Cutting from one angle to another is as easy as watching your options in the Source monitor while the Sequence Timeline plays and pressing the corresponding button to the camera you want.

The great thing about editing multicamera footage in this way is that everything is on a single layer and all transitions, timings, and even camera angles can easily be changed, just like any normal edit can be trimmed or segments replaced. Once you're out of Multicam mode, all of the traditional editing tools of the Avid system are at your disposal, and you can access any of the other "hidden" angles at any time. Trimming these shots is done just as any other trim is done. Accessing the other angles can be done with the same keys that you used to call up the angles in Multicam mode, or you can use the Next in Group and Previous in Group buttons to toggle through the angles. To change an entire segment to a new angle of the multicam group, park your Timeline locator at the beginning of the segment. To change the angle at some point in the middle of a segment, just park at that point and use the multicam angle buttons or the Next/ Previous in Group buttons, and the angle will change at that point in the Timeline until the end of the segment.

For Allyson Johnson, multicam was invaluable in culling through the huge amount of coverage in Smash's signature song-and-dance numbers. Often the coverage included in excess of eighteen different takes and angles per song. Because the songs were performed to a locked audio track, the assistant would group not just the angles of a specific take but also multiple takes into a single multigroup.

Because of the number of angles, Johnson worked directly with her assistants to determine how the multiple takes and angles would be grouped together. Since footage from only nine of the cameras can be seen at a time, they grouped similar angles together in order to compare them. Johnson explains, "We'd group all of the close-ups in one multigroup and the other shots in another group. But depending on the choreography, we'd sit down beforehand and I would choose which way to group things."

Though she would sometimes rough-cut dialog using the real-time editing and playback capabilities of Multicam mode, more often, especially with the big dance numbers, she was more deliberate in using the multicam groups from the Source monitor instead.

"In Avid if you're in Multigroup on the record side, it will cut when you make an edit," Johnson explains, "so I'm stopping at a point when I want to cut to another camera and then looking through the choices. Then the choreographer would come in and sometimes say 'I had this particular dancer do something interesting' and they'd want that on camera." In those instances, they could quickly find the coverage of that dancer or action within the multigroup.

ScriptSync

For off-line editing, especially of scripted drama and comedy, ScriptSync is one of the things that truly set Avid apart from the competition. Please refer to chapter 1 (on page 84, and also check out the video tutorial on ScriptSync on the companion website if you haven't watched it) for specifics on how to set-up your production to use this optional tool.

One of the people I interviewed for this book who relies on this feature heavily in the off-line editing phase is Peter Chakos. Peter has been cutting comedy as far back as *Seinfeld* and *Cheers* but is most recently known for Chuck Lorre's hit on CBS, *The Big Bang Theory*.

When I asked him about the bin and project organization for *Big Bang* he told me that he never even opens the bins for the show. He goes straight to the ScriptSync script that his assistant has prepared. The show is cut entirely from that.

The editing schedule for the show is very tight. On Monday, they do a preshoot day to accommodate guest star schedules, outdoor scenes, and anything else that can't be shot in front of the live studio audience for one reason or another. That material is digitized from HDCam tapes and edited in time to be rolled in for the studio audience that arrives for the main shoot on Tuesday.

GLOSSARY

Line Cut: A recording taken from the switcher, including multiple cameras switched during production.

GLOSSARY

ISO deck: A recording deck on a multicamera shoot that only records the output of a single camera.

The Tuesday shoot is done with four cameras, each recording to its own ISO deck, with no line cut. Peter sits in on the Monday and Tuesday tapings and takes handwritten notes about favorite takes. Peter says that the preshoot days include about three to six takes of each scene, while the Tuesday audience shoot usually has no more than three takes for each scene. The tapes from each ISO recording are digitized by Peter's assistant, Todd Morris, who prepares the bins and the ScriptSync script. The cameras are also grouped for multicam editing.

As Peter cuts, using his notes and the ScriptSync script window, he finds that about 80 percent of his intuition on set is correct, and that determining which camera angle to choose at any given moment is simply a matter of instinct. According to Peter, "I'll experiment to see what makes me laugh more."

Despite the perception that the laughs are "goosed up" in editing comedies, Peter says that usually the audience at the show laughs longer than television can tolerate. "On the stage the laughter is communal," explains Peter, "so all the laughs get shortened, usually by 50 percent or more." Peter attributes his timing as an editor to the fact that he, like so many editors I know, is a musician. "Timing is all rhythmic and musical," says Peter.

The show's actual content is just about twenty-one minutes per episode. Peter initially cuts the show together loosely, then does a separate tightening pass, so that the whole show tightens uniformly from beginning to end, instead of just near the end when it becomes apparent that time needs to be cut.

By Thursday night Peter delivers the first pass of the edit to the director. "I'll get shot changes—maybe three to five notes—on Friday morning. With those changes completed, Peter delivers the show Friday afternoon to get a final sign-off from executive producers Chuck Lorre and Steve Molaro. According to Peter, "Usually I'll get minimal notes from [Lorre]." From there the show goes to the studio and network, where it's rare to have additional notes or changes.

Peter's assistant handles the hand-off of the show's approved edit to Technicolor, where Tony Teresi gets a single Avid bin with the final edit and completes the on-line in Avid Symphony.

A major proponent of ScriptSync since its earliest days is David Helfand, who has cut *Friends, Weeds,* and many other shows. When we spoke, he was in the midst of cutting *The Mindy Project.*

Despite being a single camera production on a TV schedule, Helfand says, "The volume of stuff that we shoot is also overwhelming—our first edit is 1.5 to 2 times the length of the episode. So the initial assembly is 35–38 minutes." All this, for the 21.5 minutes that they have to end up with. Helfand reveals that the initial pass at the 2012 season finale was 46 or 47 minutes. Helfand explains, "The producers really like to see the ad libs, and then you have to figure out how to get in to and out of these comedic detours. They riff and do hilarious stuff and you want to keep snippets of it and that's the great challenge and to keep the pace and drive of the episode flowing. So there are a lot of tough choices. How do you keep it relevant to the scene? So we have to show them everything so that they can decide what to keep. Every episode is a juggling match. It's a lot of on-the-spot juggling. It's not the most efficient way to work. As you work through something, you can tighten it too much. It's a very tightly woven fabric, and you have to make sure to pull out just the right thread and that's where the judgment and skill come into it."

Key to finding and organizing the various riffs and adlibs is ScriptSync. "I've seen people use ScriptSync in a lot of different ways or approaches," explains Helfand. "One of the complicating things is that with film, the production would be very selective about what they shot, but with digital they let it run and run and they'll reset without breaking. But with the way they shoot now, you might have fifteen or twenty different passes between the different coverage and angles. So when you try to wade through various camera angles, I'll have my assistants drag lines for every single utterance of a line. When they reset because they flub or whatever, I'll have the assistants drag out the number of times they go over the line. I colorize the lines to indicate various passes like cuts to the scene. So my scripts look more like Richter scales instead of scripts [see Figure 3.23]. So there's kind of a creative formatting issue with how to get the text so you can see it. It becomes a very detailed roadmap that helps you dig in to the material."

Though Helfand has been an unofficial evangelist for ScriptSync among his peers, he still wishes for improvements to make the tool more useful, including the ability to edit text directly in the script inside of Avid and the ability to include sticky notes. Coloring various clips different colors is critical to his workflow, and the methodology for doing this is not great.

Adding the same color over and over is easy. In the Script tool, select the lines of text and the vertical line that represents a specific take of that line. Then at the top of the script, the button that looks like the capital letter *I* lets you assign a color to the line. But changing the color means going to the menu bar, Script > Color > (pick one of six colors). By selecting specific script lines and specific takes, it's also possible to change the coloring of the line with multiple colors (see Figures 3.24 and 3.25).

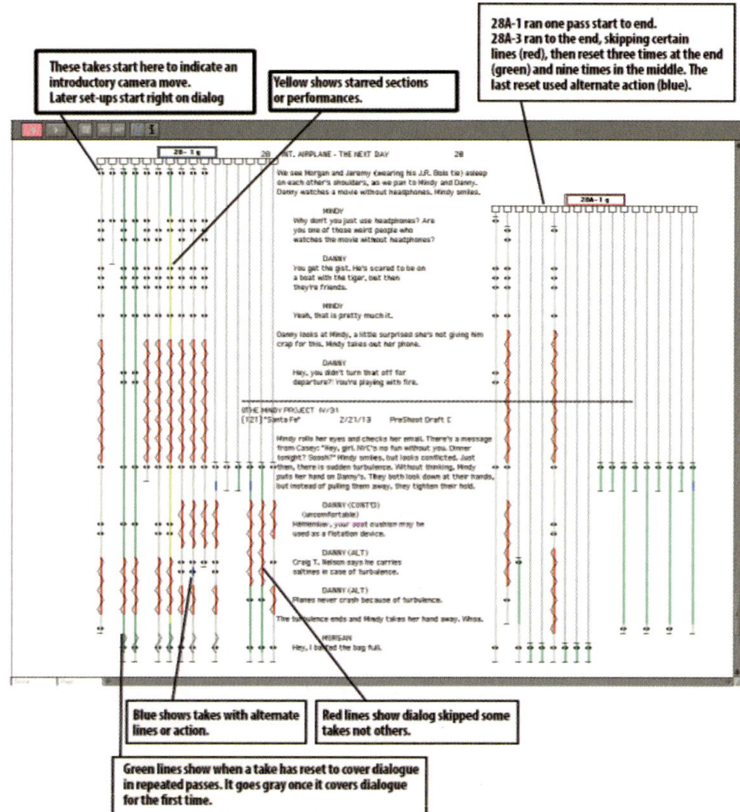

Fig. 3.23 This is a heavily notated page from *The Mindy Show*. Image Courtesy of David Helfand.

Despite a few improvements that he'd like to see, Helfand has many features that he couldn't do without. One in particular helps him judge performances that are obviously critical to editing. Helfand explains, "There's a feature where you can loop and play the various angles of a line. If your script is properly set up, you can compare all the takes of a single line by highlighting horizontally across the line of dialog so that you select all of the coverage across that range and hit play in the Script tool and it'll play all those lines. It's very powerful to be able to compare all the nuances. And I play a lot with nuance and performance, if there's a particular nuance of the inflection someone says something that I love, I'll dialog edit that inflection so that it's under another take visually. I've also helped use that in dialog editing so if a line is slurred or something, I find a similar word in a script to be able to use dialog that I normally wouldn't be able to use...."

Helfand points out that the media links to the scripts don't even have to be active. "On *Weeds*, I could go back through scripts that were two or three years old and pop them back up in my project and have lines

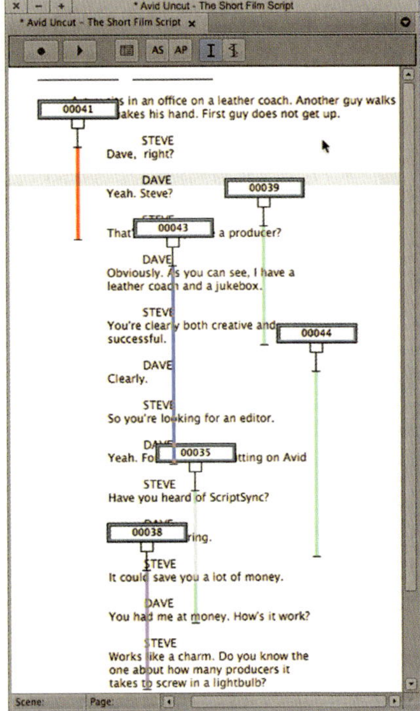

Fig. 3.24 ScriptSync colors.

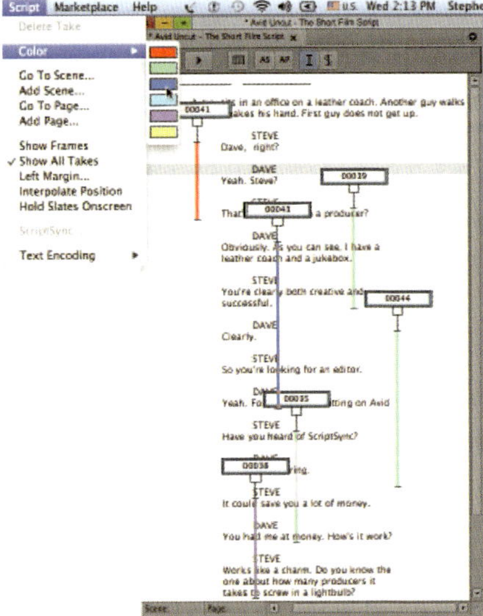

Fig. 3.25 ScriptSync Color menu.

redigitized from an episode that's years old so I can find a replacement word to manufacture things. The script doesn't have to still be linked."

There are also some features Helfand prefers to turn off, "When I use ScriptSync, I don't bother showing the frames because they may not be representative of the entire length of the clip." They also take up space on the script and may clutter highly complex scripts.

Feature editors are also very fond of ScriptSync. Wayne Wahrman (editor of *I am Legend*) relies on it. According to Wahrman, "The script [in Avid] is my primary source as I start a scene. It allows me to get to any take that I need at any time."

Even during revisions, Wahrman finds that ScriptSync has uses: "Everybody shoots a lot of film these days, and the director will say, 'I really want to use that scene where he said 'Ouch,' so then you can look at the script and find it."

And, like Helfand, Wahrman finds ScriptSync invaluable for reviewing, too. "You play all the takes of the script on a certain line back to back and find the best performance. Editing from the script has allowed me to be more thorough and sped me up a little, though editing can only go so fast because it has to go through your head."

Using PhraseFind

Nexidia's technology powers the phonetic matching capability of ScriptSync. It also powers Avid's PhraseFind. If you used the third-party program called Get with Final Cut Pro, it's the same technology and concept. PhraseFind simply lets an editor search for specific words or phrases in the audio of the clips in a project.

For decades, speech recognition software has been notoriously inaccurate. Most of the reason for this is because understanding speech is deeply dependent on understanding the context that surrounds a specific sound. The most obvious example of this is simply the number of homophones—words that sound exactly the same but that have different meanings, for example, *their/there/they're, too/two/to,* and *oar/or/ore.*

However, PhraseFind and ScriptSync are very accurate and useful because neither one depends on context. ScriptSync matches sounds with a script that already provides the context and the exact words that are trying to be matched phonetically. Because ScriptSync knows what words it's looking for, based on the search criteria that are typed in, the accuracy is nearly perfect. And it delivers its results in order of the software's confidence that it has found an appropriate match. In my experience working with the software, anything under 50 percent confidence is usually not a match.

As with ScriptSync, there are several ways to install or activate it, depending on where you purchased it or whether you're running a temporary license. For the basic install see the ScriptSync section toward the end of chapter 1.

With ScriptSync installed and some clips with intelligible speech in the program, simply hit Command-F (Mac) or Control-F (PC). This calls up a dialog that is used jointly for PhraseFind, regular text searches in Bin columns, and ScriptSync searches.

For ScriptSync to work, it needs to index all of the audio on your system (including AMA clips). You can attempt to do searches while this indexing is taking place, but it won't be a complete search until the little PhraseFind index button at the bottom of the window is completely green. In this example (Figure 3.26), you can see that I did a successful search even though the indexing was not complete, as indicated by the half-green PhraseFind Index button.

After indexing your imported or AMA clips, there are a few choices you need to get PhraseFind working. First is to type in your search in the Find text entry box. Then select the Clips and Sequence" tab (not the Script Text, Timeline and Monitors, or Settings tab). At the bottom, choose the appropriate language that the talent spoke from the pulldown menu. In Figure 3.27, I chose North American English. Then click on the PhraseFind button in the top right corner of the window.

TIP

If you're having problems finding a word that has unusual phonetics – like maybe a foreign word in an otherwise English text–try spelling the word phonetically, so "Voila" would be "wuh-lah."

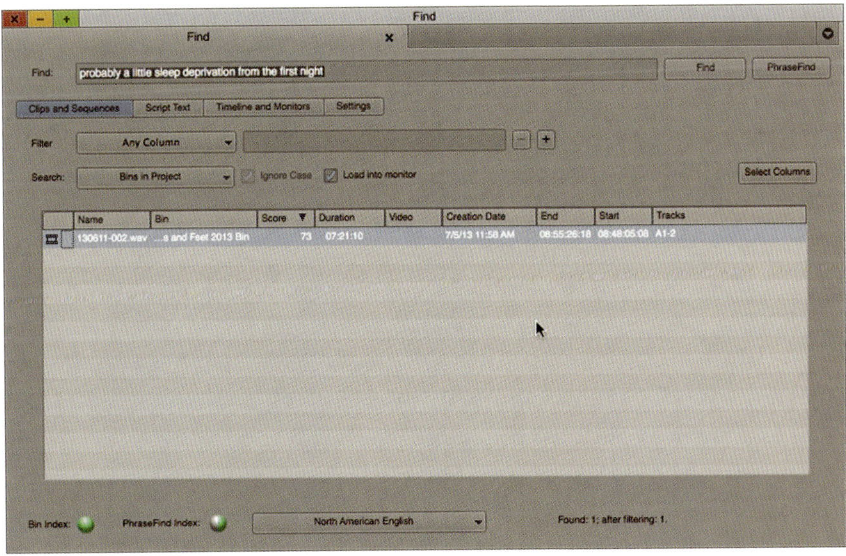

Fig. 3.26 Command-F (Mac) or Control-F (PC) calls up the Find window, which includes the ability to find text in bins, clips, scripts, or, through PhraseFind, audio.

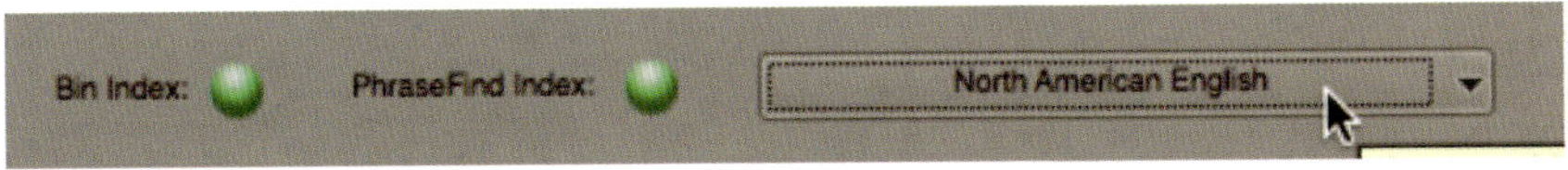

Fig. 3.27 To set up PhraseFind, you must assign a spoken language. Here, I chose North American English.

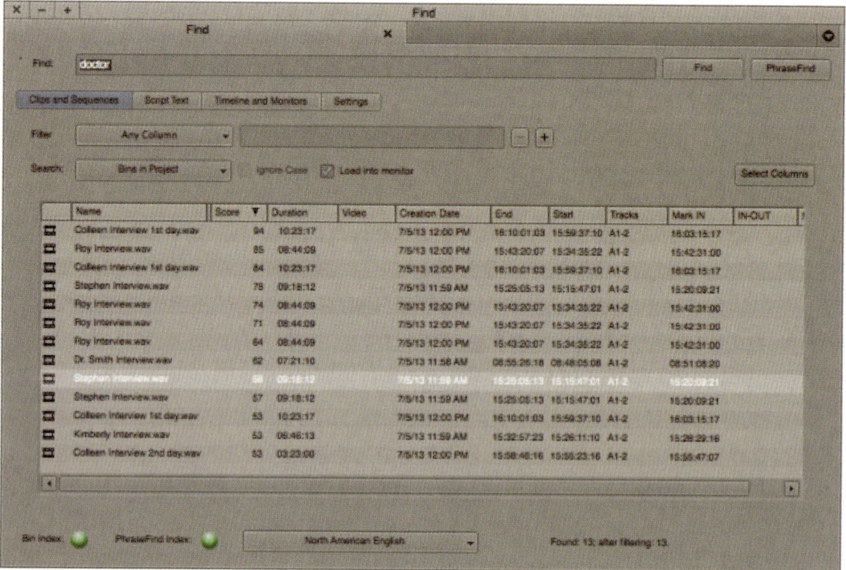

Fig. 3.28 PhraseFind will list the possible "hits" and rate each with a probability that it has found the word or phrase on which you searched.

In Figure 3.28, the word *doctor* came up in thirteen possible sound bites. I can audition these by double-clicking the clip in the PhraseFind window, which loads the clip into the Source monitor and places the Timeline indicator and an in point at the precise beginning of the phrase. Notice in the list of potential clips that each one is ranked from high to low in the Score column. This is a probability score that indicates how certain PhraseFind is that it has found the correct sound bite. All of the hits in this particular search above 60 percent were correct. The five clips with a score in the fifties did not have the word *doctor*.

By using PhraseFind to search for a more complete phrase, like "The doctors here are very well trained," narrowed the search results down to a single, correct option that had a probability score of sixty. The fewer words you search on, the more potential hits you will have.

Exploiting the Latest Features in MC6 and MC7

It's easy to get complacent as an editor—to rest on your current skillset. But as Avid adds powerful features, it's worthwhile to take the time to try out the new features and figure out how to incorporate them. Editing is similar to learning a sport or a musical instrument: you need to use a feature multiple times to "get it under your fingers" and acquire some amount of muscle memory and comfort level.

It's impossible to go back through Avid's entire history of new features, but I do want to point out some of the important ones that you may have decided to ignore from the last couple of releases. I'll start with "new" 5.0 features and work towards 7.0.

Smart Tool is the big one, as I've already discussed in this chapter. My initial reaction was pure disdain, but when I ended up mapping the four main Smart Tool buttons, I found that Smart Tool is an incredibly powerful way to work. For FCP users, the Smart Tool functions should almost come as second nature. To old, curmudgeonly Avid editors, the two trimming Smart Tool functions are obviously like the standard "red" and "yellow" trim tools, and the "red and yellow" arrow tools are similar to their old "segment editing" counterparts that used to sit at the bottom left of the Timeline. My suggestion to the curmudgeons is that before you dismiss the use of Smart Tool, map the two trim tools and two segment editing tools to your keyboard in the same order that they are in the Timeline: red segment, yellow segment, red trim, yellow trim. I personally have these mapped to the ZXCV keys. Spend some time toggling these off and on from the keyboard as you do Timeline editing. Give yourself an hour or two of experimentation before you give up on them. I think you'll find that the keyboard mapping and toggling really adds to the functionality and usability.

The rest of the functionality added in the 5.0 update included additional Timeline palette tools that mirror some of the functionality of FCP7 (see Figure 3.29). These include the Link Selection button at the top of the new Timeline tool palette. Link Selection works similarly to the link button on the top right of the FCP7 Timeline. It allows you to select a single track and have all of the other tracks that are linked to it by being from the same clip. For example, selecting a segment of video in the Timeline would also select the audio tracks associated with that video segment. Link Selection can be enabled and disabled from the keyboard using shift-L.

The ability to adjust keyframes in the Timeline can be enabled and disabled from the Timeline palette using the Keyframe Selection button.

Also in the "new" 5.0 Timeline palette is the ability to directly manipulate the length of transitions (dissolves, wipes, etc.) on either side of the transition by enabling the Transition Manipulation button (see Figure 3.29) and then clicking on the small handles that appear in the Timeline at the end of the transition effect icon (see Figure 3.30).

Version 5.0 introduced some other FCP-like functionality that allows for selecting multiple segments in very helpful ways. If you haven't played with them, and never cut on FCP7, check out the Select Left button, Select Right button, and Select In/Out button (see Figure 3.31). Select Right is one I find particularly helpful when I want to open up a complicated sequence and a ripple trim won't really get me where I want to go. Usually this happens when I want to insert something into a

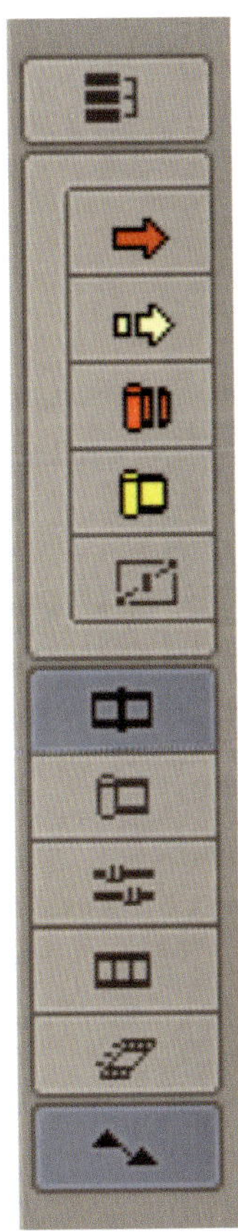

Fig. 3.29 These are the main tools added in the version 5.0 release. Collectively, they're known as the Smart Tool Timeline palette.

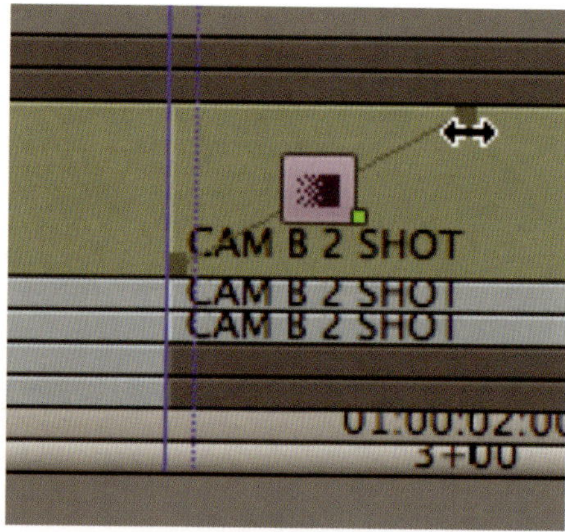

Fig. 3.30 Transition Manipulation allows for dragging handles on transitions in the Timeline to adjust the length and position of the transition.

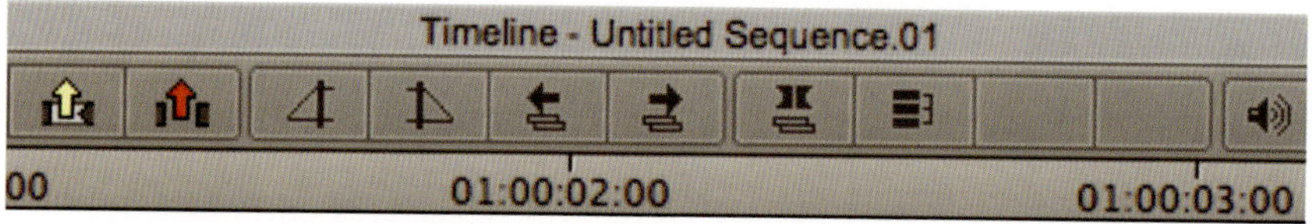

Fig. 3.31 The Select Left, Select Right, and Select Between Marks tools are very useful. In the default UI, they are available at the top of the Timeline.

complicated edit, but I don't yet know the length of the footage I want to insert. When the button is clicked, it selects all segments on the Timeline on tracks that are enabled and are at or to the right of the Timeline position indicator. The Select Left button does the same thing, but obviously selects segments to the left instead of to the right. The Select In/Out button selects all segments on enabled tracks between or "under" the in and out buttons.

Version 5.0 introduced the ability to have stereo-specific tracks. Using the keyboard to add a single mono audio track was always Command-U (Mac) or Control-U (PC). In this release, came the added keyboard command of shift-Command-U (Mac) or shift-Control-U for a stereo audio track (see Figure 3.32). The stereo track is represented by a single track in the Timeline indicated with a small double-speaker icon. Note that the source track feeding the stereo track in the Timeline is also a single track with a double-speaker icon, indicating that the source is stereo.

In order to have stereo sources when you import stereo or other multichannel audio, you must select the Options in the Import dialog and then the Audio tab of the Import Settings. At the top of this tab is the choice for Multichannel audio. Click the Edit button (see Figure 3.33), and in the resulting Set Multichannel Audio window, under A1 and A2,

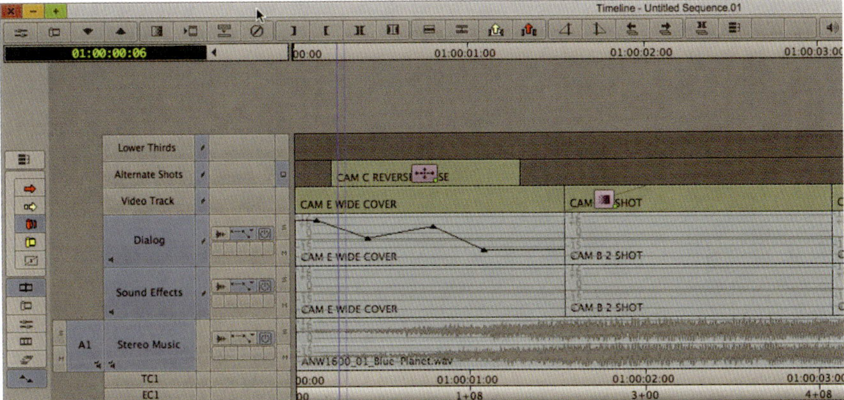

Fig. 3.32 Prior to 5.0, stereo was always represented in the Avid Timeline as dual mono tracks panned left and right. With the release of 5.0, you could create true stereo tracks, indicated by a small double-speaker icon, instead of the standard single-speaker icon.

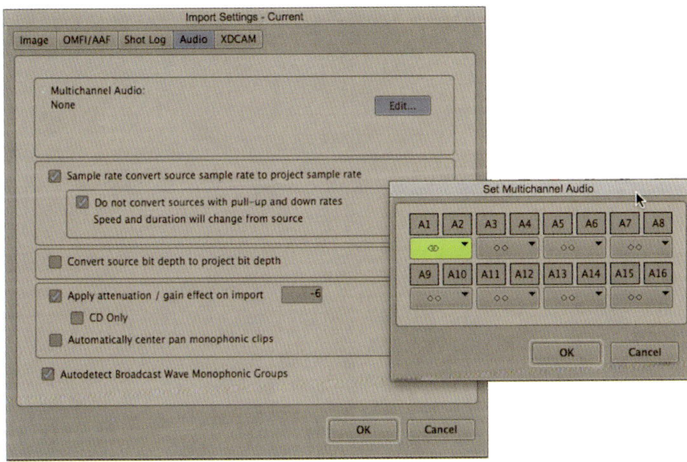

Fig. 3.33 Audio Import setting allowing you to import multi-track audio as mono, stereo or surround. In this figure, channels 1 and 2 are being imported as stereo, indicated by the linked circles in the green-highlighted button. Note that the circles under tracks 3–16 are not linked. They would be imported as mono if there was audio on those tracks.

use the small pulldown menu to select the choice for Stereo. Otherwise, your audio will be imported as dual mono.

If you accidentally create a stereo track that really should be dual mono (like a camera recorded with a boom mic and a lav mic on separate channels), you right-click the stereo track and select Split Track to Mono from the contextual menu. You can also right-click a stereo clip in a bin and do the same thing.

In 6.0, the big changes were 64-bit and the new UI, including changes to keyframe defaults and behaviors.

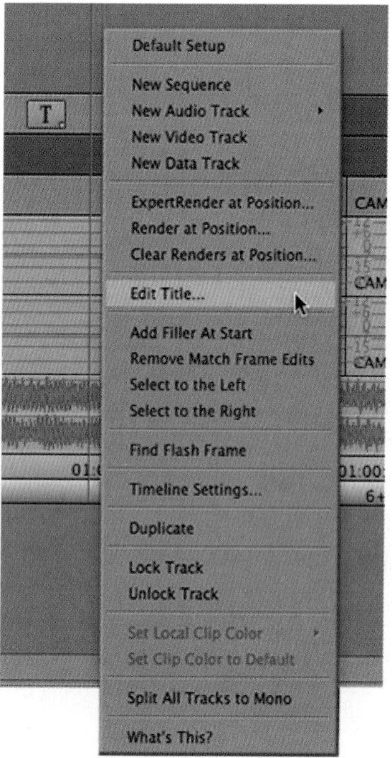

Fig. 3.34 Edit Title in the Timeline.

Previously, the default interpolation between keyframes was spline. After 6.0, the default became linear. And when keyframes are added with Advanced keyframing, you can add start and end keyframes by right-clicking in the Effect Editor and selecting Add Start and End Keyframes. Keyframes can be added universally or to individual parameters or groups of parameters.

Version 6.5 allows you to revise titles in the sequence without going into Effect mode (see Figure 3.34). Just place the cursor over the title in the sequence and right-click. From the contextual menu that appears, choose Edit Title, revise, and save as normal.

Version 6.5 also introduced the ability to control sliders in effects and color correction with the mouse slider.

In 6.5, you can copy and paste entire regions of audio keyframes. Sometimes I download watermarked stock music to see if it will work in the sequence and whether the client will like it. And if I spend time rubberbanding the audio to "duck" the audio around sound bites, I don't want to lose the work I put into that when I pay for the unwatermarked version. To reuse the keyframes, either select the audio segment or set in and out points around enabled tracks, Command-C (Mac) or Control-C (PC).

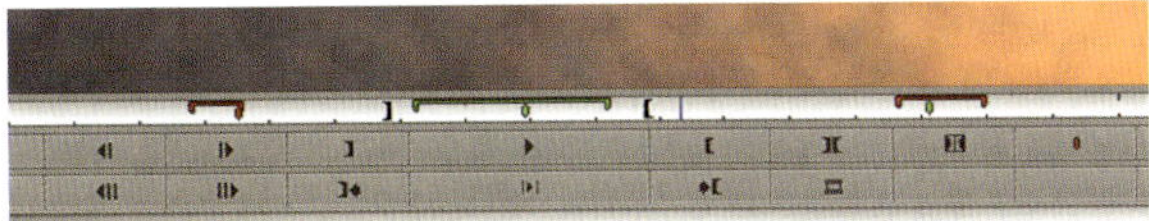

Fig. 3.35 Version 7 added the ability to create spanned markers. These define marked regions instead of the singular moments in time indicated by regular markers.

Then move to the new tracks and use Shift-Command-V (Mac) or Shift-Control-V (PC) to paste audio frames. A dialog box will appear allowing you to select Automation Gain or Automation Pan. To paste the volume keyframes, choose Automation Gain.

If you want to bump the volume of a keyframe up or down, use Shift-Command (Mac) or shift-control (PC) along with the up and down arrow keys.

Video Tutorial

To see some of these audio features in action, check out the video tutorial on the companion website (www.focalpress.com/cw/hullfish). For many editors, this may have snuck in under the radar, but this is a great opportunity to add some functionality to your editing arsenal.

In Version 7, there is a very cool feature called Spanned Markers (see Figure 3.35). Basically this allows you to set an in point and an out point in the Source or Record monitor and hold down the Option key (Mac) or the Alt key (PC) while clicking on any color marker to create a spanned marker of the same color that spans between the in point and the out point that you set. At its initial release, there was no way to convert these spanned markers back to in and out marks at a later time. This is great for clips that have several useable segments, like an interview, for example. Each useful answer from the interview clip can be marked with a spanned marker. Spanned markers cannot overlap each other, though they can overlap regular markers. Avid implemented this feature because it is essential to their development of the Advanced Media Workflow Association (AMWA) AS-11 specification. AS-11 is an OP1A MXF file specification for delivery of finished programming to broadcasters. The specification requires program segment markers, which is where Spanned Markers comes in: to define the program segments. If you deliver in this format, you'll need to use spanned markers to define each segment of your program.

One of the biggest features added to MC7, in my opinion, is the ability to import video resolutions higher than HD while continuing to utilize their full resolution throughout the edit process. Previously, you could import higher resolutions, but they would be converted to HD, so any blow-ups or repositioning would be only from the HD media, not the original 2K, 4K, or higher resolution. With the release of MC7, you can now import that high-resolution media and access full resolution throughout the editing process, as was explained in chapter 1 starting on page 66.

Dynamic Media Folders are a new addition as of MC7. A complete description of how these work and how to set them up is also in chapter 1 starting on page 61.

The ability to add LUTs and do advanced color management were added in MC7 and were discussed in chapter 1 starting on page 72. Background transcoding and consolidating were also added in this version, allowing you to continue editing while these functions churned away simultaneously.

Audio

Audio gets the short end of the stick in most video editing apps. Luckily, Avid's sister app is the hands-down worldwide leader in audio editing and recording: ProTools. And as the two programs have grown up side-by-side, more and more of the audio power of ProTools has been incorporated inside Media Composer.

One of the last ProTools innovations to make it to Media Composer has been a long time coming: cached audio waveforms in the Timeline, which finally arrived in version 7. Final Cut Pro had cached waveforms years before Avid, but to me, the audio waveforms in FCP were not that useful because they didn't sync well with the actual audio. With Avid's waveforms in the Timeline, it's possible to actually set marks based on the visual cues in the waveforms. With FCP, the waveforms in the Timeline were often off from the actual audio by several frames.

Mixing

There are a number of audio manipulation tools and methods. The most obvious to discuss is clearly the Audio Mixer. And my first word of advice is to use the Menu to Button mapping function (see chapter 2, page 133) to map the Audio Mixer button to a handy keyboard button or location in your editing UI. Mine is mapped to F12.

Going through the Audio Mixer's (see Figure 3.36) functionality from top to bottom, left to right, you will find the following:

Sequence Mix Format: This button allows you to set the sequence in the Timeline as a stereo, 5.1, or 7.1 mix. Click on the button to see a pulldown menu with the choices. When the Timeline is in stereo format, there are stereo pan controls above each track. But in 5.1 or 7.1 mix format, you get these cool, surround mixing "graphs" that allow you to move the position of the audio on that clip or track to a specific surround location in the "room." It's possible to manipulate the surround location directly in this small graph, but obviously the precision with which you can place a sound in the surround field is limited at this size. Directly below the graph is a small green dot inside a green box. Clicking on that icon will call up a larger and more full-featured surround mix interface. When you are done placing the sound, choose the "Close" button.

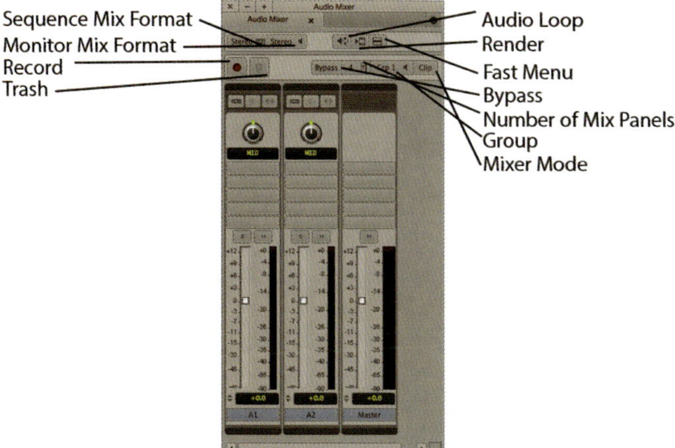

Fig. 3.36 This is the Version 7 Audio Mixer. This is displaying on a two track Timeline. Your mixer may look different.

Monitor Mix Format: The next button to the right is the Monitor Mix Format button. If options are grayed out, it's because you may not have the audio hardware to monitor a surround mix. The audio hardware is probably part of your video card, though it could also be part of an audio interface box. This allows you to choose to monitor your mix in a number of different ways, including direct tracks out. You can monitor your audio with the same settings as your Sequence Mix Format or with different settings, such as monitoring a 5.1 sequence as stereo if you are only using headphones or don't have the hardware to mix to surround speakers.

Audio Loop: This button allows you to listen to a specific section or clip over and over as you finesse your mix, edit, EQ, or audio effect.

Render: Next is the Render button for rendering effects.

Audio Mixer Fast Menu: Accessible by Control-clicking (Mac) or right-clicking (Mac or PC) anywhere in the Mix Tool, this calls up options that allow you to institute the clip's audio tweaks (filters, pans, volume levels) across the entire track (globally) or between an in and an out point. This is very useful and time saving. You can also remove gain or pan adjustments globally. As of MC7, the Fast Menu also gives you access to the Mix Tool Display Options which lets you customize the look and functionality of the Mix Tool. With it you can Show Group/Link/Mirror buttons, Show Pan Controls, Show Effects Buttons, Show Mute/Solo buttons, Show Faders, Show Meters, and Show Legends. The default is for all of these to be on. Personally, that's the way

I'd leave them, but every editor may have reasons for disabling some of these.

Record: In the next row of buttons, the Record button allows you to record mixer level changes in what is known as Auto Gain mode, which must be enabled using the Audio Mixer Mode button, which is the last one in this row. So, to enable the record function, you must first toggle the Audio Mixer mode button from Clip (the default) to Auto. Then, you can set an in and an out point for where you want to record Auto Clip Gain volume changes, or just set the Timeline play indicator to the area you want to adjust. After pressing the Record button, any changes you make to the mixer during real-time playback of the audio are recorded and will affect the audio levels on subsequent playback. These recorded level changes can be seen in the Timeline as Auto Gain level changes. If you want to mix your levels in real time, this is the way to do it.

Trash: The next button, the trash can, allows you to delete your recorded level changes.

Bypass: Next is the Bypass button, which clears the effect of all of the Audio Mixer level settings. You may want to do this if you are sending a mix out directly to a real hardware-based external audio mixer that will be controlled live, or if you want to lay something to tape without having your "temp" audio mix influence the layoff.

Number of Mix Panels: The next button allows you to toggle the Audio Mixer from 4 to 8 to 16 tracks of faders. Obviously, there's no need to waste precious screen real estate on 8 or 16 channels of faders if you only have 4 tracks of audio. If you have even more audio tracks in your sequence, you'll need the next button, which allows you to choose which set of audio tracks you're displaying sliders for—again saving screen real estate.

Group: The Group button allows you to look at tracks in groups that extend beyond the number of tracks in the mix panel, so if you are viewing four tracks with the Number of Mix Panels button, but you have six tracks in your Timeline, the Group button allows you to see the other two tracks in a second group. Similarly, if you are monitoring sixteen tracks in your Mix Panel but you have additional tracks in your sequence, the Group button would provide access to those additional tracks.

Audio Mixer Mode: The final button in the row is the Audio Mixer Mode button, which allows you to mix the clip, auto or live. Clip Mode allows you to set a level across an entire clip. Auto allows you to execute Auto Gain volume and pan changes while listening in real time. (See under Record.) The Live setting allows you to control the mix levels of each channel live as the Avid Media Composer output is sent to an external source. These levels are not recorded or stored. The Live mode also overrides any existing volume and pan automation settings.

Below these buttons are the audio sliders themselves. Across the top of the sliders are three buttons: Gang, Stereo Link, and Stereo Mirror (see Figure 3.37).

Gang: Enabling the Gang button on multiple tracks allows you to control all of those "ganged" tracks with a single slider.

Stereo Link and Stereo Mirror: The Stereo Link and Stereo Mirror buttons are active only with stereo audio tracks, either tracks that were imported as stereo tracks in the source monitor or tracks that were created as stereo tracks in the Timeline.

The Stereo Link button allows you to "gang" the pan controls on a stereo track in a stereo sequence so that the pan controls move in a parallel direction together. The Stereo Mirror does the same thing, except the pan controls are ganged to move by equal amounts in the mirrored, or opposite, direction. Stereo Mirroring has to be used in conjunction with Stereo Linking. To experiment with this, try starting with both pan controls set to MID. With Stereo Link on, moving one pan control to R35 will also move the other pan control to R35. But with Stereo Mirroring on, moving one to R35 will move the other to L35.

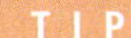
TIP

To create a mono track in the Timeline using a keyboard Command, use Command-U (Mac) or Control-U (PC). To create a stereo track in the Timeline, use Shift-Command-U (Mac) or Shift-Control-U (PC). Stereo tracks in the Timeline have a double-speaker icon indication below the track number in the Timeline.

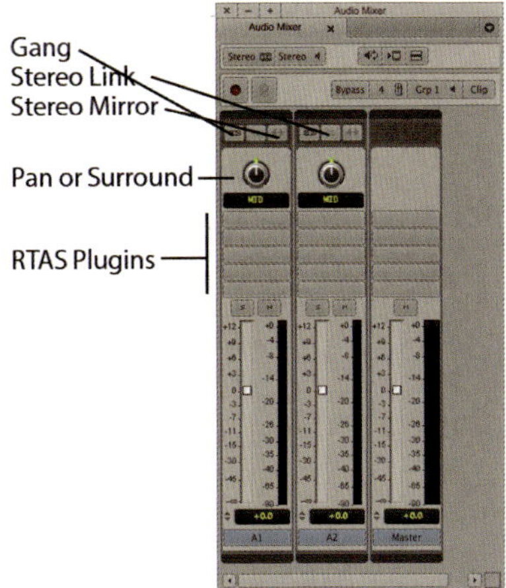

Fig. 3.37 Audio Mixer Panel detailing the tools in the slider section of the panel.

Pan or Surround: Below those buttons is a space for either the surround controls or the pan controls, depending on whether the Sequence Mix Format is set to stereo or 5.1/7.1. The controls will also depend on whether the track is a stereo or mono track. For mono tracks, a single pan control will be in the space. For stereo tracks, two pan controls will be available. For 5.1 or 7.1 surround, there will be a single surround control interface on each track or a dual control for stereo tracks. Below the surround control is a green button in a green box. This allows access to more detailed control of the surround sound, including larger graphical displays for positioning the sound for the track and specific pan controls for each of the five or seven channels of surround. Use the Close button to switch back to the regular Audio Mixer controls.

RTAS Plug-ins: Below the pan and surround controls is a series of five blank gray rectangles. These are used for enabling, patching, and controlling the RTAS (Real-Time AudioSuite) plug-ins (see Figure 3.38). The order of the plug-ins matters. Plug-ins that are upstream will alter the sound of plug-ins downstream. The streaming order goes from top to bottom. RTAS plug-ins can be dragged and dropped to new positions in the stream. Clicking on one of the blank RTAS rectangles calls up the RTAS Tool. Clicking

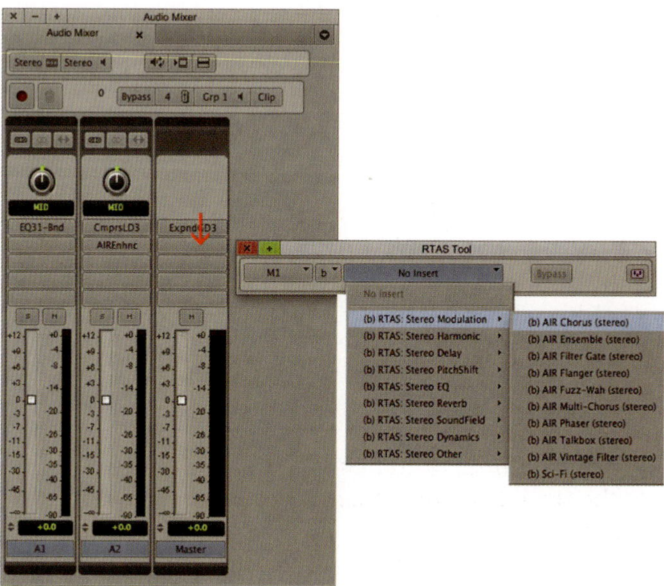

Fig. 3.38 The gray rectangles in the Audio Mixer are for RTAS (Real-Time AudioSuite) audio effects.

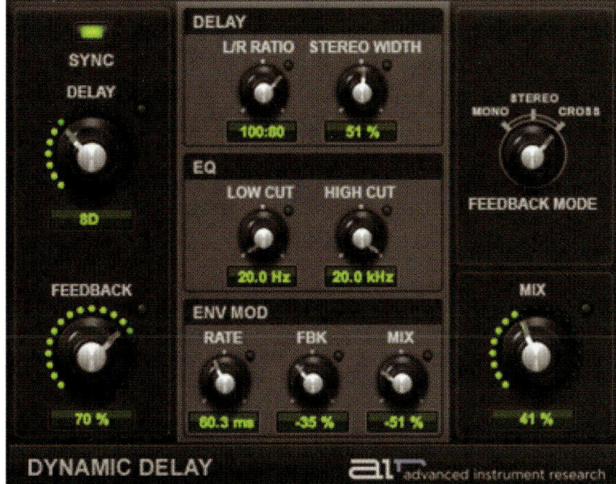

Fig. 3.39 This is the control window for just one RTAS effect, Dynamic Delay.

on the pulldown (No Insert) shows the menu of options. Selecting an option enables the controls for that option (see Figure 3.39). Clicking on an enabled RTAS rectangle (one with an effect in it) also enables the controls for that effect. Dragging a rectangle with an effect onto another rectangle with an effect will replace that effect.

Solo and Mute: Below the RTAS effects are the Solo and Mute buttons. It is possible to have multiple tracks soloed. Also, soloing automatically mutes the nonsoloed tracks.

Below the Solo and Mute buttons is the slider for the track. These can be controlled by manipulating the small square in the slider (the slider handle) or numerically using the number keypad or QWERTY numbers. You can also manipulate the sliders using the left, right, up, and down arrow buttons on the keyboard. Holding the shift key down at the same time as the arrow keys increases the speed of the changes. The numeric readouts for the levels can be read in the volume level display below each slider.

Finally, at the bottom of each track in the mixer is a track selector button (with the number of the audio track) that allows you to enable and disable the track from the mixer. Note that if you rename a track in the Timeline, the renamed track maintains its original track number in the Audio Mixer. Also, reordering tracks in the Timeline has no effect on the order of the tracks in the Audio Mixer.

In MC7, there is the addition of a Master slider for control of the overall level. This was possible in previous releases, but the control was

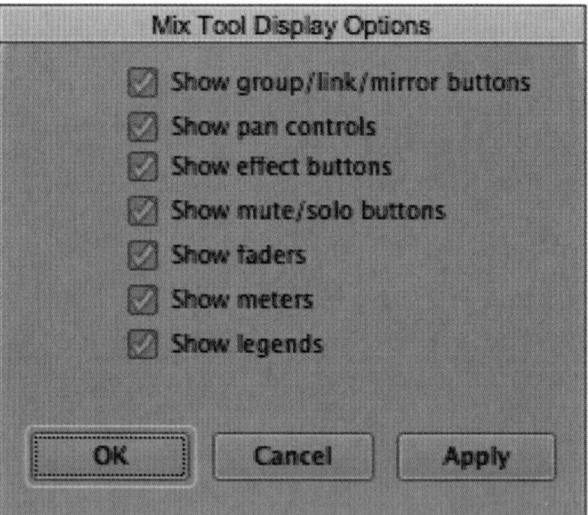

Fig. 3.40 There are myriad new options for displaying the Audio Mixer. From the Fast Menu in the Mixer Tool, choose Set Display Options to call up this window. Or right-click on the Mixer itself for the same menu pulldown.

hidden fairly deeply inside the Audio Tool. In MC7, the Master slider is easy to access from within the Audio Mixer. It also provides a global mute switch. The Master fader also allows you to see a "VU meter" for the full stereo or surround mix channels beside it. To me this means that I rarely have to open both the Audio Tool and the Audio Mixer. With the addition of the Master fader comes the ability to add RTAS effects to the overall audio mix instead of just individual channels.

A few changes to the mixer in version 7 (released in Summer 2013) included the following.

The mixer doesn't bother showing you faders for tracks that don't exist. This is just a screen real estate saver.

The Audio Mixer also gets the full "Avid customizability" treatment. Instead of providing a one-size-fits-all solution, Avid allows you to right-click on the Mixer and toggle a number of displays to save screen real estate or visual clutter (see Figure 3.40). Avid lets you take this to the extreme, even eliminating sliders (see Figure 3.41), which would seem to be the heart of any mixer, but there are some editors who prefer to set levels numerically or perhaps with an external physical fader device. So if you're on a laptop and need fader control in a compact package, you can turn it all off and still have some kind of control.

AudioTool

The Audio Tool looks like it's just VU meters, but it actually hides quite a bit of functionality within. The key to turning Clark Kent into Superman

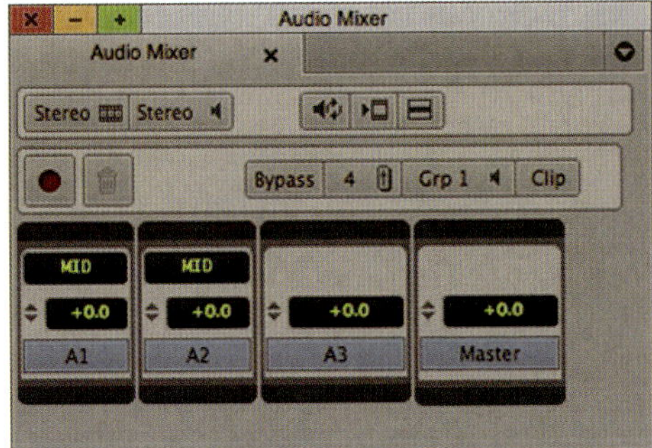

Fig. 3.41 By eliminating most or all options in the Mix Tool Display Options window, you can end up with a Mixer with no sliders. Many editors are comfortable setting levels numerically, so this can save screen real estate if you don't need the sliders.

is the seemingly innocuous PH button in the top right-hand corner of the Audio Tool. PH stands for Peak Hold, but it actually is a pulldown menu that gives access to Input Settings, Output Settings, Peak and Infinite Hold, Play Calibration Tone, Calibrate, Set Reference Level, and Set Calibration Tone (see Figure 3.42).

Input and Output Settings: These are the same settings that are accessible through the Audio Project Setting; they allow you to choose the inputs and outputs and routing of your video card or audio device.

Peak and Infinite Hold: These are used to determine the highest peaks in your audio. Peak Hold is the default, but with Infinite Hold, the VU meters "hold" the highest peak audio levels in the display until playback is paused or stopped. This is a useful QC for audio. With Peak Hold, the highest peak is held only for a moment.

Play Calibration Tone: Use this option to play tone across all channels.

Calibration: This option allows you to make fine adjustments to the incoming levels of audio. This is usually done when setting up the system using incoming tone to set a universal level on the Avid editing system that matches the source, such as a mixer feeding the inputs.

Set Reference Level: This option allows you to determine the reference level in dB compared to 0VU. The default is –20 dB, but some shops prefer –14 dB. This is a personal preference. If you aren't sure what to do here, stick with the default. If you regularly work with an outside audio post house, ask them where you should set this. Note that at its default, the –20 dB mark on the left of the

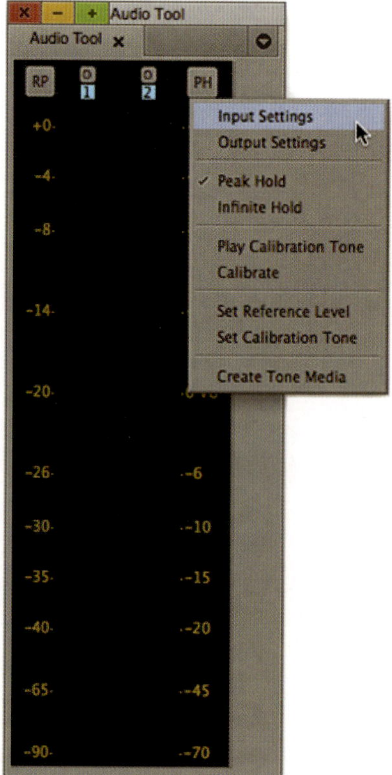

Fig. 3.42 The Media Composer Audio Tool with the PH menu displayed.

VU meters in the Audio Tool is at the same level as the 0VU mark on the right side. If you change this to −14 dB, then the −14 dB mark will be equal to the 0VU mark and the −20 dB mark will be above 0VU.

Set Calibration Tone: The default is −20 dB and 1,000 Hz (1 kHz), but you can set this to whatever you want, though this is pretty much the industry standard. For example, for all of you band geeks out there, you can set the Hz to 440, and you'll have a perfect concert-pitch A. Setting it to −777 gives you a funky wave, like a siren. Typing in 0 gives you white noise (or maybe it's pink noise), so if you want to generate "static," this is the way to do it. This is a great way to do a hearing test. Sadly, my hearing topped out between 14 kHz and 15 kHz.

Audio Project Settings

Audio project settings were touched on in chapter 2.

Rubberbanding

Editors are very set in their ways of whether they prefer doing level changes within a clip by rubberbanding or by breaking the clip into pieces and adjusting the levels of each piece. It's very easy to Add Edits to audio clips and adjust the relative levels of the clip on either side of the Add Edit. If that's the way you like to work, then stick with it. Many editors who work with outside audio houses, or in-house ProTools colleagues, claim that rubberbanding in Avid causes havoc when collaborating. I would disagree with this. It could be that information that *used* to be true is no longer true. The audio post houses I contacted had no preference for rubberbanding or Add Edits. Most audio post houses prefer an AAF export and can strip or use any audio mixing metadata from that file as they see fit. For me, it would seem that breaking up the clips with Add Edits would be worse for the end audio mixer because it's easier to strip rubberbanded audio levels than to delete hundreds of Add Edits. Probably the best thing to do, if you are sending your audio mix out for a final mix, is to call the audio post house ahead of time and find out what they'd prefer. If you're doing the final mix yourself within Avid, then it's simply a matter of personal preference. For me, my personal preference is to rubberband any levels that need to change within a clip.

> **TIP**
>
> If you know you are going to use an audio clip, or video clip with audio, multiple times in a sequence, set the level in the Source monitor before cutting it into the sequence. That way, each subsequent time you cut the source into the sequence, you'll at least be starting at the same audio level that you used before. It may still need tweaking, but if the levels are very low or very high, at least the source-side audio tweak will get you closer to the ballpark for all future level changes.

Rubberbanding functionality changed a little back around version 5 with the introduction of Smart Tool. This was necessary because Smart Tool meant that there was a lot more clicking and moving of elements in the Timeline that could cause havoc with rubberbanding.

In the latest version, rubberbanding requires enabling Volume in either the Track Control Panel on each track, or from the Timeline Fast Menu in the Audio Data/Volume pulldown menu selection. With that enabled, you can see any previous rubberbanding that was done in a track, but you can't manipulate it. You can also add keyframes to the audio tracks using the Add Keyframe button, but you can't change the levels. To enable manipulation of the Volume levels in the Timeline, you have to enable the Keyframe button in the far left of the Timeline, under the Smart Tool and Editing Mode buttons.

With that button enabled (indicated as blue), hovering over a keyframe in the Timeline will cause the cursor to change to a hand. Individual keyframes can be dragged up or down to change volume level. You can monitor the specific dB changes as you drag by looking at the bottom of the Timeline, just to the left of the Scale Bar. To make dragging volume keyframes easier or more accurate, you can enlarge the audio tracks you're working on by selecting the tracks you want to enlarge and pressing Command-L (Mac) or Control-L (PC).

You can also Option-drag (Mac) or Alt-drag (PC) keyframes to move them horizontally (forward or backward in time) in the Timeline.

As I mentioned earlier in this chapter, there are a number of ways to manipulate groups of keyframes. You can shift-select multiple keyframes and then manipulate just one of the selected keyframes to move them all. Or you can mark in and out points around a series of keyframes on selected tracks—even the entire Timeline—and manipulate all of the keyframes between the marks by moving just one of them. This actually can lead to a common mistake. If you have selected tracks and have left a mark in and out from a previous edit operation, you can accidentally move multiple audio keyframes when you meant to manipulate only one. If this happens, undo, then clear your mark in and mark out.

A keyframe can be removed by selecting it and hitting the Delete button. Multiple keyframes can be Shift-selected and deleted with this method.

AudioSuite and RTAS

AudioSuite and RTAS (Real-Time AudioSuite) plug-ins are worthy of a book of their own, if not a book for each individual filter. These are specialized audio filters (effects) that came from the DigiDesign/ProTools side of the Avid family. Some of these are pretty simple to understand and some of them really require some serious audio engineering chops to utilize properly. Because I don't have those chops and because teaching them to you even if I had them would be beyond the scope of this book, I'll show you how to apply AudioSuite plug-ins and I'll walk you through the parameters for a few of the basic and useful AuduiSuite plug-ins that don't require a degree from MIT or to have Sir George Martin to explain them. Actually, I'm quite certain that George Martin never used AudioSuite plug-ins on the Beatles albums, but you get the idea.

To figure out where to start with AudioSuite, I fall back on the most critical aspect I consider when trying to teach or learn something, and that is, what is the application? In other words, what is the problem I'm trying to solve? The things you may want to do to your audio with an AudioSuite plug-in include some pretty useful items: time compression or expansion (slomo or speed up), pitchshift, reverse audio, gain (good for audio recorded at low volume), complex EQ, Compression, Expansion, and De-essing. There are others, but these are the ones I have used in the past to good effect.

To give a simple example of an application for an effect, what if you wanted to have a piece of music end at a specific point in the track sounding as if it was a record player that had been turned off? This is usually

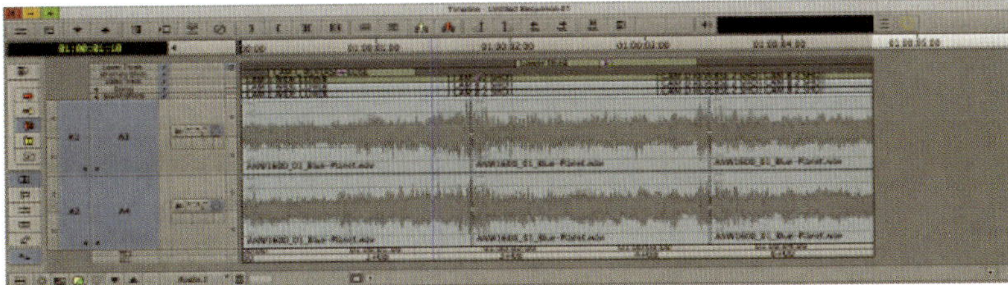

Fig. 3.43 This Timeline has music as dual mono on tracks 3 and 4. The first Add Edit is on a beat of music where I want the slow-down effect to start. The second Add Edit is where I want it to end.

done as a comic underscore or sometimes as a dramatic finish to something bad happening.

The answer to that effect is VariFi.

Drop in a music track on a sequence. VariFi is a mono-only effect, so your music will need to be dual mono, not stereo. Find a spot in the song that you want to have end with the "record player stop effect." Place an Add Edit on the beat where you want the music to stop. You also need to add an Add Edit a second or so later (see Figure 3.43). The key is that the length of the segment you create becomes the duration for the slow-down effect, so I added my second Add Edit on the fly at about the time I imagined I wanted the slow-down to stop. In my case, the segment I created between the Add Edits was 1:12 (42 frames). You could try shorter or longer. You will probably have to experiment a bit.

Once you have the segment created, park the Timeline locator somewhere inside the segment. Then choose AudioSuite from the Tools Menu. The AudioSuite Window will come up. Make sure that the text bar at the bottom says "Connected to DAE." Above that is a pulldown for where the effect will render. You can probably leave that where it is, but you may want to direct the renders to your RAID or media drive. Above that is a pulldown menu that says "None." Click on that pulldown menu and you'll see a pulldown with about thirty choices of filters (see Figure 3.44). These are all either free filters that came with your system or they are demo filters that you can use for a short trial period, or you may have purchased some AudioSuite plug-ins and installed them already. Go down to what is probably the bottom of the list and select VariFi.

This applies the effect in the Timeline to the segment you were parked on and to the tracks you had selected (see Figure 3.45). To be able to edit the parameters of the effect, click on the purple plug-in icon just to the left of the pulldown menu you were just using (see Figure 3.46). That will call up the "effects editor" for the selected AudioSuite plug-in (see Figure 3.47). One of the reasons I chose this effect is because it has only two options: slow down and speed up. The slow down option is the default, I think, and it's what we want for our "record stop" effect. If you

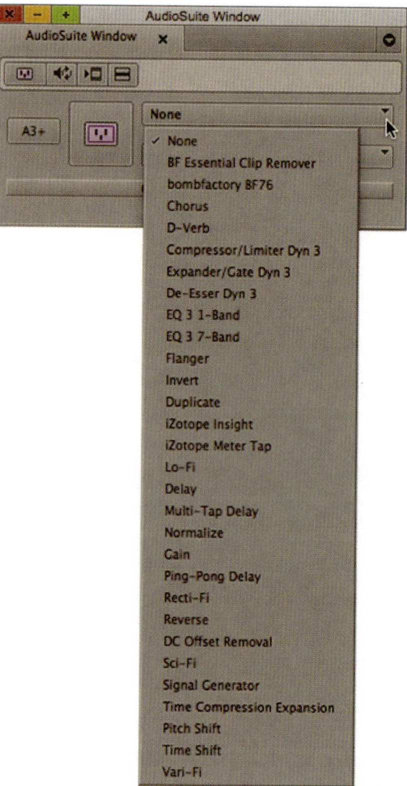

Fig. 3.44 AudioSuite plug-ins are applied from the pulldown menu in the AudioSuite Tool. VariFi should be at or near the bottom of the list.

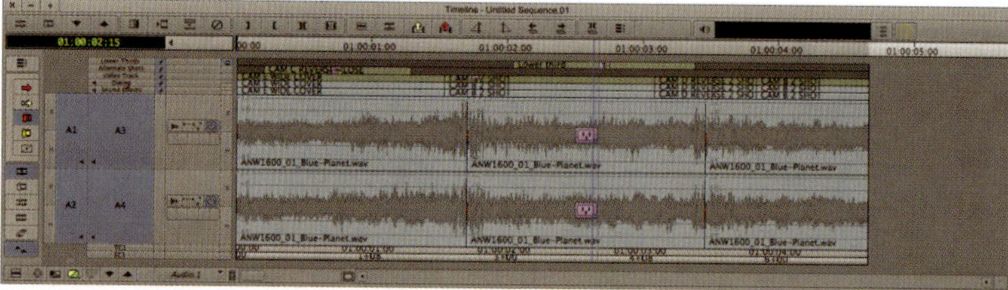

Fig. 3.45 As soon as the plug-in is chosen from the AudioSuite Tool, it is applied to the enabled tracks at the point of the Timeline locator.

want the beginning of a track to sound as if the record player was just starting up, then you'd choose the "speed up" effect. You can preview it to hear what it sounds like. You can also use the options to render or cancel the effect. If you want to, trim the length of your segment to lengthen or shorten the "slow down" effect. Then render the segment.

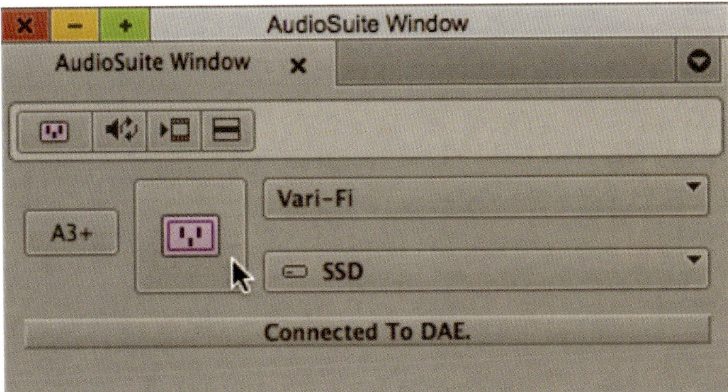

Fig. 3.46 In the AudioSuite Tool, choose the Activate Current Plug-in button, which looks like a purple electrical outlet.

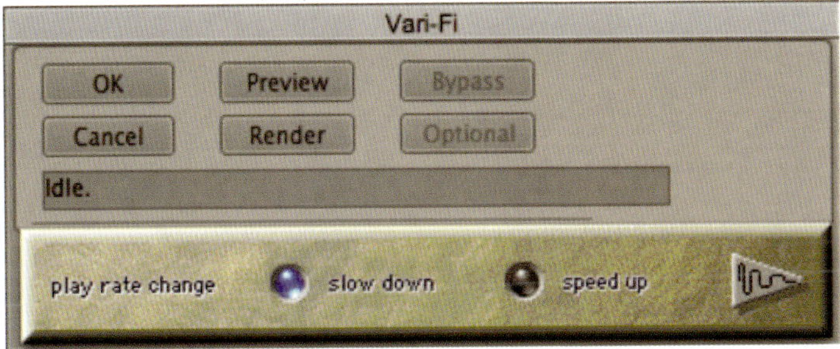

Fig. 3.47 The VariFi AudioSuite plug-in is probably the simplest effect possible. Two choices are available: "slow down" and "speed up." Choose "slow down."

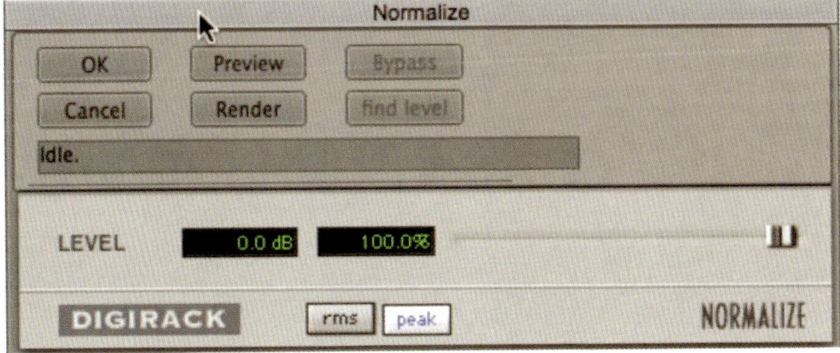

Fig. 3.48 This is the UI for the Normalize AudioSuite effect.

A great use of AudioSuite plug-ins that I employ fairly regularly is the Normalize function (see Figure 3.48), which is sometimes confused with Compression or even Gain. What Normalize does is increase (or sometimes decrease) the volume of the overall clip to a specific level. The default is 100 percent. That means that the plug-in makes the loudest part of the clip 100 percent of the volume it can be without distorting.

I try to set Normalize to something around 80 percent. This is different from Compression, which will attempt to alter the dynamics of the audio within the clip, making soft sections louder and loud sections softer, so that everything is kind of at the same level. Normalize makes the loudest parts of each clip the same level. This is great for most interviews. If you have a number of speakers talking in a sequence, and not all of them spoke at the same level or were recorded at the same level, if you drop a Normalize effect on each clip, they will all sound about the same level, and you will not have to listen to them all, watching levels the whole time. Normalize will preserve the dynamics of speech during each clip while still making all of them sound about the same level.

I've also used the Gain effect on tracks that were simply not recorded with enough volume. If you can't get the audio level loud enough by cranking Clip Gain up to +12 in the mixer and cranking up the rubber-banded audio to +12, the answer is AudioSuite Gain, which can raise levels by impressive amounts (see Figure 3.49).

Another good AudioSuite plug-in for everyday use is the BombFactory Peak Limiter (see Figure 3.50). This is very good for talent who tends to

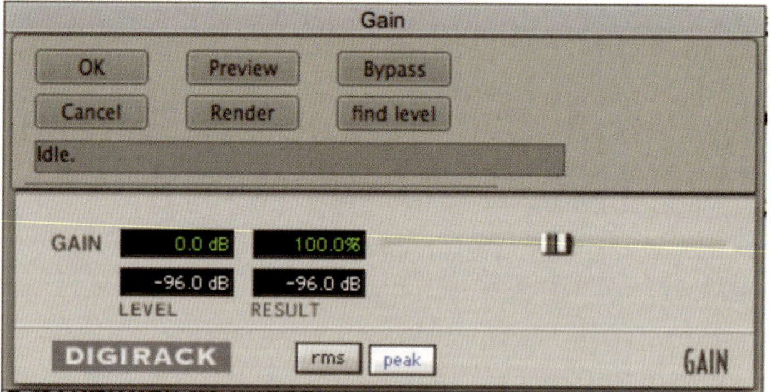

Fig. 3.49 This is the UI for the AudioSuite Gain effect.

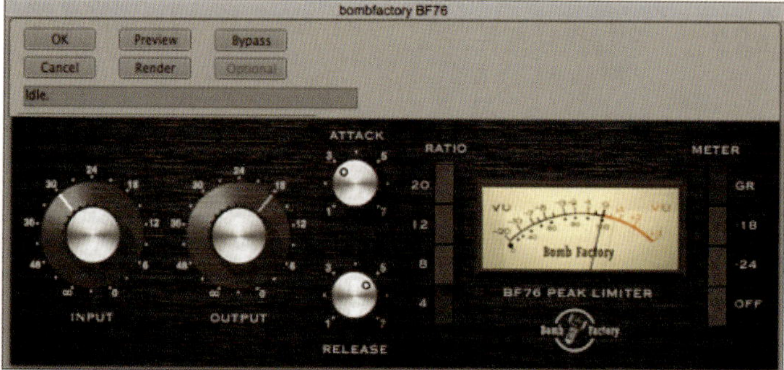

Fig. 3.50 This is the UI for the AudioSuite bombfactory effect.

speak at a given level most of the time, but occasionally hits a word or syllable incredibly hard. The Peak Limiter can allow you to set a nice general audio level for most of the speech using the mixer, but then drop the Peak Limiter effect on top to cap the louder syllables. That allows you a louder overall volume for the rest of the speaking.

TIP

When adjusting AudioSuite plug-ins, it's often highly useful to set in and out points relatively close within the clip. This sets the parameters for the looping of the audio as the preview of the effect happens. Having a short in–out period lets you hear your adjustments more quickly as you tweak parameters. I usually do about three to five seconds.

EQ

There are many ways to do EQ inside the Avid editing interface. The most straightforward way is with the very basic Audio EQ Tool in the Tools menu (see Figure 3.51). There are more advanced EQ tools in AudioSuite and RTAS (see Figure 3.52).

Occasionally, I'll get in to some EQing of tracks, but usually it's limited to simply rolling off some low-end rumble or trying to attenuate some high-end hissiness. How much you play with the EQ options will depend on your skill, your deadlines, and your need to fix something.

One of the cool things with the basic EQ Tool is that there are a number of preset EQ settings that you can play with if you just want to do some drag-and-drop experimentation. This book is supposed to be an advanced book, so dragging and dropping presets may not seem to be

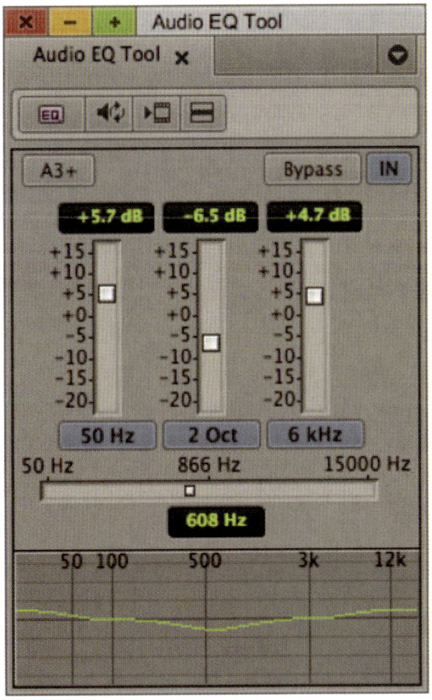

Fig. 3.51 Basic EQ can be done from the Audio EQ Tool.

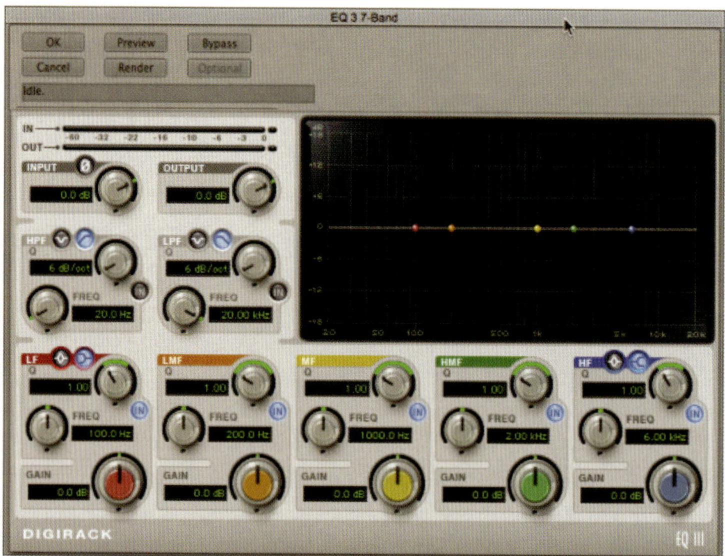

Fig. 3.52 More advanced EQ is possible with any one of a number of AudioSuite plug-ins.

on track with that goal, but in reality, most high-end editors do not have skillsets that extend into every nook and cranny of audio and video engineering, so it's possible that some simple presets can save the bacon of a high-end editor.

To apply these presets, park on a clip in the Timeline to which you would like to add an EQ preset and open the EQ Tool from the Tools menu. In the Audio EQ Tool, there is a fast menu near the top of the window with presets for things like Female Voice with Presence, Male Voice with Presence, Music High Boost, Music Low Boost, NTSC Hum Buster, and even Telephone EQ, to give voices the sonic presence of coming through a telephone (see Figure 3.53).

The features of the basic Audio EQ Tool follow:

EQ Effect: Along the top, there is a purple EQ effect icon. This effect icon can be dragged and dropped either directly to a Timeline clip or to a bin. If the icon is dragged to a bin, it can be renamed to provide an idea of what it was used for, and then it can be used on clips in this project or even on other projects as well as shared with other editors in the same way any bin elements can be shared.

Loop: There is a loop button that can be used to preview the area between in and out points on the Timeline. This is useful when trying to dial in an effect.

Render: Next is the Render button, followed by the Fast Menu that contains the various preset EQ effects that we just discussed.

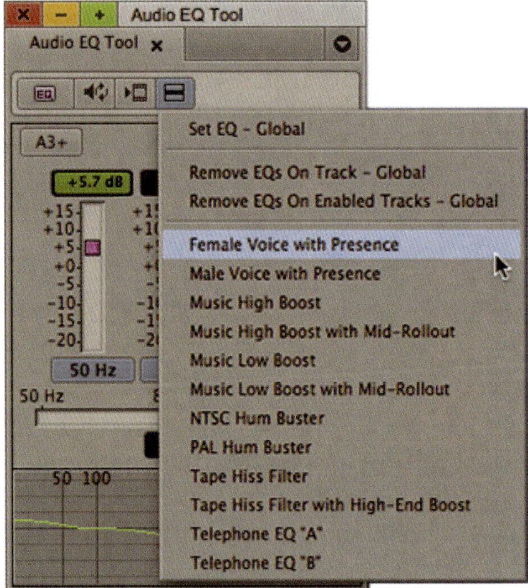

Fig. 3.53 The EQ Tool has a number of preset options that can be accessed from the Fast Menu.

Track Selector, Bypass, Disable: Below those buttons to the left is a Track Selector button that allows you to add the effect to various tracks without having to go to the Timeline to enable them. To the right are the Bypass and Disable buttons. Bypass temporarily bypasses the EQ effect, which is good for A/B listening comparisons, while the Enable/Disable button actually disables the effect. Option-clicking (Mac) or Alt-clicking (PC) on the Enable button actually resets the effect to its factory default state.

There are three faders that can be manipulated with the mouse, the numeric keypad or the arrow keys on the keyboard. The three sliders represent the three bands of EQ: low, middle, and high. Below the faders, are three blue EQ band pulldowns that allow you to adjust the specific Hertz that the band will control. The low band options are 50, 80, 120, and 240. The middle slider allows you to set either a broad band for the midtones or a 1/4 octave "notch." The spectrum of the middle band is controlled by the horizontal slider beneath it. To see and hear what this does to the sound, first raise or lower the center slider by about 10 dB. Then, as you are listening to a loop of sound, slide the horizontal slider slowly to the left and right. The slider to the right is for the high-end EQ and can be set to 6 kHz, 8 kHz, 12 kHz, and 15 kHz. The graph at the bottom of the tool gives you a visual representation of the changes you make with the sliders and pulldown frequency ranges. Remember that when you find an EQ that you like, you can drag the purple EQ icon from the tool to a bin to save it. Make sure to name it something descriptive.

Note

The RTAS effects that I set up in the new Audio Mixer also appear in the RTAS "boxes" in the expanded Track Panel. And, be careful with RTAS effects in the master track. I dropped them on as a test and forgot about them, and as I added a bunch of new cuts to my sequence, I was trying to figure out why everything had an "echoey" delay to it. It took a while to realize that the problem was the RTAS effect I'd added to the master track!

Timeline Audio Changes in MC7

One of the other cool things added to the audio functionality of MC7 is the ability to adjust levels in the Timeline with much more precision, thanks to the addition of a little "minislider" that can be viewed on every clip in the Timeline.

To turn on this capability, first expand the Track Control Panel using the little right-facing arrow next to the timecode number in the Timeline.

Then, in the expanded Track Panel, click on the Clip Volume/Pan control (looks like rubberbanded keyframes). Choose Clip Gain from the pulldown menu. (see Figure 3.54). This adds little minislider icons to all of the clips in that track.

You can now click on the minislider icon on any segment in the Timeline to control the level using the fader that appears next to it, or you can click and control the volume with a numeric entry that can include decimal points for incredibly precise control (see Figure 3.55).

Clearly, this is one of those things that you probably will want to set up a Timeline View for because the visual clutter of all those little sliders is simply not necessary most of the time. If you've never saved a Timeline view, let me walk you through that.

With all of your sliders (Clip Gain) turned on your tracks, go to the little pulldown menu at the bottom of the Timeline, near the left side.

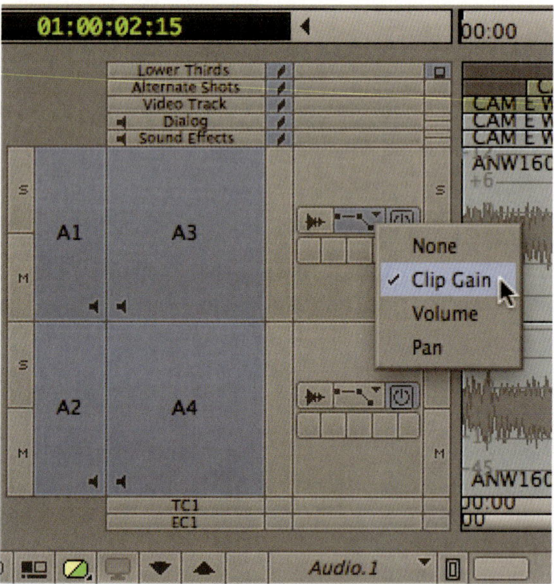

Fig. 3.54 In the Track Control Panel of the Timeline, click on the Clip Volume/Pan button, just to the right of the Waveform button. From the pulldown menu, choose Clip Gain to add the minisliders in the Timeline.

It will either say "Untitled" (if you haven't added any Timeline views yet in the current User Setting) or it will have the name of the current Timeline View.

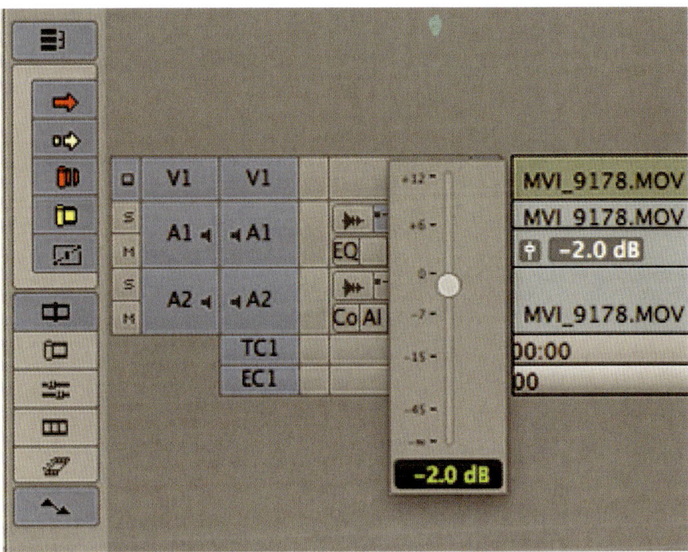

Fig. 3.55 Click on the small white slider icon in each segment in the Timeline to reveal a mixer in the Timeline. Levels can be set by dragging the slider or by typing a number and pressing Enter.

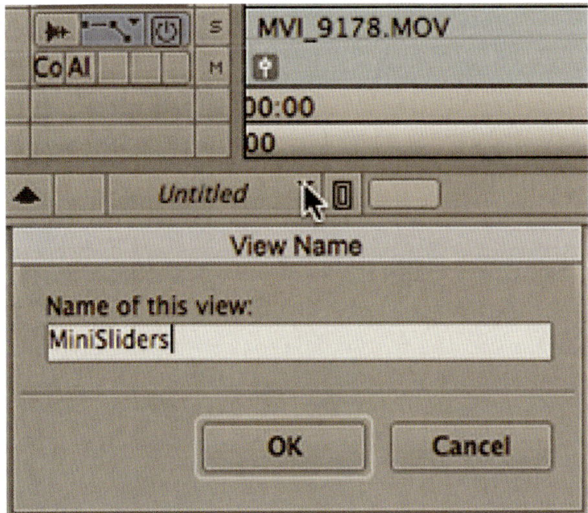

Fig. 3.56 With so many options to add in the Timeline, saving specific views, like MiniSliders, is a good idea.

Name it something you'll remember. You could use Avid's terminology for the minisliders, which is Clip Gain, or you could use my more memorable option, "MiniSliders" (see Figure 3.56). To be able to toggle the minisliders off to reduce the visual clutter, you'll need to go back to the Track Panel and turn all of the Clip Gain off. With the minisliders hidden, go back to the Timeline View pulldown and choose the Save As option. When the View Name dialog comes up, save the view as "No minisliders," or something like that. Now you can switch between functionalities at the click of the Timeline View button!

Avid also made some changes to the Audio Settings. The monitor volume used to be in the Audio Project Settings Output tab, but it's been moved to Audio Settings. It also now retains whatever level you set it at even after shutting down the app. The bad news for almost everyone is that this new functionality does not apply to software mode or if you are using a third-party I/O device, like a KONA card.

3D—Stereoscopic

Video Tutorial

"If you're interested in experimenting with stereoscopic editing as you work through the rest of this chapter, make sure to download a few stereoscopic clips, from the companion website (provide same URL as rest of the book). These clips are courtesy of ArtBeats, who has a great collection of stereoscopic and other stock footage, available at www.artbeats.com.

The first thing to know about editing stereoscopically is that without Avid Nitris hardware it has to be done on an NVIDIA 3D Vision Home bundle, not the Pro bundle, which must include a Windows 7 PC (no Mac support), and it has to be done on an NVIDIA card (HPz800 with Quadro Fermi 4000 PCI-e or better or a Z400 with Quadro Fermi 2000 or better). You'll also need NVIDIA 3D vision infrared emitter and glasses and, of course, a video monitor capable of 3D. The bundle will also come with an emitter hub that plugs into the computer and syncs with the glasses.

As I have pointed out in several places, a full description of stereoscopic editing and preparation in this book is impossible because the basic Avid stereoscopic editing document alone is nearly 120 pages.

With a Nitris, there are several options, one of which requires third-party 3D HD converter hardware. A standard Nitris must be upgraded to a stereoscopic-capable Nitris, which requires an additional DNxHD codec board to access full-frame DNxHD capture and output capabilities for 3D.

You'll hear two terms to describe the stereoscopic image: *frame compatible* and *full frame*. Full frame can be frame compatible and can be transmitted to the monitor via a single SDI, which is only half-resolution per eye. Full frame can be delivered in two ways: separate and sequential. With separate, the left and right images are treated separately and fed to separate outputs or files. With sequential—or packed files—the left and right images are sent to the display device at twice the display rate and the display device switches between them. This requires a 3D monitor capable of 120 Hz. (When viewing mono images, only 60 Hz is required.) Frame compatible combines both "eyes" into one image. There are several methods for displaying frame-compatible stereoscopic clips: Side by side, Over/Under, Checkerboard, Interlaced, and Anaglyph, where the left and right images are fed to separate RGB channels.

To achieve full-frame media output, two SDI ports between Nitris and the monitor are required to achieve full resolution on both eyes. For HDMI, frame-compatible media is delivered at half-resolution for both eyes, but for full-frame media, only the leading eye is sent. Frame-compatible media is often stored "squished" side-by-side in a single frame. Obviously, the best quality is achieved with full frame, but just as with standard 2D off-line, you can use a lower quality (frame compatible) for off-line and then use the full-frame media for on-line finishing.

You'll need to set up stereoscopic options when creating a new project that is set to be stereoscopic (see Figure 3.57). There are a number of options: Off, Leading Eye, Left Eye Only, Right Eye Only, Side by side, Over/Under, and Full. The setting determines how editing functions and renders will be processed, but it is possible to use and monitor other types of stereoscopic footage. You can also switch this stereoscopic project setting throughout the project. One value of this is to display only one of the "eyes" while in the rough cutting stage to speed review and editing of material. But it's important to create the project as stereoscopic from the

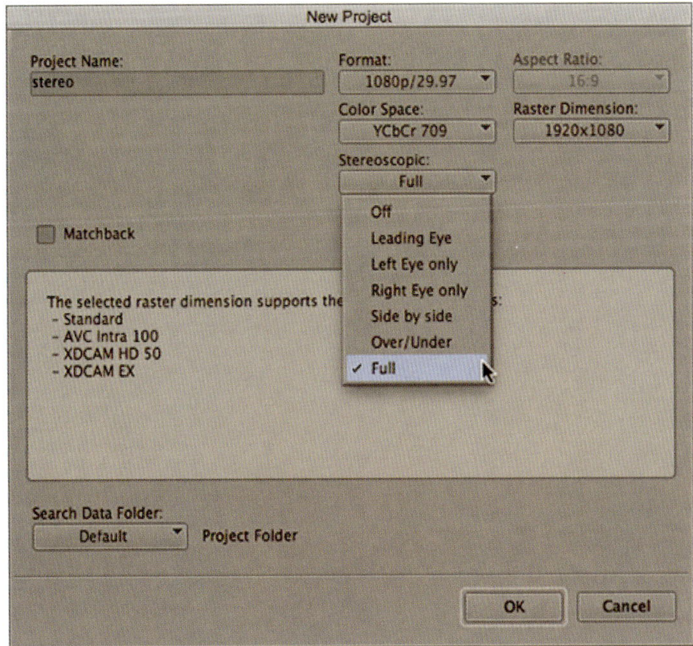

Fig. 3.57 The start of any stereoscopic project begins in the New Project window, setting the preliminary stereoscopic settings.

beginning. Without a stereoscopic process, all of the options, tools and interface discussed for the rest of the chapter will not be available.

The Composer window is obviously where you're viewing a lot of your stereoscopic footage, and it's possible to set the viewing modes on the Composer to be the same as or different from the stereoscopic project settings. These viewing modes can be set for each side of the Composer window (Source/Record) by using the contextual, right-click menu on each of the monitors. The most common method is to screen and rough-cut using primarily the leading eye (see Figure 3.58), switching to full frame to better analyze the stereoscopic imaging.

Considerations for Ingest

The considerations and directions for capturing 3D from tape are lengthy and complex and probably require a $100K deck. With the predominance of file-based capture, I'm simply going to ignore this option for the sake of using the space for other more pressing concerns.

Avid strongly recommends using AMA for ingest of stereoscopic instead of importing. They also recommend not AMA-ing from the cards, but only after transferring the material to a RAID. Like any other AMA

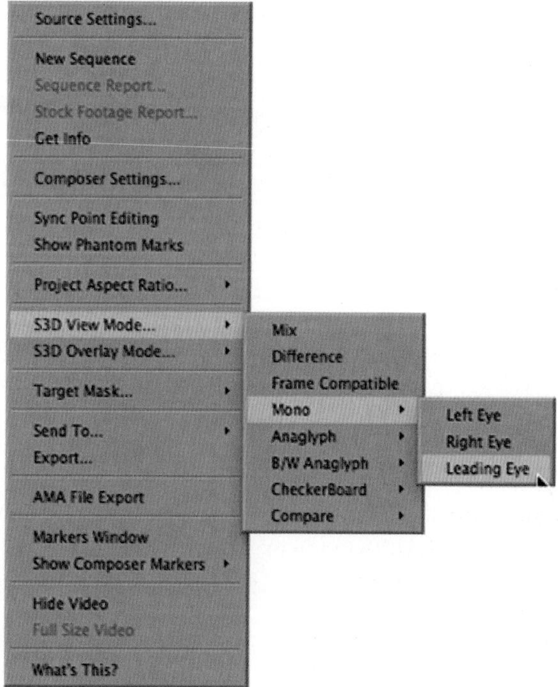

Fig. 3.58 Most of the off-line on stereoscopic projects is edited by setting the S3D view mode to Leading Eye.

footage, from the menu bar, select File > AMA Link. Browse to the top level folder containing any subfolders with the left and right eye files. Avid recommends that you use the Single Bin Based on Selected Folder option so that the master clips all populate a single bin. At this point, it's important to get certain information into the appropriate columns of the clips to create the stereoscopic clips. If these files are coming from a dedicated stereoscopic dailies workflow, there will be an AAF or ALE file that accompanies the media that will populate the columns properly when imported, though with ALE, it may be necessary to relink.

To use an accompanying AAF or ALE, create or open a bin in which you want to create the master clips and right-click the bin to get the contextual menu. Choose Import with your AAF or ALE file selected. Sometimes, it is also necessary to import data from the 3D camera rig as well to create the proper stereoscopic files with the proper settings and parameters.

At this point, your stereoscopic clip isn't actually a stereoscopic clip, at least as far as Media Composer is concerned. If you note the look of the clip icon, it looks just like all of the other clips in your other bins. To work in stereoscopic mode, you need to generate a stereoscopic clip from the master clip or clips that you've ingested. But before you can do that you have to make sure that you have set the S3D channel information and the S3D Group Name in the correct bin column for the master clips.

To enter this information, you first have to have the correct S3D bins columns in your bins. To add the S3D columns, with your bin selected, from the menu bar select Bin > Columns. Add at least the S3D Group and S3D Channel columns, but you might as well add all of the S3D columns while you're at it. Just click on them to select them (selected columns are overlaid with blue). Changing the settings in these is possible by right-clicking in the S3D columns of the clips. In order to use the stereoscopic functions in your Avid editing application, these master clips must be referenced from a stereoscopic clip.

To create a stereoscopic clip, certain S3D columns for the contributing master clips must be defined: the S3D Channel of each master clip must be set based on its type of stereoscopic media (for example, left, right, over/under, side/side, or monoscopic).

You've also got to assign an S3D Group Name for each master clip that will be part of a stereoscopic clip, with all master clips forming a stereoscopic clip having the same group name. And you've got to use a unique group name for each stereoscopic clip. It's possible to assign these group names automatically. With clips selected in the bin, right-click to call up the contextual menu and choose Populate S3D Group Names Automatically.

Only after all of the master clips have been given the correct stereoscopic attributes can you create the actual stereoscopic clips. Again, select the clips that you want to transform in the bin and right-click on one of them. This calls up the contextual menu. Selecting Create Stereoscopic

TIP

If you sorted the clips so that there is a separate bin for each eye, it's easy to grab all of the clips and set the S3D channel to that eye. Then sort by timecode to keep the clips in the same order. It's possible to shift-select or lasso or select all of the clips and change one column in all of the clips to the same information.

clips calls up a menu. In it you are provided with several options. Avid suggests creating a new bin for the stereoscopic clips (see Figure 3.59). Most of the default choices can be left the way they are unless you understand what you're doing. The sole exception might be the Destination bin for created stereoscopic clips option. It is suggested that you save the true stereoscopic clips separately from the source clips. When the proper stereoscopic files have been created, you'll see that the clip icon for those clips looks like a double or "layered" clip icon (see Figure 3.61).

The Command Palette has numerous specialized buttons that can be mapped to your keyboard or UI for stereoscopic (see Figure 3.60). When you add the Stereo Menu button to the viewer, it provides a menu of stereoscopic view commands. The menu is accessed from the contextual menu by right-clicking or Option-clicking (Mac).

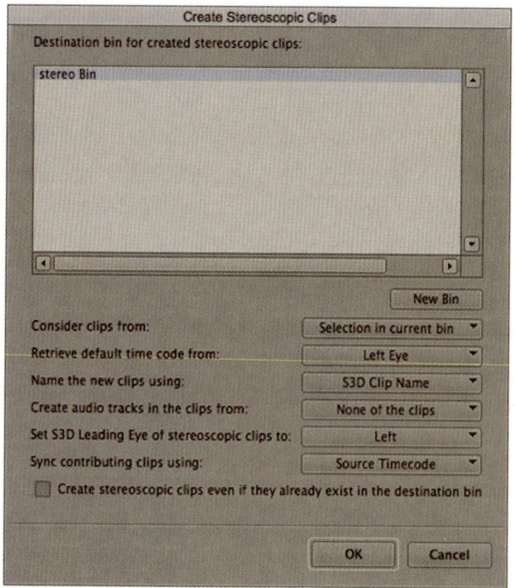

Fig. 3.59 When creating stereoscopic clips, it is suggested to place these clips in their own bin, separate from the original source clips.

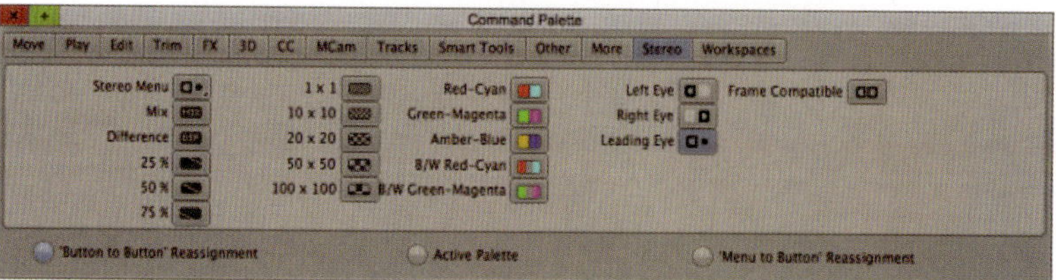

Fig. 3.60 The Command Palette has a full tab of buttons that you can use to customize your keyboard and UI when you are editing stereoscopically.

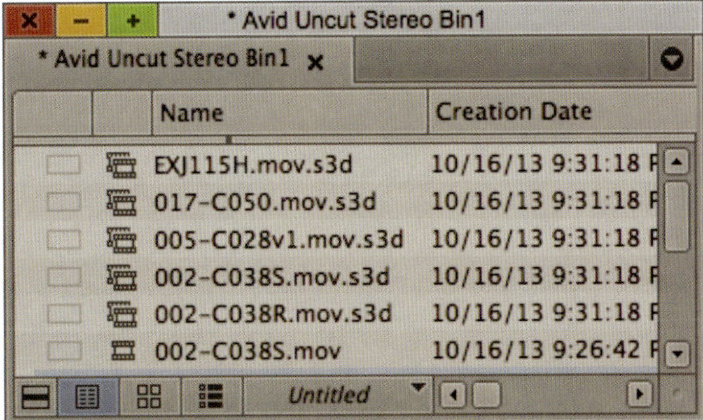

Fig. 3.61 This bin shows five stereoscopic clips at the top and a single regular clip that was one of the source clips for one of the stereoscopic clips above it.

Fixing Stereoscopic Errors

For most footage prepped in production or dailies for stereoscopic use, there's no reason to make any changes, but if corrections are needed, there are a number of options for five specific errors: (1) the alignment or disparity between eyes needs correction, (2) the color between the two eye views needs to be matched or adjusted, (3) one of the eyes is not in temporal sync with the other eye, (4) the eyes are swapped (left for right or right for left), and (5) if the image from the camera is inverted (horizontally, vertically, or both). These corrections are applied on the source side of the stereoscopic clip, not in the Timeline.

Load a stereoscopic clip into a stereoscopic project and toggle on Source/Record in the Timeline so that the Timeline displays the source instead of the sequence. You'll know you are monitoring the source in the Timeline because the Timeline indicator is green instead of blue. Click on the Stereo Correction mode button at the bottom of the mode buttons, below the Smart Tool buttons on the left side of the Timeline (see Figure 3.62). The button is visible and active only in stereoscopic projects. Similar to the standard color correction mode, where there are three viewers in the Composer instead of the standard two in editing and trim modes, the Stereoscopic Correction mode has Left, Output, and Right viewers. The Output viewer in the middle shows both eyes close together for easy comparison (Figure 3.63).

This gives you access to both the Color Correction effect and the Spatial Alignment effect buttons under the Output monitor. Color Correction is the default because you should apply color corrections before you apply spatial corrections.

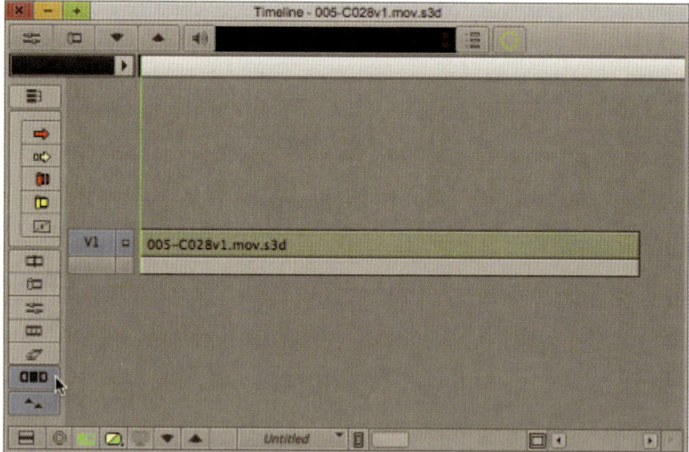

Fig. 3.62 A stereoscopic clip is loaded in the Source monitor and the Timeline has been switched to show the Source monitor. The Stereo Correction button is at the bottom left of the Timeline. The green vertical timeline indicator shows that you are viewing the Source side timeline, otherwise, it would be blue.

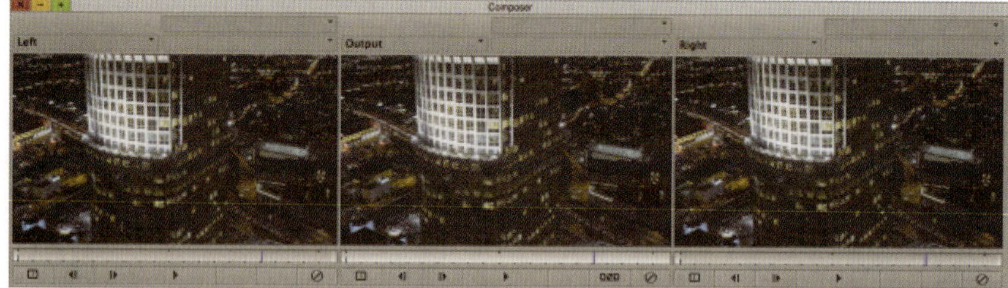

Fig. 3.63 The Composer viewing a stereoscopic clip in Stereo correction mode.

Color Correcting Stereoscopic Clips

Stereoscopic images are usually recorded by two distinct cameras. Sometimes those cameras are set up so that the recorded images are color corrected slightly differently, just as some multicamera shoots sometimes have slightly different looks from camera to camera if not properly "shaded" during production. Also, stereoscopic images are sometimes shot with a "beam-splitter," which allows one of the cameras to shoot *reflected* from a semi-reflective glass while the other camera shoots *through* the semi-reflective glass. The difference in the reflected or transparent glass can also cause color shifts. These fixes are done on the Source side from the Stereo Correction mode. Pick an eye and use the color wheels to make your adjustments. When done, exit Stereo Correction mode by toggling the Source/Record in the Timeline, switching the Timeline back to the sequence side.

Fixing Spatial Alignment

Fixing spatial alignment can be done in the source side using the same Stereo Correction mode used to get to the stereo color correction tools, or in the sequence by applying it to clips already edited into the Timeline. To view the spatial alignment in the Source monitor in a system without Avid hardware, from the Settings tab of the Project window, double-click Full Screen Playback. That displays a list with options for S3D Viewing (see Figure 3.64). One of the items is the S3D Overlay list. The S3D Overlay list includes options to display the minimum and maximum limits for where objects are represented, either in front of, or behind, the screen plane. Those limits are called Parallax Near (showing green guides where images appear close to the viewer) and Parallax Far (showing blue guides for images further away from the viewer). Also, there are Parallax Near/Far (showing both green and blue guides) and Follow S/R, which displays stereoscopic footage based on the overlay chosen for the Source/Record viewers.

In the sequence, to apply an effect, open the Effect Palette: Command-8 (Mac) or Control-8 (PC). Drag effects from the effect palette to the clip in the Timeline. Open the Effect Editor using the Effect Mode button.

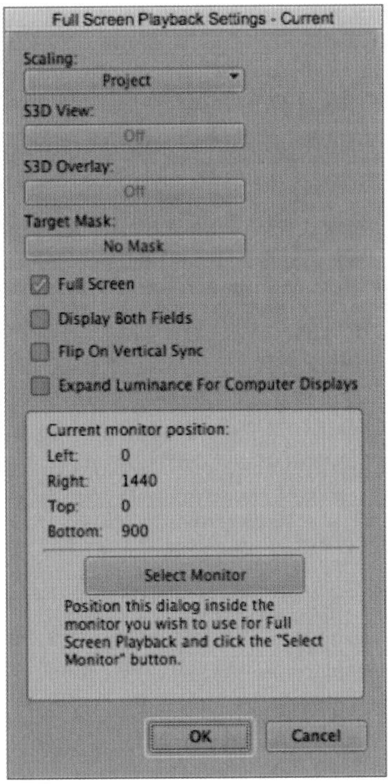

Fig. 3.64 Full screen playback settings.

(See chapter 4, starting on page 256, for more information on this and on the multiple ways that effects can be applied to individual clips, transitions or groups of clips and transitions.) In stereoscopic projects, a special group of stereoscopic parameters will be available. For full-frame stereoscopic clips, two render files per effect are actually created, one for each eye. In the Effect Editor, expand the S3D Spatial Alignment group and expand the Vergence group if you want to converge or diverge the two eyes (see Figure 3.65). If this results in black around the edges or the frame, use the Zoom group or Auto-Zoom to zoom in so that the black edges are no longer visible.

Considerations When Editing Stereoscopic

While doing a rough cut, you can choose to view only a single eye in the Composer window. This method will feel very similar to standard nonstereoscopic editing. To choose which camera or "eye" to use, select the Format tab of the Project Window. There is a pulldown under Stereoscopic that allows you to choose the leading eye (see Figure 3.66). Also, it's possible to edit nonstereoscopic footage into a stereoscopic project, as well as to mix frame rates.

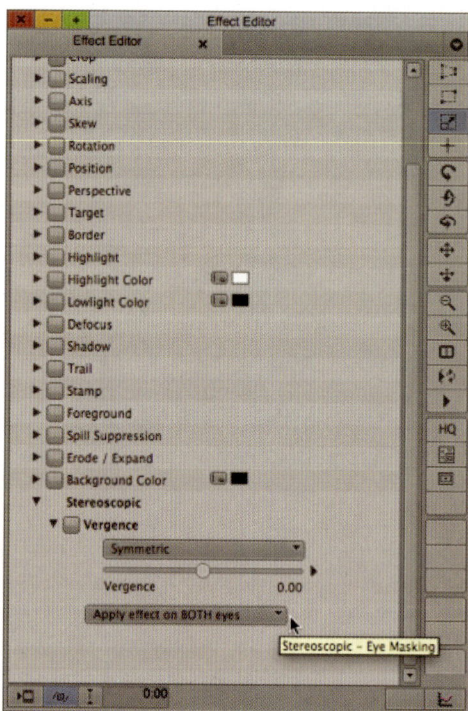

Fig. 3.65 Effects mode with a stereoscopic clip.

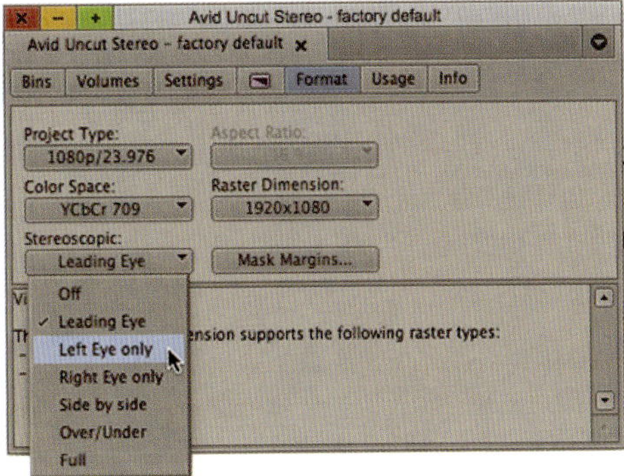

Fig. 3.66 The Format tab of the Project window.

With audio, if you are editing full frame, each camera or eye will have its own audio. The left eye will be displayed first, and the right eye second. When you consolidate or decompose stereoscopic sequences, Media Composer references all of the master clips that make up the stereoscopic clip and copies (consolidates) those master clips as well. When you delete a stereoscopic clip, you'll be presented with the option to also delete the media from the contributing master clips. The master clips themselves aren't deleted so that they can be referenced again if needed. As with regular nonstereo clips, you can choose Reveal File, which shows you the associated media files for the contributing master clips.

Editing Effects

Even tools like the Title Tool have specific stereoscopic settings that allow you to adjust the parallax between two eyes. In stereoscopic projects, the Effect Palette will have an S3D group which allows you to set the Stereo Floating window, which allows you to crop out edge violations. You also have access to Stereo Vergence, which allows you to set the disparity between eyes. Also, Stereo Depth Transition, which adjust the depth on the outgoing and incoming clips for a seamless transition from one shot to another. And finally, Stereo Spatial Alignment, which allows you to align the 3D spatial manipulation of the two "eyes" so that they're aligned (see Figure 3.67).

When you adjust this alignment, the S3D Overlays (green and blue vertical lines) allow you to judge how far apart the left and right eyes can be separated so as not to make the viewer uncomfortable. To see these Overlays, right-click on the Record or FSP viewer and set the S3D mode to Mix, Difference, or Anaglyph (Figure 3.68). Then set the S3D Overlay mode to Parallax Near, Parallax Far, or both. The parallax (amount of separation) between the left and right eyes when they are superimposed over each other needs to not exceed the distance between

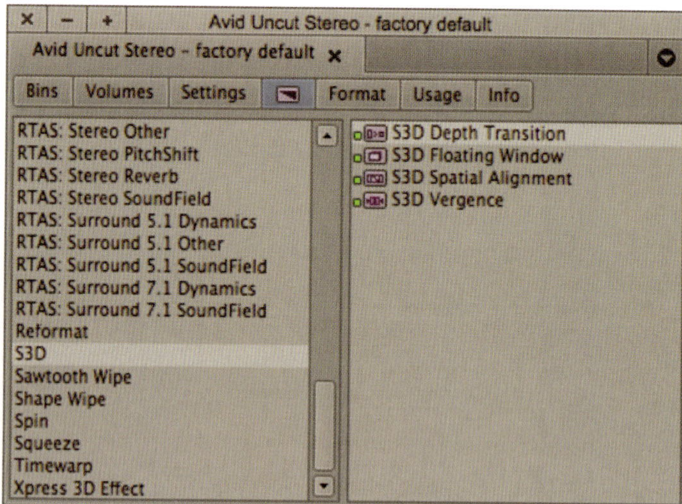

Fig. 3.67 The Effect Palette has an S3D group of effects with stereoscopic specific capabilities.

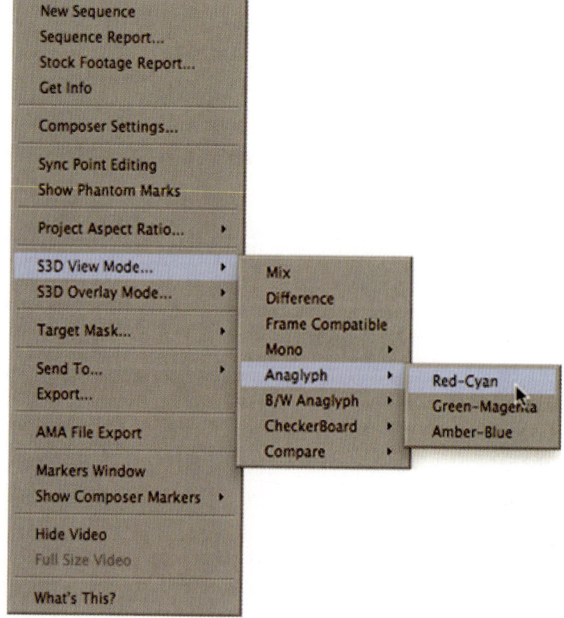

Fig. 3.68 Stereoscopic contextual menu in the Record monitor.

the Overlay lines (see Figure 3.69). The distance between these Overlay lines is set in what is called the depth budget for your project. The depth budget calculates how far in front or behind the screen your 3D objects can appear based on elements such as the size of the screen and the average distance the viewer is from the screen. The depth budget is set in the Settings tab of the Project window. Double-click the S3D setting. In that setting you can set the Unit System (feet or metric), Viewing Distance, Screen Width, Raster Resolution, Eye Separation (stick with the default), as well as other measurements (see Figure 3.70).

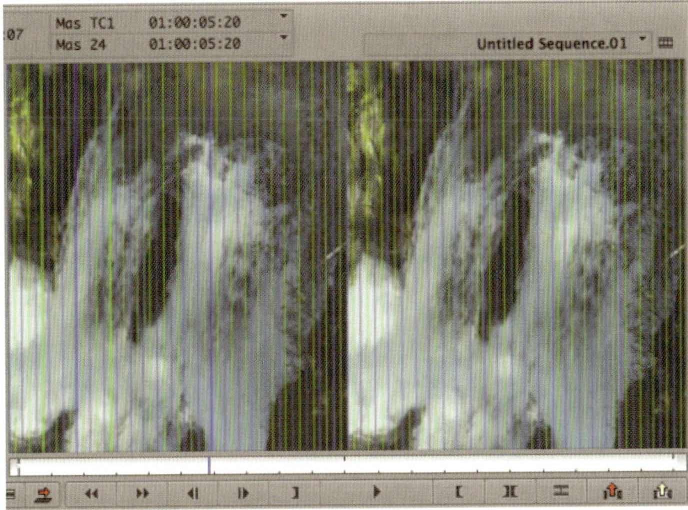

Fig. 3.69 Parallax Overlay shows your "budget depth" using blue and green vertical lines.

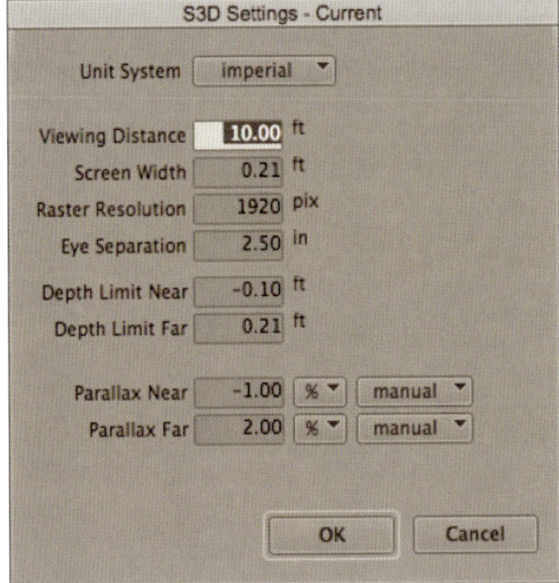

Fig. 3.70 The S3D Settings window.

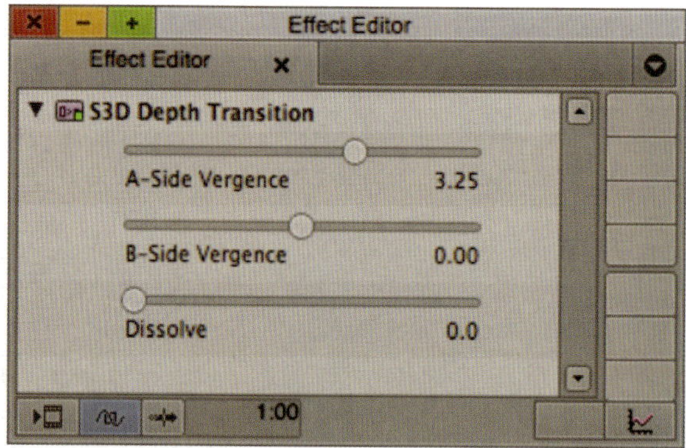

Fig. 3.71 The Depth Transition Effect smoothes depth differences between shots.

One of the critical elements on a successful stereoscopic edit is keeping a consistent depth from one shot to the next. Objects in the shot need to be kept at the same relative distance so they don't jump forward or backward from the viewer between shots. To assist with making a smooth transition between shots, select the transition point between two clips in the Timeline, open the Effects Palette—Command-8 (Mac) or Control-8 (PC)—from the S3D group of effects, and double-click the Depth Transition effect or drag it to a transition. Set the S3D View Mode in the Record monitor to Difference. This allows you to see a blend of left and right eyes, highlighting the difference in depth, represented by dark gray pixels. Adjust the vergence of the outgoing shot with the A-Side slider, setting it to a neutral parallax (see Figure 3.71).

Avid effects in a stereoscopic project can either be stereo-aware, stereo-agnostic, or stereo-specific. Not all third-party effects (AVX) may have stereoscopic functionality.

Adding Titles

Titles are a great place to be able to add a nice sense of stereoscopic depth. To manipulate a stereoscopic title, add the title as you normally would. (See the Title Tool section of chapter 4, starting at page 283.) Open the Effect Editor and move the Vergence slider to adjust the distance between the two eyes, creating a title that either pops out from the screen or recedes back behind the screen. Titles can also be manipulated in 3D space by promoting the title to 3D using the small 3D Promote button that looks like a small 3D cube in the bottom right corner of the Effect Editor.

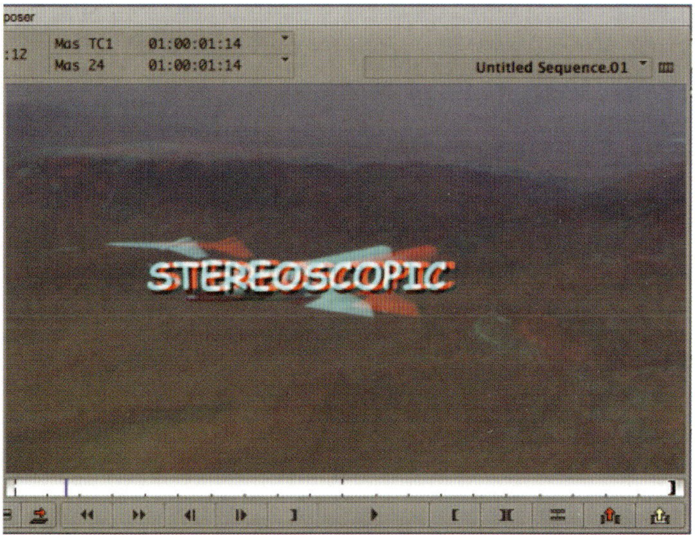

Fig. 3.72 A title from Title Tool viewed in red/cyan anaglyph with the Vergence changed in Effects mode to make the title appear "3D".

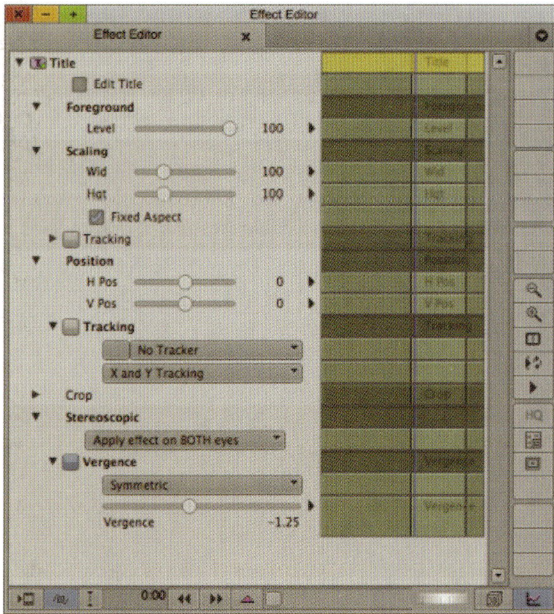

Fig. 3.73 In Effect mode, the Vergence slider affects the 3D positioning of the title.

Outputting Stereoscopic Sequences

There are three ways to output your sequence. Few people will have the resources to lay off a stereoscopic project to tape, but it can be done. Because of the few people to whom this applies, I'll just advise you to refer to the Stereo3D workflow guide put out by Avid. The other method is to export the sequence to some file-based media.

You can export stereoscopic projects to file-based media in four different modes:

1. Full, which uses the full frame for both left and right eyes, creating two separate files
2. Side by side, which creates a single file, using side-by-side half-resolution images
3. Over/under, which places the left and right eyes over and under each other instead of side by side in a single file
4. Leading eye/left eye only/right eye only, which creates a single file that is not stereoscopic at all, but a monoscopic display of the stereoscopic project

These options are chosen when you choose Export from the File menu and decide to export a QuickTime file, AVI, MPEG, HDV, or DV Stream.

The third method is to prepare your sequence for some outside stereoscopic on-line company to complete by exporting an AAF or an EDL. These are chosen from the Export menu as well. From the options, select AAF or EDL and work with your on-line company to choose the proper options that allow you to deliver a working file to them.

CHAPTER 4

Effects

Adding effects in Avid is a Timeline task. It happens only on the Sequence side in Avid (see Figure 4.1). You can't place effects on a source or an original clip. Effects work begins with the Effect Palette—the list of all your effects, similar to FCP's Effect menu—which can be called up from the keyboard by pressing Command-8 (Mac) or Control-8 (PC); from the menu: Tools > Effect Palette; or from the Effect Palette tab in the Project window.

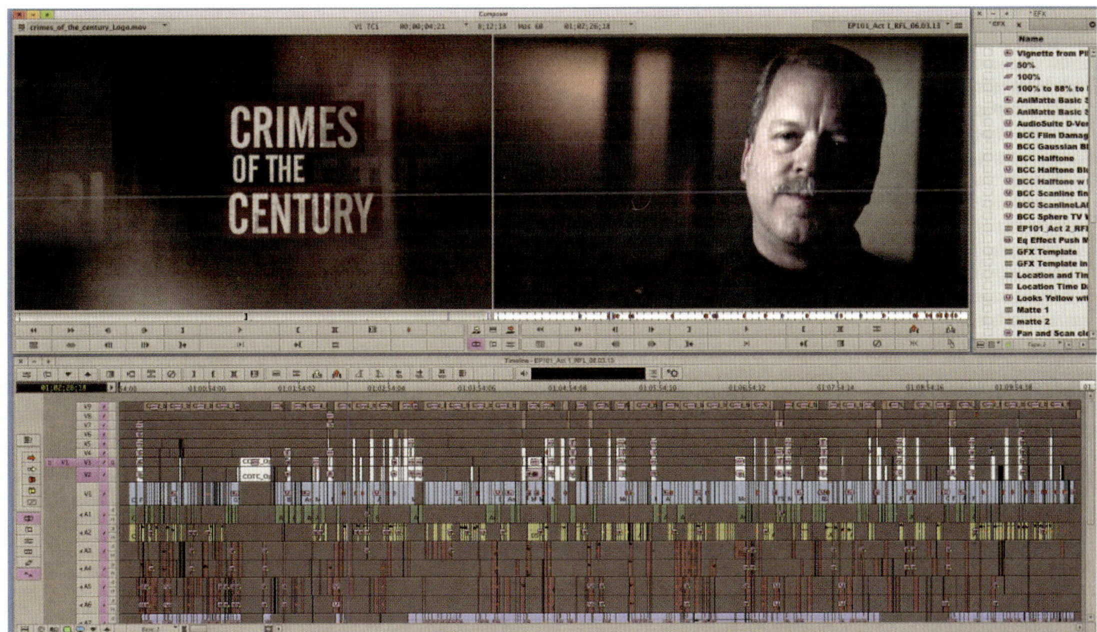

Fig. 4.1 This section of Robert Landau's Timeline for *Crimes of the Century* on CNN is packed with more effects per square inch than you're ever likely to see anywhere else. Image courtesy of Robert Landau, Scott Free Productions and CNN.

The Effect Palette is identical to the Effect tab of the Project window. Here, (Figure 4.2), the effects in the Image category, which is another common category of effects, are shown.

To add an effect, drag and drop an effect from the Effect Palette onto the segment or transition in the timeline, or select the segment or transition and double-click an effect.

The Effect Palette has two columns. The column on the left is a list of categories, or groups of effects. The column on the right shows the individual effects in the selected category or group (see Figures 4.2 and 4.3).

From years of cutting on Avid, the two big categories that I almost always use are Blend and Image plus a good deal of Reformat and Key (see Figure 4.4). Beyond that, I mostly use third-party plug-ins, usually GenArts Sapphire and Boris Continuum (BCC). These third-party effects work in Avid through a portal called AVX (Avid Visual fX). AVX allows third-party developers to interact with the Media Composer software. AVX effects are indicated by a small, purple, three-pronged "plug" icon (see Figure 4.5). Sometimes AVX effects launch their own independent

TIP

At the very bottom of the left side you can directly access the effects that are inside any open bin. The names of the bins will appear to the left, and the effects inside any selected bin will appear to the right. So, if you have a bin of favorite effects, you can open it and hide it and still have access to those effects from the Effect Palette.

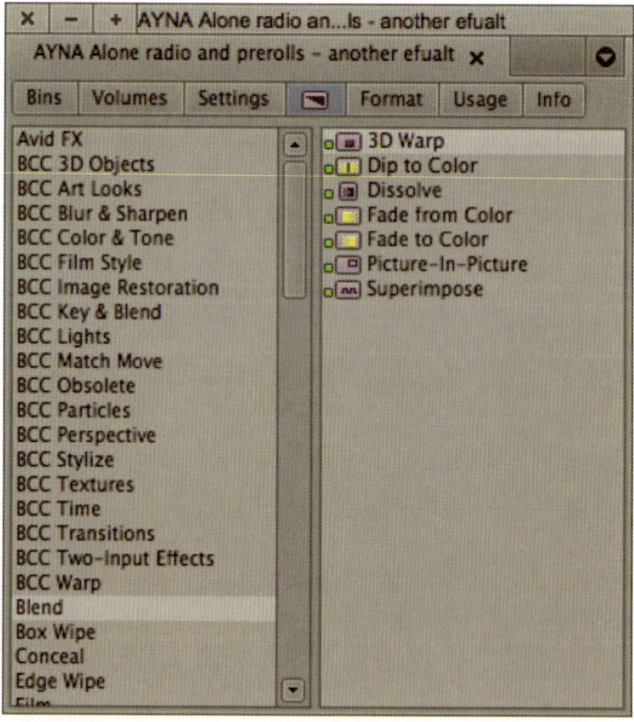

Fig. 4.2 Effect Palette in the Effect tab of the Project window. This shows the effects from the Blend category, which are commonly used.

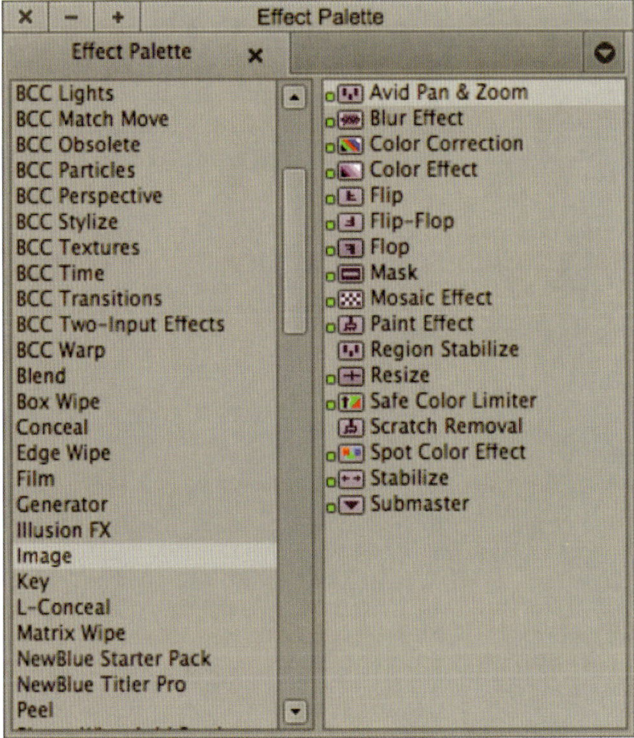

Fig. 4.3 Effect Palette in the Effect Tab of the Project Window. This shows the effects from the Blend category, which are commonly used.

user interfaces and sometimes they operate using the standard Avid effects user interface.

General Effects Concepts

Instead of dwelling on specific effects, the most valuable use of my time and yours is in understanding effects concepts that apply to all or most effects and that will save you time and effort.

When considering the layering or stacking of effects in the Timeline there are a few important concepts to understand. Oftentimes, the ways effects work in Avid are fundamentally different from the way they work in FCP. In Avid, the layers are viewed from the top (highest number track) down (through to the lowest numbered track). If you have a key layer, it goes on the track above the fill layer, for example, because Avid considers that it is seeing down through the key to the filler below. Similar to FCP and Premiere, if you want a title keyed on top of some video, the title needs to be added to the track above the video (see Figure 4.6).

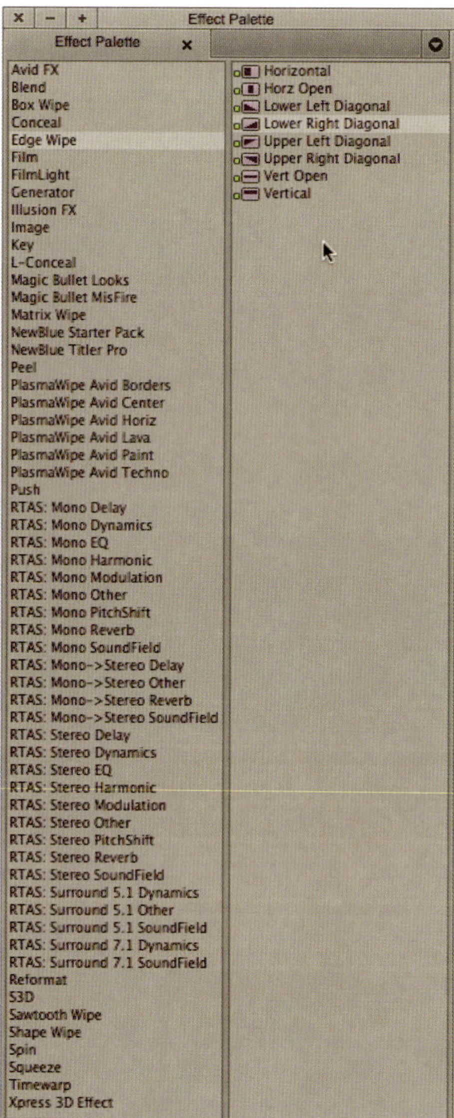

Fig. 4.4 This is the Effect Palette. Yours may look different because of third-party effects that you've added or because Avid may have added some new effects categories in a newer version of Media Composer.

Application

Applying effects, as discussed a few paragraphs earlier, is pretty straightforward and relatively intuitive. But there are methods and situations in which some specialized knowledge can come in handy.

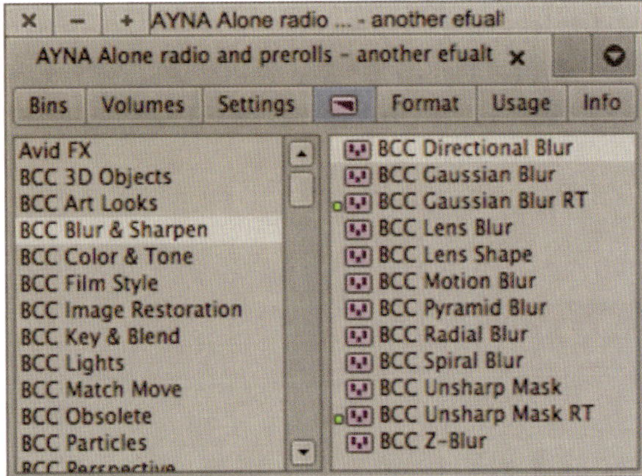

Fig. 4.5 AVX plug-ins (third-party effects, generally) are indicated by a purple icon that looks like a three-pronged electrical plug. BCC plug-ins are common third-party AVX effects on Avid systems.

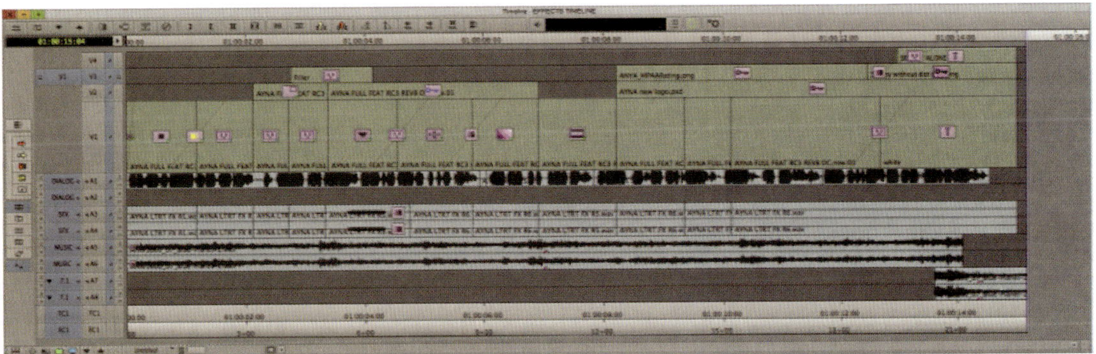

Fig. 4.6 This is a Media Composer Timeline with effects of numerous types on virtually every single clip and transition. Effect icons are purple and have a variety of icons, depending on the type of effect.

For example, you may want to apply the same effect to multiple transitions. To do this, you have to have both the Effect Palette open and be in Effect mode. Go into Effect mode with either of the Effect mode buttons on the left side of the Timeline, or by using the menu Tools > Effect Editor. Or, if you have mapped the Effect mode button to your keyboard, as described in chapter 2, then just give that a quick tap.

With the Effect mode (see Figure 4.7) active, select either the first or last transition in the group; then lasso the entire series of transitions, dragging from above the top track downward and either left or right until you've selected the group of transitions to which you want to add the effects. This will not work from the beginning of the first cut in a sequence or the end of the last cut in a sequence. You can Shift-select transitions to add them or subtract them from the group, including the beginning of the first clip in the sequence or the end of the last clip in the sequence. The selected transitions will be indicated in blue (see Figure 4.8).

With the transitions selected, you can double-click on any transition effect in the Effect Palette, or even double-click transition effects that you have customized and saved in bins. This will apply that effect to all of the transitions.

Applying effects to multiple segments is similar and somewhat easier. For small numbers of segments, the easiest way is to use one of the SmartTool Segment mode buttons (red or yellow arrows to the left of the Timeline) and simply click and then shift-click on subsequent segments.

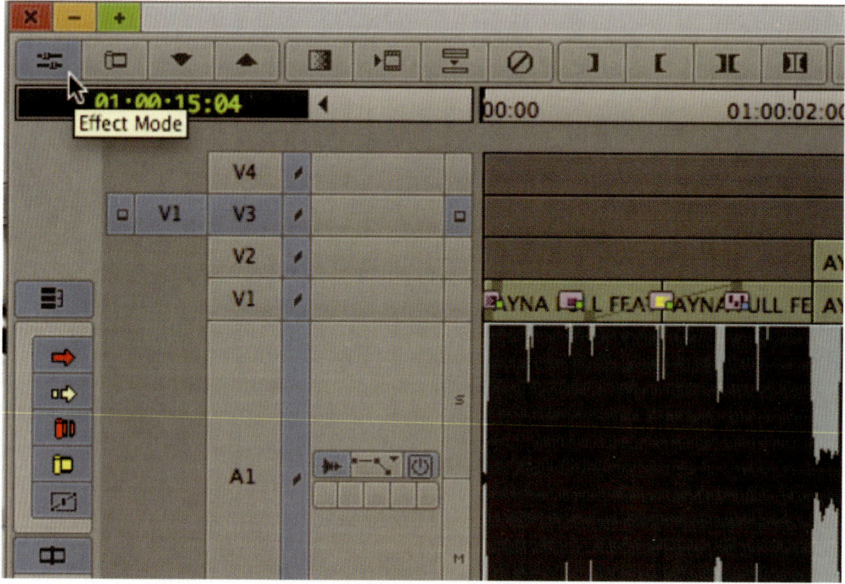

Fig. 4.7 The Effect mode icon is in numerous locations throughout the user interface, including in the top left corner of the Timeline (as shown here), as well as in the Timeline under the SmartTool, and in the Composer window below and between the Source and Record monitors.

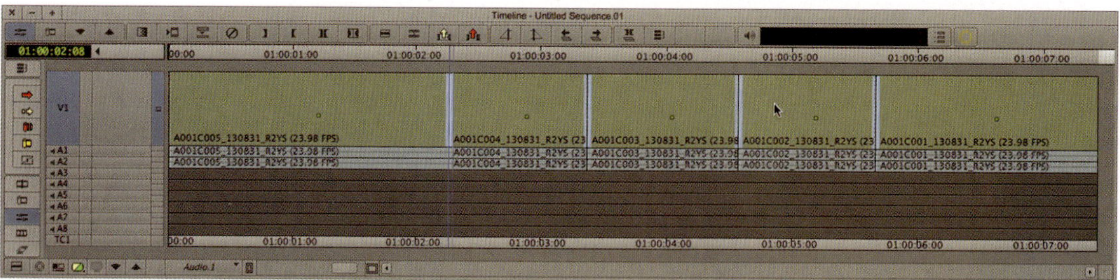

Fig. 4.8 Avid Timeline with four transitions selected. The selected transitions are indicated in light blue.

With your segments selected, double-clicking on an effect in the Effect Palette, or on an effect in a bin, will apply it to all of the segments. Unlike with transitions, you do not have to be in Effect mode to have this work.

However, if you want to use the lasso method of selecting multiple segments, first go into Effect mode and then lasso the multiple clips that you want the effect applied to. Shift-selecting segments will add or remove them from the selected segments, which are indicated in blue.

Effects in Filler

Another way to add an effect across multiple clips or to a region of a sequence is to add the effect to the Filler in a track above the region that you want to affect.

Using this technique is especially useful for things like adding an Anamorphic Mask or an overall color correction effect, such as making an entire sequence or a section of a sequence black and white. This is also the best way to add the Timecode Overlay effect if you want to lay off a sequence with "timecode burn-in" (see Figure 4.9).

To do this, simply create a video track above the track or tracks that you want to apply the effect to and drag and drop the effect onto that track.

This method of adding effects to filler can also be used to create transitions. The easiest way to do a nice glow dissolve is to use something like GenArts Sapphire Glow Dissolve and drop it onto a transition, or there are also some options for doing this with Avid FX transitions. But the free way to do basically the same effect is with a series of small, stacked filler segments on either side of a transition, each with effects on them.

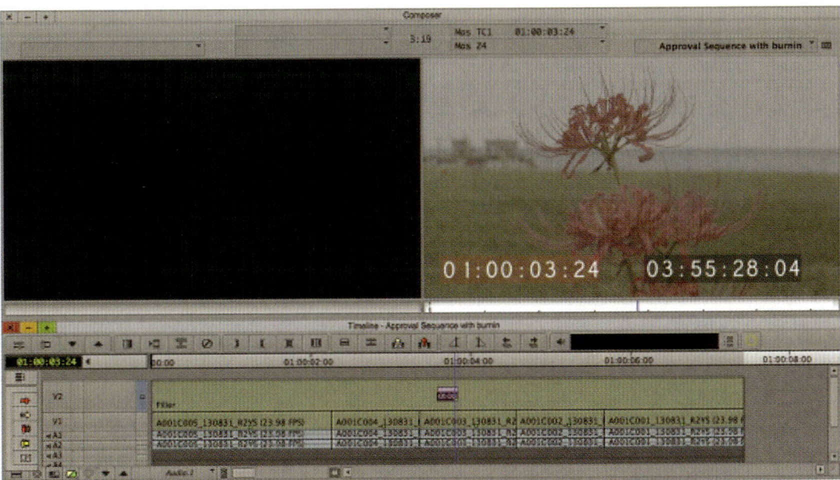

Fig. 4.9 From the Generator Effects category, the Timecode Burn-in Effect has been dropped on to the Filler on video track 2 above the video clips on video track 1.

Video Tutorial

There are files to download on the companion website so that you can follow along with this tutorial. (www.focalpress.com/cw/hullfish) Look for a bin called "Glow Transition Example."

To try this yourself, look in the Project files for a bin called GlowTransition Example. Relink the media (to the 4K Artbeats footage) or replace the two clips in the Timeline on V1 with any other footage. Because the effects are on the tracks above the video, you can change the video below and the effect remains the same but with new media. If you like this effect, you can use the subsequence called GlowTransition.sub01 and edit it onto any two tracks above any transition in any other sequence, eight frames before the transition. (You may have issues using this with frame rates other than 29.97, but you should be able to convert it.)

Though you can use the GlowTransition subsequence without knowing how it works, understanding the "recipe" for the subsequence provides an understanding for how you can build and save subsequences like this that can be applied over and over again.

Place two shots sequentially in V1 and add two video tracks above them using the keyboard command: Cmd-Y (Mac) or Cntrl-Y (PC). Then park the timeline locator on the transition by Command-dragging (Mac) or Control-dragging (PC) near the transition so that the timeline locator is "magnetically" locked on the transition as you drag near it. At the transition between the two shots, add an edit to the filler above the edit by holding down the Option key (Mac) or Alt key (PC) while pressing the Add Edit button. That modified adds edits to *filler only* at the point of the timeline locator regardless of which tracks are enabled. (This is a useful capability for trimming and maintaining sync.)

Next, place an Add Edit in just the filler, eight frames earlier. There are a few ways to accomplish this. One is to tap the left arrow key eight times. Another is to use the numeric keypad to enter −8 (minus 8). Then use the Option key (Mac) or Alt Key (PC) while pressing the Add Edit button to add edits to the filler. Use the same method to place Add Edits in the filler 16 tracks after your V1 transition (24 frames from your current position). When you are done, your sequence should look like Figure 4.10.

A Glow Dissolve is basically two different effects stacked on top of one other. The main idea of a Glow Dissolve is that the picture gets brighter at the end of the first segment and the brightness fades back to normal after the transition to the second shot. But just applying a ramp in brightness and back looks very harsh, so the other thing that's needed is to ramp a blur from normal (0 blur) to blurry at the end of the first clip and then ramp it back to normal from the beginning of the second clip.

In Avid, effects above affect tracks below, so the order of the stacking matters. I wasn't sure which would look better, so I tried it both ways, and I discovered that having the Color Effect above the Blur worked better than the other way around.

Drag the Blur Effect from the Image category of the Effect Palette to the two segments of filler on either side of the transition on V2. Then drag the Color Effect effect onto the two segments of filler on either side of the

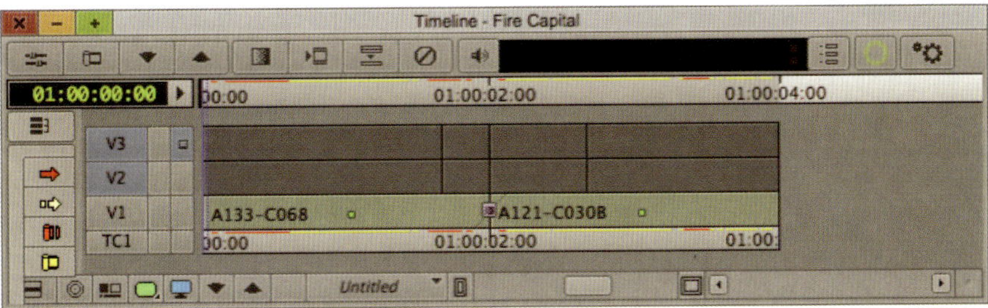

Fig. 4.10 Add Edits in filler on V2 and V3 create segments where the effects will be added.

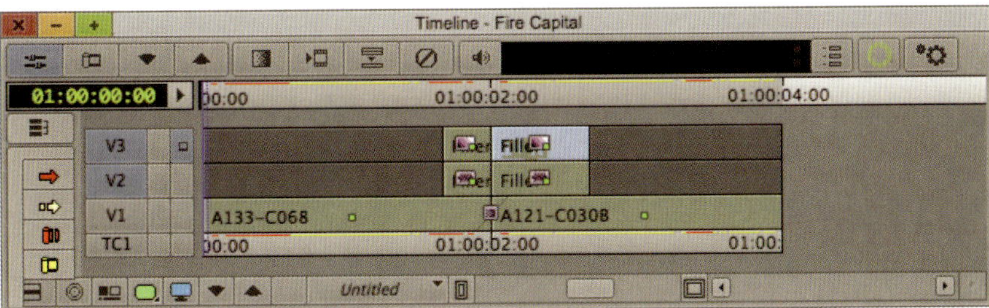

Fig. 4.11 This is the final effect in the Timeline. It creates a Glow Transition, as if a flashbulb went off, but fairly organic and better than a boring dip to white.

transition on V3. To help soften the transition a bit, place an eight-frame dissolve on the transition on V1 by clicking the default keyboard command for Dissolve, which is the forward slash key above the Return or Enter key, and typing 8 into the Duration Text box. Your sequence should look like Figure 4.11.

To edit the Blur Effect, I went to Effect mode and clicked on the first Blur Effect (see Figure 4.12). The Blur Effect is designed to blur portions of the image (for example, hiding the identity of a person or obscuring a license plate). I wanted to blur the entire image, so I used the Rectangle tool to select the entire image. To make sure that you are able to select the entire image, first click on the sequence monitor and hit Command-K (Mac) or Control-K (PC) to shrink the image in the window, which allows you to draw a rectangle around the entire image without missing any pixels on the edge. Then click on the small rectangle-shaped tool in the toolbar to the right of the Effect Editor and lasso the entire frame in the Sequence monitor. With the entire image selected, the Blur HOR and VERT sliders become active. Under the Sequence monitor, go to the end of the Timeline and add a keyframe, using the purple triangle under the sequence monitor Timeline. Change the HOR and VERT sliders in Blur to about 20. Move to the beginning of the sequence Timeline and add another keyframe. Move the HOR and VERT sliders to 0. This ramps the blur up from nothing to blurry over the course of eight frames.

TIP

There is also a blur effect built in to the Paint Effect tool, so if you want to blur a specific custom shape or to blur based on brightness or a color (using a matte), then use the blur functionality in the Effect Palette under Image > Paint Effect.

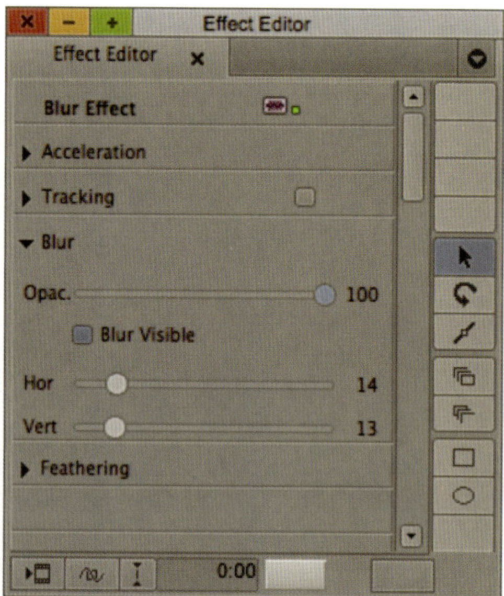

Fig. 4.12 This is the Effect Editor on the Blur Effect. Yours may look slightly different because I resized the Effect Editor window so that it would have less "white space" for the book.

Click on the second Blur Effect in the Timeline and go through the same procedure, except that you want the first keyframe of the second Blur effect to be set to 20 and the last keyframe set to 0. This ramps the blur from blurry at the beginning of the second effect to sharp at the end.

At this point the effect is simply a blur dissolve. To complete the "glow," we need to change the W Point (White point) slider in the Luma Range parameter of the Color Effect. Again, we want to have this ramp up on the first effect and then ramp down on the second effect.

Select the first of the two Color Effects in the Timeline and go into Effects mode. Add a keyframe to the first frame of the effect and don't change any of the parameters because you want the effect to start as "normal." Then go to the last frame of the Timeline under the Sequence monitor and add another keyframe. On the second keyframe adjust the W Point slider from 235 to 16. This ramps the brightness up from nothing at the beginning to nearly full white at the end. Then click in the second Color Effect in the main Timeline, set the first keyframe to 16, and then add another at the end of the effect and set that to 235. This ramps the brightness from fully bright to normal by the end (see Figure 4.13).

That seems like a lot of work for a simple transition. The cool thing is that because all of this was done in the filler tracks, you can save the four effects on tracks V2 and V3 as a subsequence of prebuilt effects and drop it on any transition in this project or any future project. (However, Effects

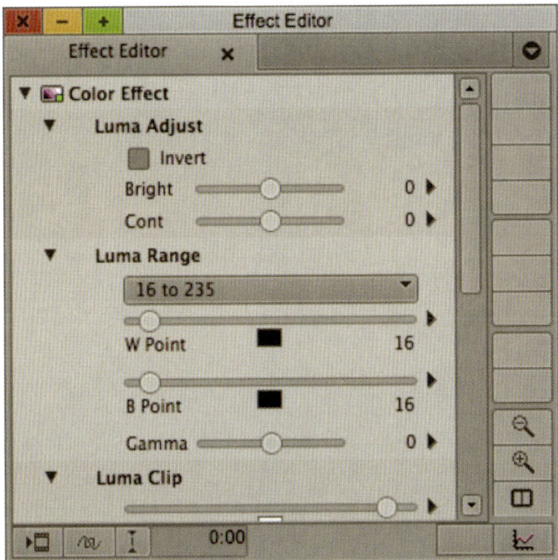

Fig. 4.13 This is the Effect Editor on the Color Effect. The White Point (W Point) slider has been moved from 235 (default) to 16 in order to create "highlights" out of any video level above black.

are sometimes finicky about the frame rates that they were originally created in, so you may want to create versions of your favorite effects in multiple frame rate projects. You can convert Effects subsequences to new frame rates, but they might act a little squirrelly.)

To save the effect, enable tracks 2 and 3 (enabled tracks are indicated by blue color in Track Panel) and make sure track 1 is disabled (gray). Mark an in point at the beginning of the effect and an out point at the end (Figure 4.14). Then, from the top right corner of the Composer window, above the Sequence monitor, drag the icon that looks like the sequence icon (see Figure 4.15) to a bin and name it something memorable and descriptive (see Figure 4.16). To reuse this effect, you simply load the subsequence into the Source monitor by dragging it there. Then make sure your tracks are patched correctly (they default to the same tracks they came from) and do an Overwrite (red arrow) edit. If you double-click the effects subsequence, it will instead load it into the Sequence monitor and Timeline track and will look like Figure 4.17.

The cool thing about this transition is that after you apply it above a new transition, you can actually use the trim controls to extend or shorten the length of the effect on either side of the transition. You could also create this transition as two effects instead of four. To do this, Add Edits on V2 and V3 a few frames before and after the V1 edit. Then add the Blur to the filler on V2 and the color correction to V3. Instead of ramping up on one side with one effect and back down on the other side with another effect,

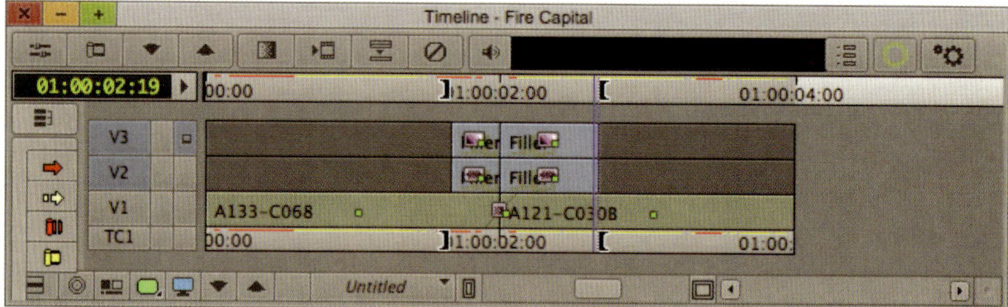

Fig. 4.14 The Glow Dissolve effect is chosen by selecting the appropriate tracks (V2 and V3 are enabled) and marking in and out points around the effect.

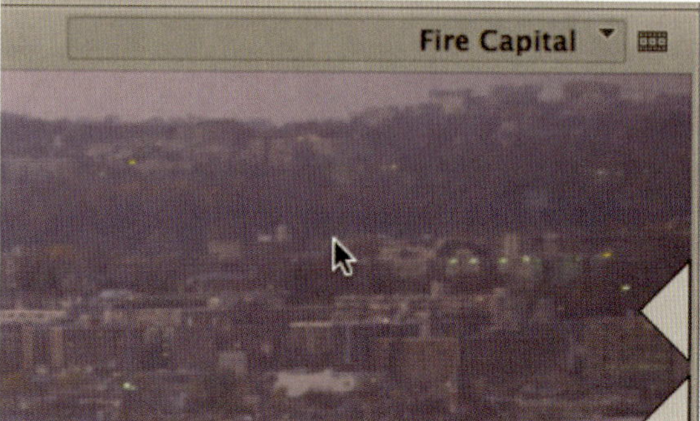

Fig. 4.15 The small icon to the far right above the Sequence monitor is like a subclip, but it "subclips" sequences. So this is called a subsequence. If you drag this icon into a bin, it will save the selected tracks and marked area of the sequence as a subsequence.

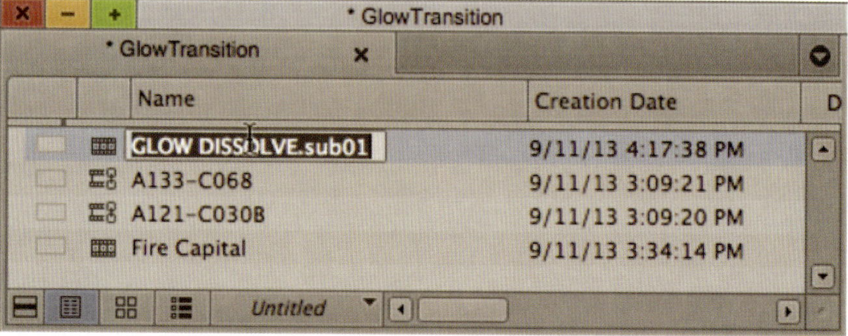

Fig. 4.16 Here, the Glow Dissolve subsequence has been dragged to a bin. The text is reversed over black to indicate that you can simply type to rename the subsequence if you want to.

simply ramp up and then back down within a single effect on each track. It might be good to place a marker in the effect at the point where the effect reaches its maximum level so that when you apply it above future transitions, you will know where to position the effect in relationship to the lower track edit point. This is one of the advantages to the original way

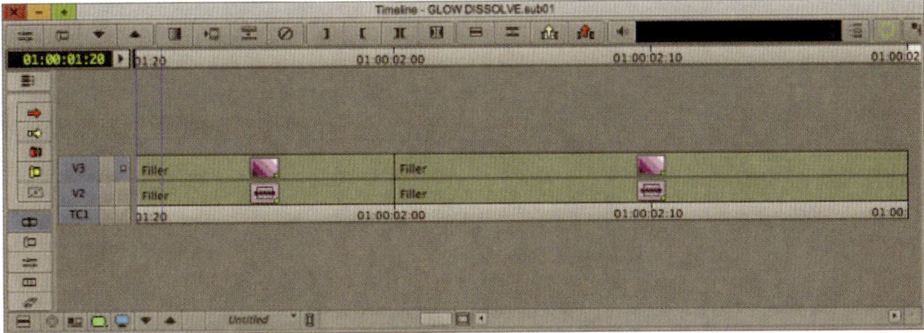

Fig. 4.17 The timeline above shows the subsequence of the glow transition. This can be edited above any transition to instantly apply the glow effect.

I created the transition: it's easy to place and trim by matching the edit points on all tracks. If you trim the alternate version that I just described, the "peak" of the effect will move depending on the lengths of the trims on either side of the effect.

Collapsing, or Nesting, and Stepping In and Stepping Out

Nesting doesn't usually get a lot of attention. There are a number of good reasons to nest layers of effects, and there are a few reasons not to nest them. But first, an explanation of nesting: if you have multiple layers of video tracks and you wish to combine them all into a single video track, use the Collapse button (which is sometimes referred to as the Nest button, especially by FCP editors). Similarly if you have a single video track and drop multiple effects onto that track, the track "autonests" the various effects into a single track with multiple effects in layers that are in a nest. In order to layer effects, you need to Option-drag or Option-double-click (Mac) or Alt-drag or Alt-double-click (PC) subsequent effects. To access the layers of video or layers of effects in a nested section, you can use the Step In button to step down through the various layers and find the section you want to fix or revise. When you want to come back out of the stack of effects or videos, you use the Step Out button to come out of the stack and back to the standard Avid UI (user interface). You can also double-click a nest in the Timeline to have it "expand" inside the Timeline.

One of the reasons for creating a nested effect is to maintain a clean Timeline that allows for easy on-line versioning, for example, keeping all of the video on track 1 so that titles can go on track 2 or allowing for the easy creation of a "textless" master by just monitoring or enabling track 1.

Another reason for creating nested effects is to allow for a clean transition from a "stack" of effects to another "stack" of effects or even a single video track. Even in the example of a simple key over a video track, if you want to fade out at the end of that "stacked" effect, it is actually not clean to just place dissolves on each track because that means that as the clip on

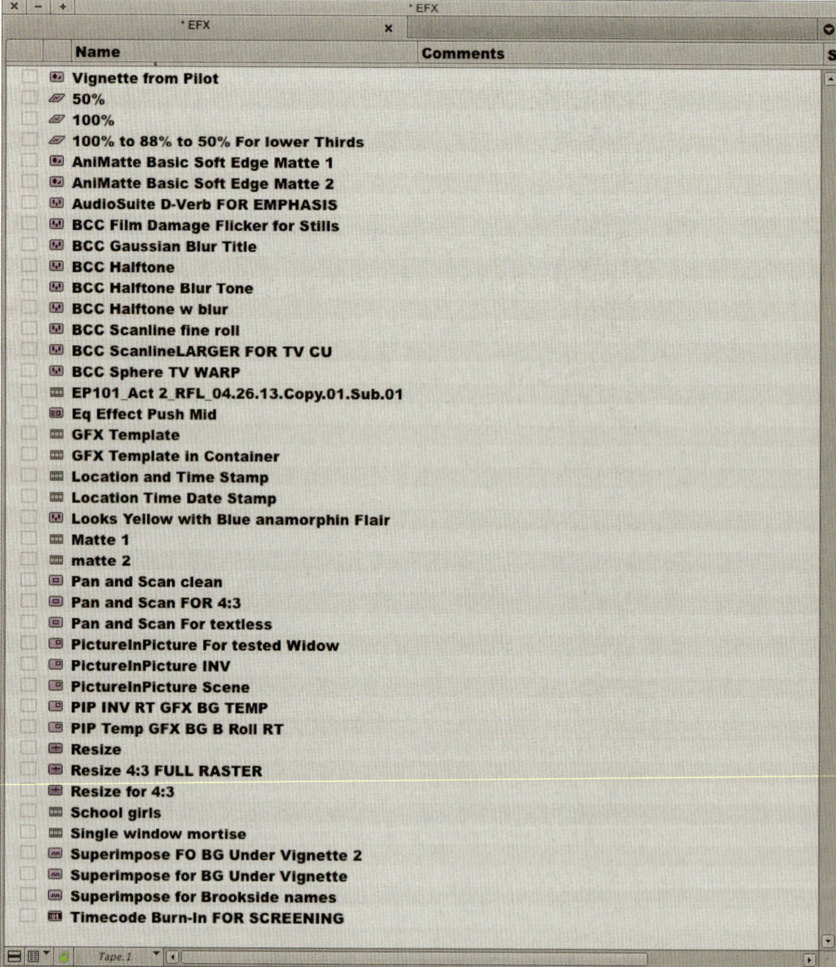

Fig. 4.18 Robert Landau's Effects bin for "Crimes of the Century." Note the descriptive names of the effects.

V1 is fading to black, the key on V2 is actually changing opacity throughout the fade. This concept is actually the same in Final Cut Pro. The analogy of how this works is with a live or taped production using a video switcher. You would not want to try to fade out one ME (mix effects bus) with one fader at the same time you faded another ME. Instead, you would run the MEs into a single fader and fade them both out with a single fader. That's kind of what nesting the stack of effects does: it feeds them all into a single ME that you can execute with a single effect, whether that's a dissolve or a fade or a wipe.

Nesting is pretty simple. With multiple tracks in the Timeline, select the part of the effect you want to nest by marking the area with an in point and an out point with selected tracks and click the Collapse button (see Figure 4.19). All segments do not have to start or end at the edit points.

TIP

If you don't already follow this practice, I would suggest creating a new project and a new bin in the new project called My Effects (or something similar) and saving effects like this to that bin each time you find something that you like or might consider using on a future project. Then each time you start a new Avid project, one of the first things you should do is to select the Bin tab in the Project window and from the menu bar, choose File > Open. Or use Command-O (Mac) or Control-O (PC). Then navigate to the My Effects bin and open it in your new project. This gives you a huge head start on any new project. The image (Figure 4.18) is of Robert Landau's Effects Bin from his *Crimes of the Century* project.

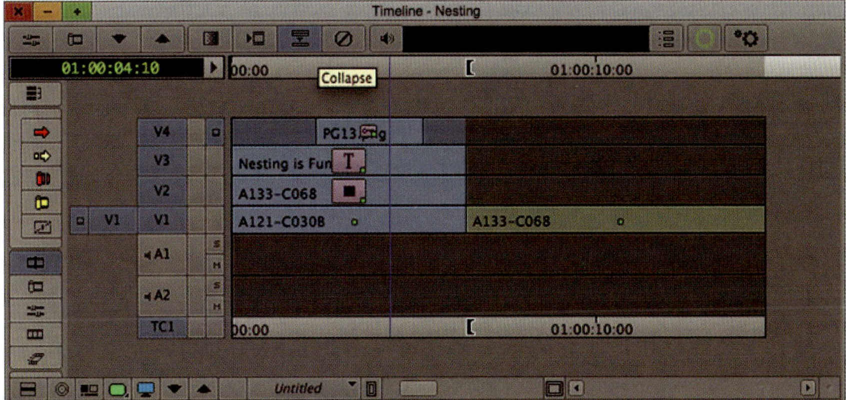

Fig. 4.19 The front portion of this sequence is about to be collapsed. The tracks that will be collapsed are selected and indicated with a blue highlight, and the portion of the sequence that will be collapsed is selected between the mark in and mark out. Notice that the Collapse icon in the Timeline is selected (blue), and the Text Help indicates that the Collapse icon has been chosen.

Once you hit the Collapse button, all selected tracks collapse to a single track. Nested Effects are represented on the Timeline with a small downward-facing arrow icon to indicate that there is a nest of collapsed tracks within (see Figure 4.20). It's also possible to segment select the tracks you want to Collapse instead of using marks.

There really is no way to "uncollapse" a nested effect. If you want to uncollapse the Nest immediately after creating it, you can simply undo it with Command-Z (Mac) or Control-Z (PC). But once you've missed the opportunity to undo, you're stuck. The work-around is to step in to the nested effect, select all tracks and copy the entire section, and then step out of the effect and paste the selected tracks back over the nested clip.

Stepping in and stepping out is done with the Step In and Step Out buttons. I use these a lot, so I have them mapped next to the Home and End keys (see Figure 4.21). When you step in to an effect, the Timeline changes from representing the entire sequence to representing only the tracks and duration of the nest.

The Step In and Step Out keys in the default UI are in the bottom left corner of the Timeline (see Figure 4.20).

You can also view and edit the collapsed section by double-clicking it in the Timeline. This behavior is actually controlled by an option in the Display tab of the Timeline Setting with the check box for "Double-clicking Shows Nesting." This behavior bothers some people. If you regularly get tripped up by having nesting displayed inside the main Timeline, then uncheck this option.

As I mentioned earlier, you can also nest by simply Option-dragging (Mac) or Alt-dragging (PC) effects on top of other effects. Each time you do this, it autonests just that effect into a nest, so that there are nests

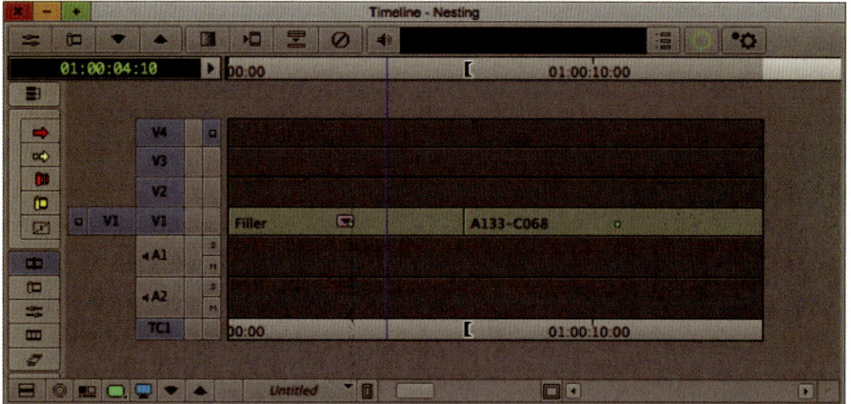

Fig. 4.20 Immediately after the Collapse button is selected, the Timeline looks like Figure 4.20. The three tracks in Figure 4.19 between the mark in and mark out have been collapsed to the single track here in Figure 4.20. The collapsed, or nested, section is indicated by a downward-facing effects triangle icon with a green dot that indicates that it will play back in real time and does not need to be rendered.

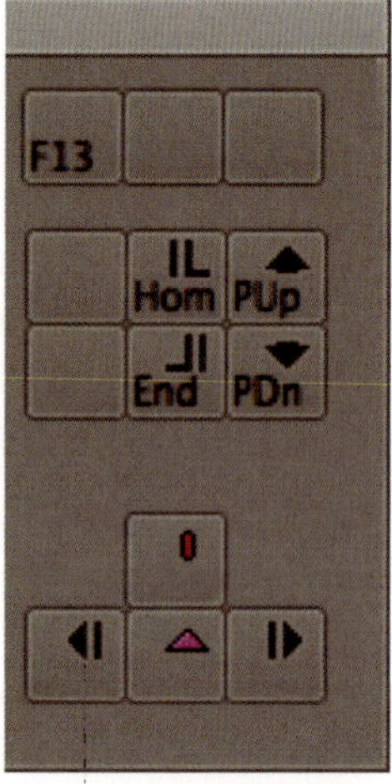

Fig. 4.21 Note that the Step In and Step Out buttons are mapped to the PUp and PDn buttons on my customized keyboard.

within nests, and you have to use Step In multiple times to step down through a series of nested effects to get down to the lowest level of the basic clip or clips that are the foundation of the set of nests. There's no indication in Media Composer how many layers of nests are at a specific location or which layer you're on. Each time you step in, you simply see a single video layer with the effect icon for that layer (see Figure 4.22). The only way to see all the layers of effects stacked is with the double-click method in the main Timeline, but you'll need to double-click once for

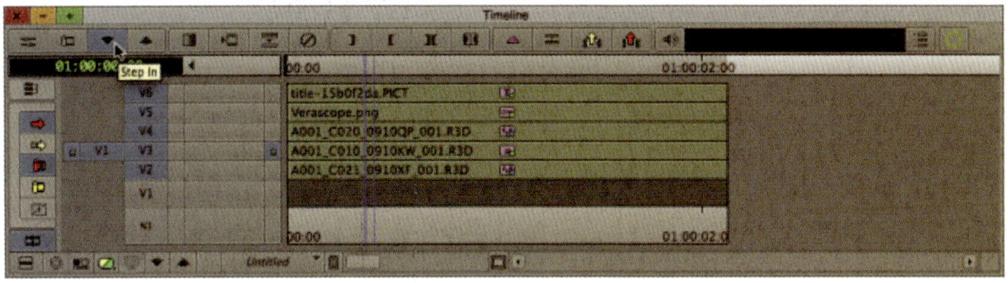

Fig. 4.22 When you Step In to an effect, this is how the Timeline displays it.

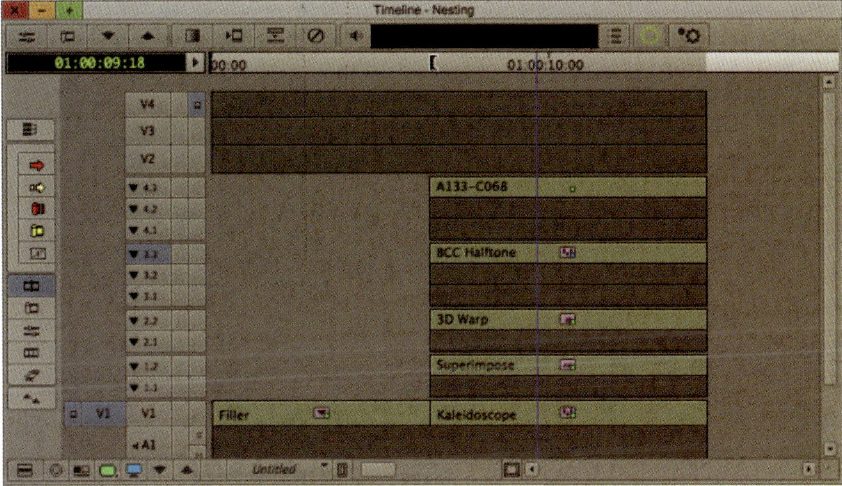

Fig. 4.23 If you double-click on a nested effect in the regular Timeline, the nest will be revealed in this "exploded track" view.

each nested effect. Strangely, though, the "stack" of nested effects shown in the main Timeline will be upside down, with the base video track on the highest video track and the last effect added on the lowest video track (Figure 4.23).

Effects Settings

Though I discuss effects settings in the chapter 2, I want to do a quick review and focus on the importance of preparing some settings specific to effects before continuing.

The first thing you should do to increase your efficiency in working with effects is to have a conveniently mapped button on your keyboard for the Keyframe button. The default keyboard setting has one mapped to the ' or " key (see Figure 4.24). This is a pretty good spot for it. My personal

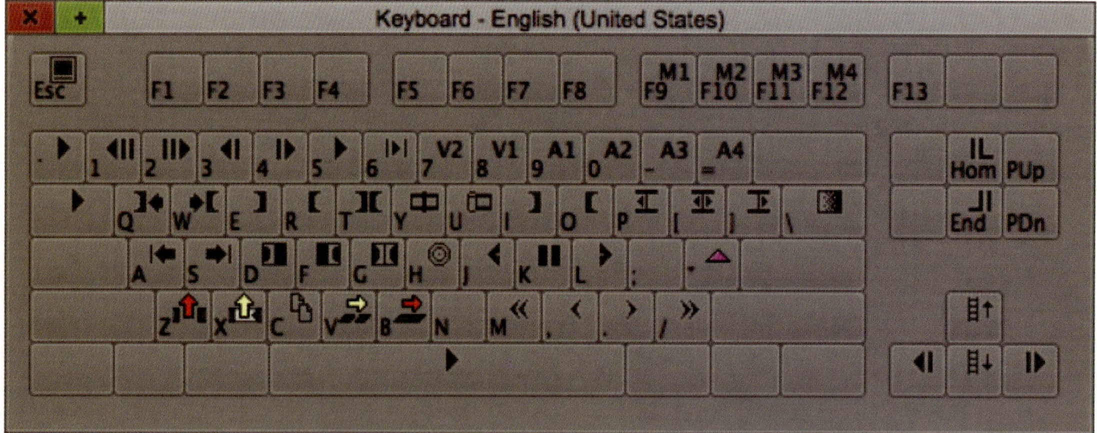

Fig. 4.24 This is the default Avid keyboard mapping. The most important effects button mapped to the default is the Keyframe button: the purple up-arrow to the left of the Return/Enter key.

keyboard setting has it on the down-arrow key. The keyframe button also gets a lot of use in doing rubberbanded audio.

Another efficiency in dealing with effects is using the Effects Editing Workspace. This is accessible from the menu bar: Windows > Workspaces > Effect Editing. Customizing this workspace even further is a good idea, but the Avid preset is a good start.

Another good thing to map for the sake of being efficient is the Effect mode button (see Figure 4.7). This is not mapped anywhere on the default keyboard. If you think you'll be going into effects mode a lot, I advise mapping this key to a convenient space on your keyboard. If you don't want to mess too much with the default setting of the keyboard, then I suggest possibly putting it on the Shift-Y or Shift-U key. The Trim and Edit modes are on the nonshifted Y and U keys, so possibly placing the Effect Editor on one and the Color Correction mode on the other would put all of the mode keys together. Another idea is that the E and R keys are the same as the I and O keys. So, switching the E key over to the Effect Editor would be an easy mnemonic device if you don't mind losing one of your In Point keys.

Of course, it's a great idea to review all of the various effects buttons in the Command Palette to see if there are other effects tools that you would like to map to specific places on your keyboard or elsewhere on the UI.

There are only so many keys to map on a QWERTY keyboard, so I try to make use of the UI buttons when possible. Usually I choose to use the UI for buttons that I know need to be used with mouse clicks anyway, like the Collapse button. Before you use Collapse, you need to mouse the selection to start, so you might as well just keep using the mouse to select Collapse instead of having a key devoted to it.

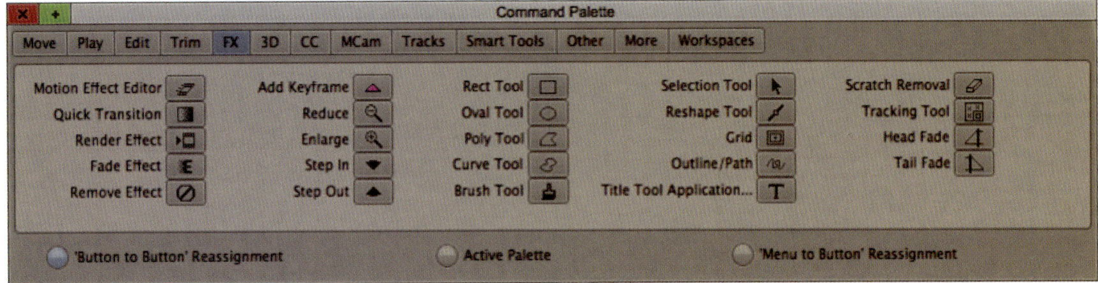

Fig. 4.25 This is the Effect tab of the Command Palette. It is accessed with Command-3 (Mac) or Control-3 (PC). If you do a lot of effects, you'll want at least some of these functions mapped to your keyboard. The other tab with effects-oriented functions is the 3D tab.

I also try to only map things to the keyboard that I use a lot. Collapse is important, but not something I use more than once or twice a week at most, so that's another reason it goes in my UI instead of on my keyboard. The Collapse button is visible in many of the Timeline screenshots from earlier in this chapter, such as Figure 4.20. The Collapse button is just to the left of the Delete Effect button at the top of the Timeline.

Finally, it's possible to map a special "effects keyboard" that you use only when you're doing a lot of effects editing. I'm not a big fan of this "multikeyboard" approach to editing, but if you're comfortable with it, it's a huge plus in Avid. You can trigger settings so that when workspaces launch, for example, the Effects Editing Workspace, other settings—such as keyboard settings or Timeline settings—will automatically switch over to match. These "automatic" settings need to be named with a common name. For example, if you want a specific Timeline, keyboard, and bin view to be paired with the Effect Workspace, they'd all have to be named "Effects" or something similar. This functionality is outlined on page 162 in Chapter 2.

Keyframes

One of the important aspects of that last Glow Dissolve tutorial was using keyframes. The keyframes we used in the Composer window are called Basic Keyframes. Basic Keyframes are manipulated under the Sequence monitor.

Adding keyframes and changing the parameters at those keyframes is pretty straightforward, but you may need to do some fine-tuning that requires a little more understanding of what is possible with the basic keyframes.

One of the first things to realize is that the behavior of keyframes is not like some other applications. For example, if you move to a new spot on the Timeline in some apps and begin to modify a parameter, the app will automatically add a new keyframe to the spot with the modified parameters. Avid can work like that, but you need to make sure that you do not have one of the other keyframes in the effect that is still active. With Basic keyframes, if you move to a new spot in the Timeline of the effect without disabling any enabled keyframes (indicated in pink/purple), then the parameter adjustments will be made to the enabled keyframes, *not* to the location where the Timeline locator has been moved. This behavior is different when using Advanced Keyframes, which we'll get to in the next section.

Some Basic Keyframe Behaviors

Shift-clicking keyframes toggles whether they are enabled or disabled.

You can select all of the keyframes in the effects Timeline using Command-A (Mac) or Control-A (PC). Shift-Command-A (Mac) and Shift-Control-A (PC) disables all keyframes in the effects Timeline.

With no keyframes enabled in the effects Timeline, if you move to a new location in the effects Timeline and make a parameter change, a new keyframe will automatically be added at that spot.

If you want to change the position of an effect keyframe, you can use the trim tools. In the default keyboard, these are on the M<>? keys. With a single keyframe enabled, use the <> keys to move the keyframe back and forth one frame at a time, or the M and ? keys to move ten keyframes at a time.

You can also hold down the Option key (Mac) or the Alt key (PC) to drag keyframes to new locations. And even though the trim keys can move only a single keyframe at a time, when multiple keyframes are selected, you can drag them together with this method.

Video Tutorial

Most of these behaviors are a little easier to explain when you see them at work, so there's a quick video tutorial in the Project media on the companion website for this book, www.focalpress.com/cw/hullfish.

If you want to jump between keyframes, you'll need to map the Fast Forward and Rewind keys to your keyboard. I have always mapped FFWD to my Tab key and REW to Shift-Tab, but this will not work to jump backwards between keyframes. For the Rewind key to jump to the previous keyframe, it can't be mapped to a shifted key.

Another useful keyframe behavior is the ability to copy and paste keyframe data. Enable a keyframe (enabled keyframes are purple/pink) and Command-C (Mac) or Control-C (PC) to copy. Enable a different keyframe and Command-V (Mac) or Control-V (PC) to paste the information onto the enabled keyframe.

Advanced Keyframes

For a lot of effects, basic keyframes will work fine. But for some effects, you need more specialized and specific control over the keyframes of every single parameter and you need to determine the way the effect is

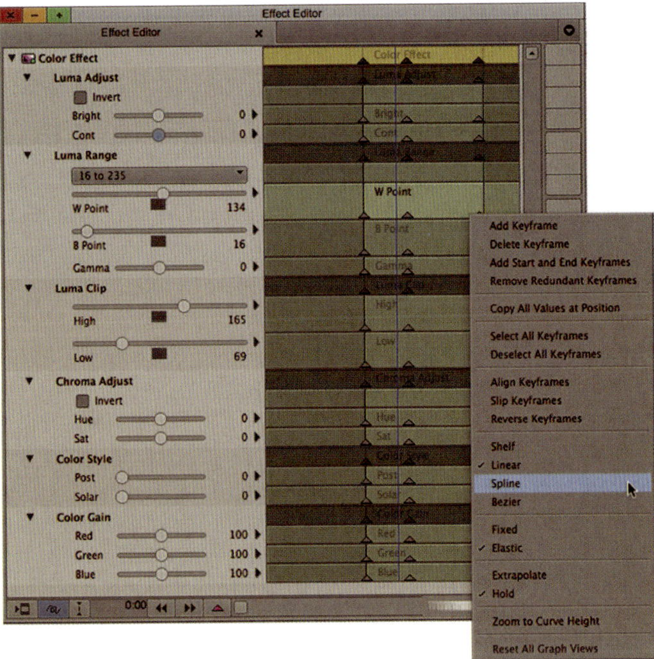

Fig. 4.26 The contextual menu of the Effect Editor allows you to choose how to manipulate Advanced Keyframes and affect their behavior.

interpolated* between those keyframes. For that kind of functionality, you need to use Advanced Keyframes, which are enabled using the small Advanced Keyframe icon at the bottom right corner of the Effects Editor.

All of the cool stuff that you can do with Advanced Keyframes is accessible from the contextual menu that you access by right-clicking in the keyframe or graph area of the Effect (see Figure 4.26).

Here's the list and a little rundown of how each menu item works and what it's used for:

Add Keyframe: This is fairly pointless, especially if you have it mapped to your keyboard.

Delete Keyframe: As with Add Keyframe, this is also relatively pointless because you can do the same thing with the Delete button.

Add Start and End Keyframes: This used to be the default for Avid, but now it's quite useful if you know you want keyframes at the beginning and end of your effect. It can save you a mouse click.

TIP

Option clicking (Mac) or Alt-clicking (PC) on any parameter box in an effect or even in a color correction will reset the parameter to the factory default. You can temporarily bypass the changes in any given parameter by simply deselecting the box (changing it from the blue "altered" status to gray "nonaltered" status. Clicking on the box again will toggle the parameter changes back on, reinstating the parameter changes at the same levels they were at before they were toggled off.

---

* Interpolation determines how an effect transitions from one keyframe to another. The default is Linear interpolation, which is a simple mathematical division of the change between parameters (for example from 0 to 100 percent opacity) equally divided over the course of the number of frames between keyframes. Linear interpolation results in a very mechanical look. For more organic moves, try Bezier or Spline. Shelf "holds" the effect until the next keyframe.

Remove Redundant Keyframes: This option is not something I've used a lot before. However, if a keyframe isn't serving a purpose, this is a way to get rid of it.

Copy All Values at Position: Essentially, this is like the copy command I described in the Basic Keyframe section.

Select All Keyframes: Does the same thing as Command-A (Mac) or Control-A (PC).

Deselect All Keyframes: Does the same thing as Shift-Command-A (Mac) or Shift-Control-A (PC).

Reverse Keyframes: This allows you to select two keyframes and have the parameters associated with the first keyframe applied to the last keyframe and vice versa.

Shelf, Linear, Spline, and Bezier: These options describe four different interpolations that can be used to get from one keyframe to the next. The default mode is Linear. This creates a straight line—a direct, even, linear, numerical change from one keyframe to another. In a five-keyframe move from a parameter set to 0 and another set to 100, the first keyframe would be 0, the next keyframe would be 25, the next 50, the next 75, and the final would be 100. Spline interpolations smooth out the incoming and outgoing transition to "ease in" and "ease out" of the transition. Bezier is similar to Spline but provides you with handles to customize the ease in and ease out amounts. Shelf holds the parameter at the level of the first effect and doesn't change to the second level until it reaches the second keyframe and then instantaneously switches to the new parameter amount. Looking at the graphs when you switch these interpolations will provide a clear visual indication of what the interpolation will look like.

Extrapolate and Hold: These options are useful if you don't have keyframes at the very beginning and end of your effect. Extrapolate and Hold dictate the behavior before the first keyframe and after the last keyframe if those keyframes are not set on the first and last frames. Extrapolate continues the trajectory of the effect before or after the keyframe and Hold maintains the parameter value of the first and/or last keyframe through to the beginning or end of the effect.

Zoom to Curve Height: This allows you to better see the curve of the graph between keyframes by filling the vertical height of the graph with the full curve height. It doesn't actually change the behavior of the effect, just the display of the graph. You can undo this command with Reset All Graph Views in this contextual menu.

The other contextual menu accessible in the Effect Editor is called up by right-clicking in the parameters area of the Effect Editor instead of in the keyframes or graph area (see Figure 4.27).

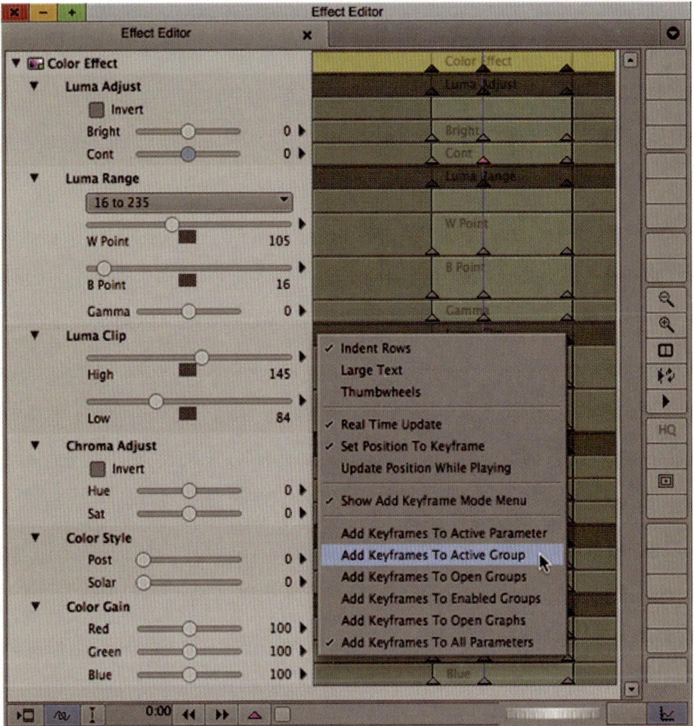

Fig. 4.27 This is the other contextual menu available in the Effects Editor by right-clicking on the left side of the Effects Editor (where the parameter names are) instead of the right side (where the keyframes and graphs are).

The options here are the following:

Indent Rows: This simply changes the alignment of the Effect Editor parameters to make it a little easier to locate parameters in a long list.

Large Text: A simple toggle between two sizes of text in the Effect Editor.

Thumbwheels: This allows you to choose between sliders and thumbwheels for controlling the parameters.

Real Time Update: Almost always, this should be left on. In some cases, with extremely complex effects stacks or with underpowered machines, you can turn it off to get more responsiveness in the movement of the effect parameter, but you can't see what you've done to the actual effect until you release the parameter.

Set Position to Keyframe: This allows you to have the Timeline indicator in the sequence follow when you select a keyframe in the Effects Editor. This allows you to see the effect of the parameter change at the keyframe. There are times when you may want to monitor another frame other than the frame that the keyframe is on while changing a parameter. In this instance you would want to toggle off Set Position to Keyframe.

Show Add Keyframe Mode Menu: Choose this to call up the options to choose where an added or deleted keyframe will appear—in an Active Parameter, Active Group, Open Groups, Enabled Groups, or Open Graphs or to all Parameters. Sometimes you may want to add keyframes to just one of these options, in which case you'd want the Mode menu to come up when you add keyframes. In other instances, you may already have this set up from a previous effect and you don't want to have to click on a mode because it's already selected or you're happy with the default behavior, in which case you want this toggled off.

Another critical issue to understand with Advanced Keyframes in Avid is that they can be elastic or fixed. This state matters when an effect is either trimmed to a different length, or when it is created on a clip of one length and then pasted or used on a clip of a different length. If the keyframes in either case are elastic, then the keyframes are placed or moved proportionally to the length of the clip. So if you have a keyframe that is in the middle of a 5-second clip and the clip gets trimmed to 4 seconds, with Elastic selected, the middle keyframe would move from 2.5 seconds to 2 seconds in. But if Fixed is chosen, then the keyframe that was originally at 2.5 seconds in would stay at 2.5 seconds if the back half of the effect was trimmed. If the front half of the effect is trimmed, the keyframe would actually change so that it still stayed associated with the same frame of video where it was originally placed.

Elastic or static keyframes are chosen in each parameter—or across them all—in the Effect Editor by right-clicking (Mac or PC) or Option-clicking (Mac) in the graphs or keyframe area. In the resulting contextual pulldown menu, Fixed and Elastic are options.

Promoting Effects to 3D

One of the concepts that may be a little less than transparent is that if you want to "fly" a title or an imported keyed graphic, you don't drop a 3D effect on top of the title, you "promote" it to 3D. In other words, to do 3D effects on a keyable graphic, you don't drop a 3D effect on it from the Blend Effects category, you go into Effects mode on the keyable graphic and click on the small 3D box icon in the bottom right corner of the Effects mode window (see Figure 4.28).

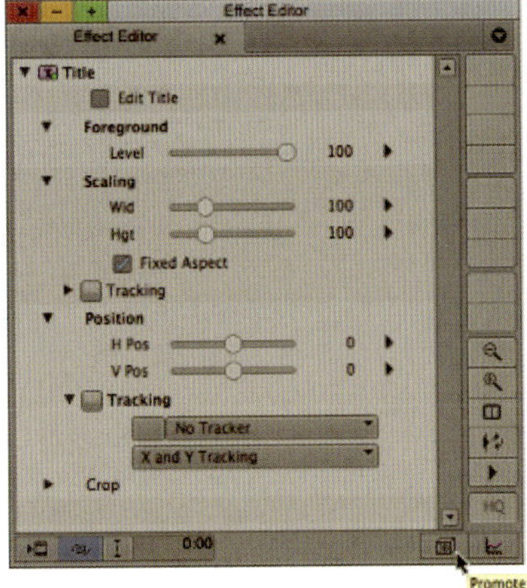

Fig. 4.28 The small 3D box icon in the bottom right corner of the Effects mode window allows you to "promote" keyable graphics and titles to 3D, allowing them to be manipulated in 3D space or to be controlled by various additional parameters, like blur, that are unavailable otherwise.

Video Tutorial

Follow along with this tutorial on promoting effects to 3D by downloading the Verascope.png file from the companion website, www.focalpress.com/cw/hullfish.

As a tutorial on how this works, cut some video on track 1 and import a graphic with an alpha channel and cut it onto V2. You can use the Verascope.png file in the project files on the companion website for this book, or one of your own images that has an alpha channel or transparency.

With the logo or keyed file cut onto track 2, go into Effect mode and select the logo on track 2. In the Effect Editor, there aren't a lot of options for manipulating the keyed file. But notice the small 3D cube icon at the bottom right corner of the Effect Editor. Clicking on that icon promotes the effect to 3D and delivers the same parameters and functionality that are in the standard 3D Effect. This technique also works with titles created in the Title Tool or in Marquee.

TIP

I like PNG file format when I create graphics for import because the quality of the file is very good and it automatically retains "transparency" when exported from a Photoshop file. Avid sees this transparency as a key.

AniMatte

AniMatte is another effect with a usefulness and application that might not be readily apparent by its name. Essentially, AniMatte allows for the creation of a rotoscoped matte of a video source.

One possible use of this would be to key a standard interview over some background footage.

As a tutorial for this, let's use the Steve MS clip from the ScriptSync project as our "interview" footage. The background is pretty ugly. Let's say we wanted to create a kind of rough rotoscope to replace the background in the clip with one of the 4K Artbeat clips or anything else you have lying around your hard drive.

The Bin with the footage links and sequence is in the project files for this chapter on the companion (www.focalpress.com/cw/hullfish) for this book. You may need to relink or re-edit the AMA footage.

Using a 4K clip, like the shot of the US Capitol building available from the Artbeats folder on this book's companion website, it's possible to create a more pleasing composition by utilizing the full resolution of the clip to reposition the dome of the Capitol to the right to make room to key the Steve MS shot to the left.

TIP

Avid added Source Settings to the Clip menu in the main menu bar, so you can map this useful command to a keystroke or to the UI, as described in chapter 2 on page 133.

To access the full resolution of the 4K clip, don't import the clip. Importing will lock the resolution to HD. Use AMA to bring the clip into an Avid bin and open it into the Source monitor by dragging it or double-clicking on it. There are a number of ways to deal with these high-resolution clips, but an easy way is to right-click on the high-resolution clip in the Source monitor to call up a contextual menu. One of the options in that contextual menu is Source Settings. In the Source Settings is a FrameFlex tab that allows you to intuitively use the clip's high resolution to zoom in and reposition and recompose the framing. Move the Capitol dome off to the right as much as possible. With that done, apply the change and cut the clip onto V1 (see Figure 4.29).

Then cut the last half of the Steve MS clip onto V2. Drag the AniMatte effect onto the Steve MS clip on V2. AniMatte is in the Key effects category (see Figure 4.30).

Because we want to be able to place rotoscoping points outside of the actual video raster, before we start to edit this effect we want to shrink the size of the video image in the Sequence side monitor. The keyboard command to do this is Command-K (Mac) or Control-K (PC). The mnemonic for remembering this keyboard command is *shrinK*. The opposite of this is to zoom bigger using Command-L (Mac) or Control-L (PC). The mnemonic to remember that is *enLarge*. This leaves space around the edges of the image so that we can place points of the rotoscope shape outside the image.

Go into Effect mode. Click on the Poly tool or the Curve tool. Personally I prefer the Poly tool because it creates fewer points for the Avid system to track. To use the Poly tool, select it in the Effect Editor (see Figure 4.31), then click on a spot in the dark gray border just outside of the video image at a point near the outside and below either arm. Then continue to click on points around the outside of Steve's body and head. You don't have to finesse these marks. I used about twelve points before clicking back on the original point to complete the "loop" around Steve's body (see Figure 4.32).

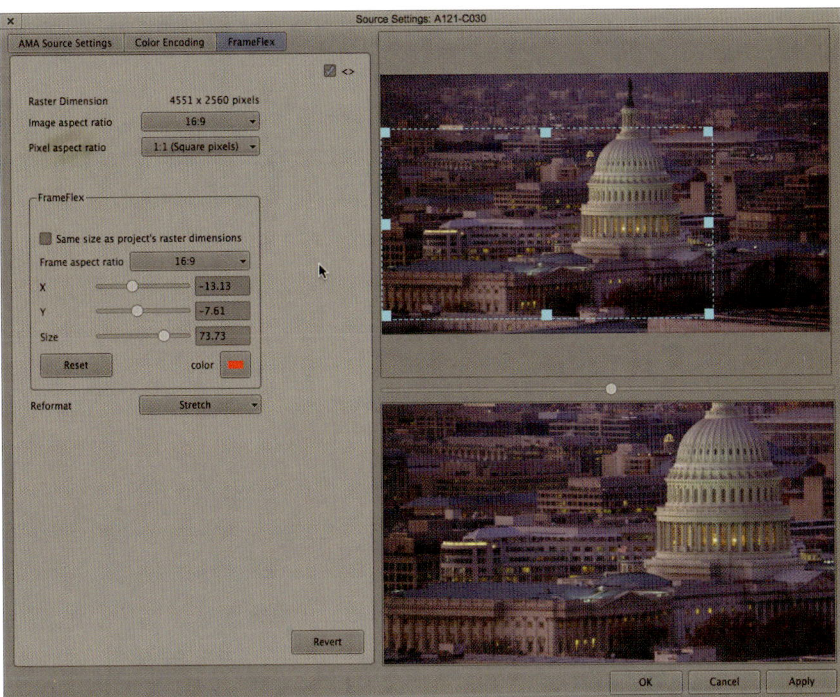

Fig. 4.29 In the Source Settings window, using the FrameFlex tab, you can utilize the full resolution of a 2K, 4K, or 5K clip to reframe the image using the native camera resolution to deliver a sharp HD image.

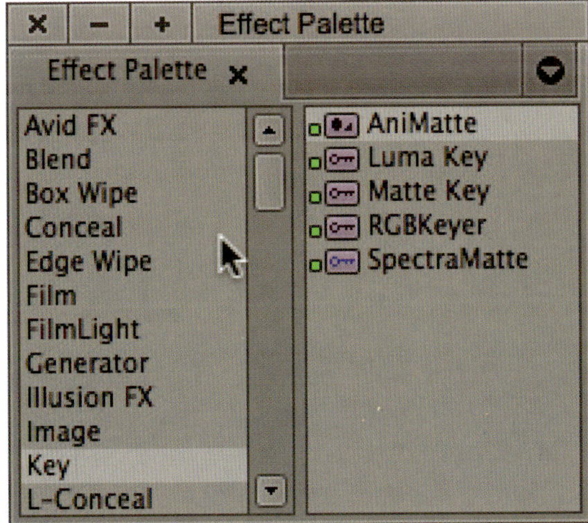

Fig. 4.30 The AniMatte effect is in the Key category of effects in the Effect Palette.

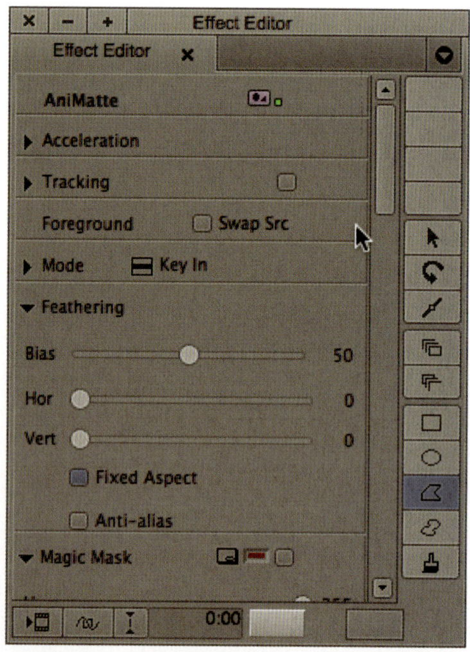

Fig. 4.31 These are the AniMatte controls in the Effect Editor. The Poly tool is selected (indicated in blue) on the right side.

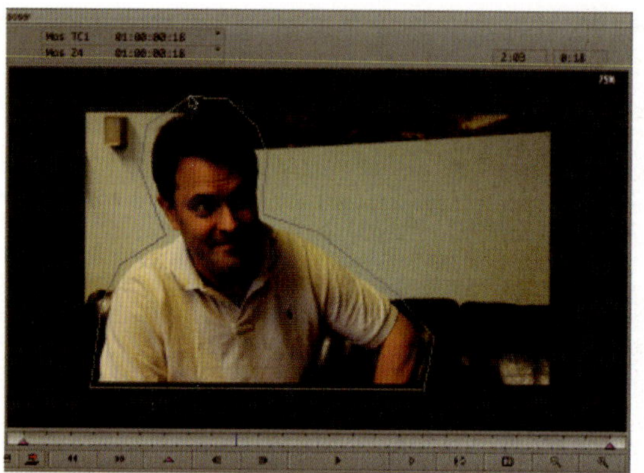

Fig. 4.32 The white outline shape in AniMatte is made by clicking on the Sequence monitor where you want to add points.

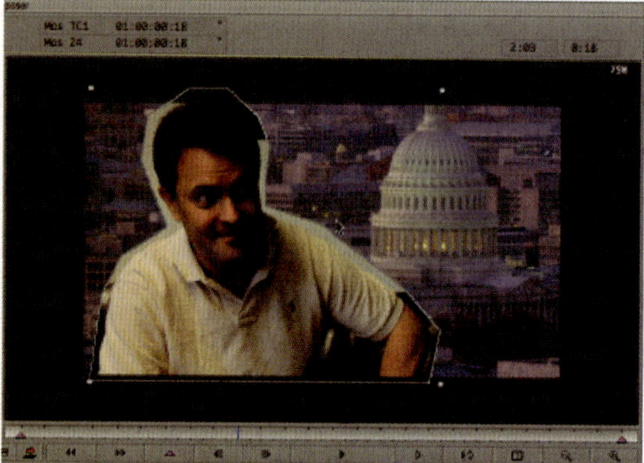

Fig. 4.33 When the last point is made to connect the "loop," or shape, the background will be visible on the outside of the shape.

As soon as the loop is closed, the background of Steve MS will key away, revealing the background image on V1 (see Figure 4.33). This is a pretty ugly key. To improve it, you can soften the hard lines into curves by clicking on the Reshape tool (see Figure 4.34; diagonal line with a point in the middle of it) then Option-click (Mac) or Alt-click (PC) on any or all of the connecting points in the loop you drew. This will turn

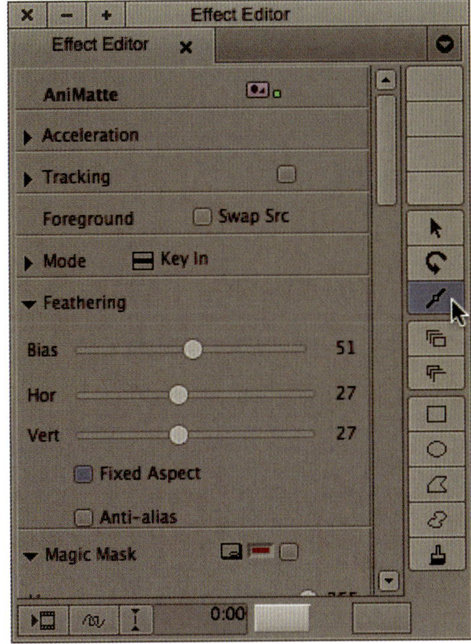

Fig. 4.34 The Reshape tool in the AniMatte Effect in the Effect Editor is indicated in blue under the arrow cursor.

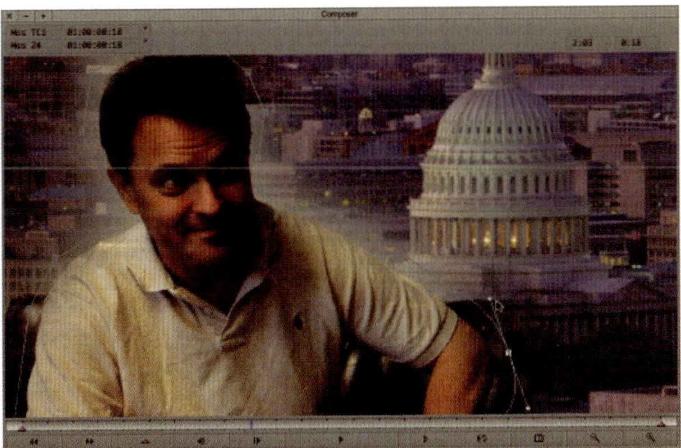

Fig. 4.35 You can turn the straight angles of a shape created with the AniMatte Poly tool into Bezier curves using handles.

the angled lines into Bezier curves, which you can manipulate using the handles that appear on either side of the points (see Figure 4.35). I point this out only to show that it can be done. In reality, you can probably leave them as hard angles, because the thing that will really improve the key is to soften the edges of the key. When the edges are softened, it won't matter much whether the angles are sharp or curved.

To soften the key, use the Selection tool (arrow icon) in the Effect Editor and click anywhere inside the loop you drew around Steve. Then with the shape selected (the outline of the shape will appear in white), move the sliders for Bias, Hor, and Vert in the Effect Editor (see Figure 4.36). Personally I like keeping the horizontal and vertical sliders approximately equal, somewhere around 20 or 30, but it will depend on the look you're trying to achieve and how well the background and foreground blend. The bias parameter controls how far the softened edge erodes the image on either side of the effect outline that you drew. A bias of 0 means that the soft edge starts at the loop and extends only outward. A bias of 100 means that the soft edge starts at the effect outline and extends only inward. It's possible that with a bias of 100 the keyed image will disappear completely, especially with a softer edge. A bias of 50 means that half of the softness goes inside the loop and half of the softness extends outside the loop.

If you prefer to use the Curve tool instead of the Poly tool, it's very similar, except instead of just clicking on various points around Steve's body, you actually draw by dragging the cursor around the outline of the body and then back to the starting point to complete the loop.

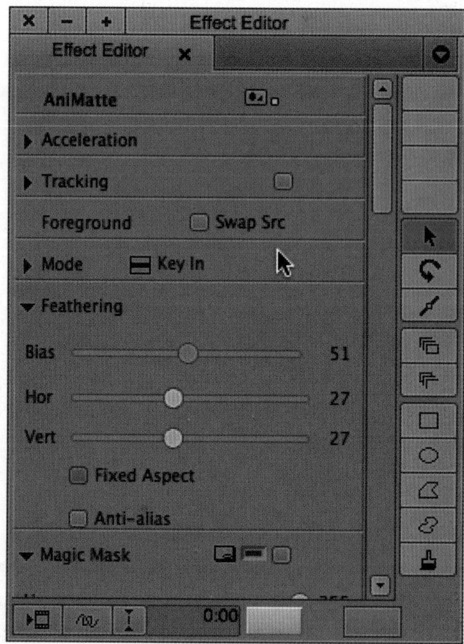

Fig. 4.36 The Feathering section in the AniMatte Effect in the Effect Editor includes the Bias, Hor, and Vert controls. The Hor and Vert controls adjust the softness of the key. Bias adjusts whether the softness extends inside or outside of the shape.

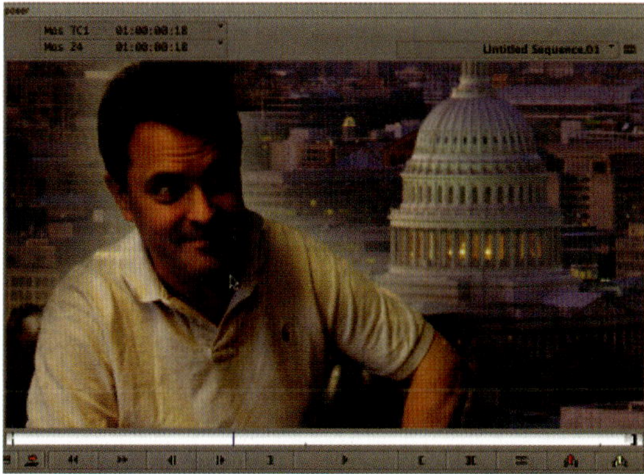

Fig. 4.37 This is the final look of the effect with the V2 effect keyed over the V1 footage cut by the shape drawn in AniMatte.

AniMatte can also be used to great effect by keying the same clip over itself and then manipulating the lower clip either by darkening it or blurring it. Then cut out either a simple large hole in the middle with the Circle tool, or cut out a subject like in the above example. This is a way to create a nice vignette, darkening the edges of a shot, or to blur the background slightly to "pop" the foreground out from the blurred background.

Title Tool

The bad news is that Avid's Title Tool hasn't been improved for over a decade. The good news is that it is a WYSIWYG text editor. In other words, it's pretty easy and intuitive to get something looking just like what you want on screen. That's a far cry from the pathetic FCP7 text generator. I'm going to cover this tool quite thoroughly because it can get you through about 90 percent of most titling needs.

Creating a text-based graphic, whether full-screen or a key, is easy and can be accomplished in at least three different ways. You can access the Title Tool from menu bar at Clip > New Title, or from Tools > Title Tool Application. You can also map or launch it from the Command Palette in the FX Tab > Title Tool Application.

When you launch the Title Tool, you first get a dialog box that asks you to choose between the simple and easy-to-use Title Tool or the more complicated and more capable Marquee titler. You are also given the option to Persist (see Figure 4.39). Choosing Persist means that until you relaunch Media Composer, you will not have to go through this dialog choice again because the choice you make now will persist until relaunch. Since we're going to explore Marquee next, do not check Persist; just click on the Title Tool button.

TIP

In an interview for the publication of *The Art and Technique of Digital Color Correction* (2nd edition), colorist Pankaj Bajpai, from Encore in LA, said that he prefers doing very organic hand-drawn vignettes instead of simple ovals to create a rich vignetted feel that doesn't scream, "Vignette here!" The organic quality of the vignette hides the obvious omnipresent oval while still accomplishing the same visual effect (see Figure 4.38).

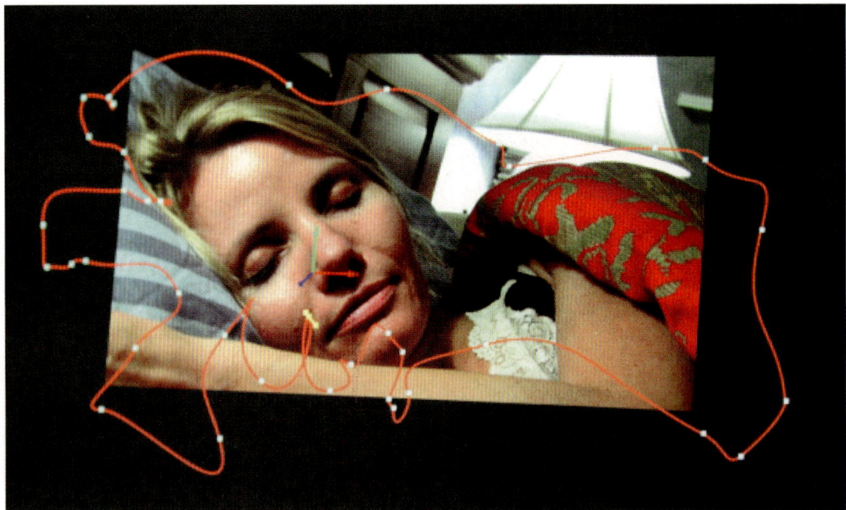

Fig. 4.38 From my color correction book, "The Art and Technique of Digital Color Correction," this image shows a mask applied to the image by Encore Hollywood's Pankaj Bagpai to vignette the image in a very organic way. The red lines are the outline of the vignette in the UI, and are not visible in the final grade. Some colorists would have done this with a simple oval. Bagpai did this to make the vignette harder to spot.

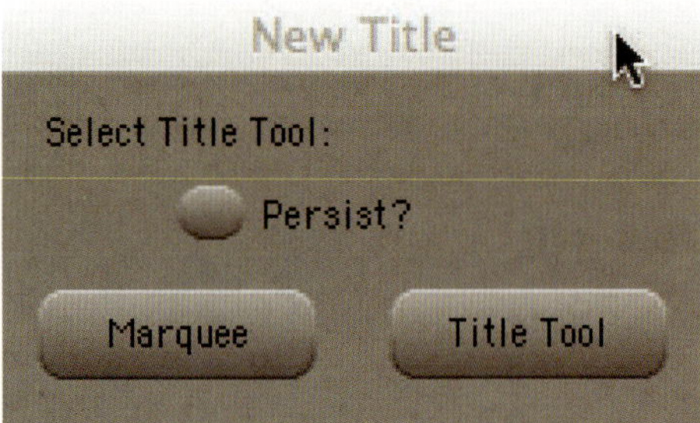

Fig. 4.39 The New Title dialog comes up when you choose to create a new title and offers the option to create the title in Marquee or the Title Tool. If you know you always like to use one or the other, then the Persist? option can be used so that this dialog doesn't come up again until Media Composer is relaunched.

This launches the Title Tool application (see Figure 4.40). It used to be run as a process inside the Avid editing application, but now it is a self-contained application on its own, with the background for the Title Tool fed by the current location of the Timeline indicator in the Timeline. Because it is a separate application, you can still navigate the Timeline if you want to see your text against a different part of the Timeline or even against a different track level. The Title Tool background won't refresh to the new image from the sequence until you click on the Title Tool.

There are several defaults at play when the Title Tool opens. The default font, which is not permanently changeable, is Geneva 48-point

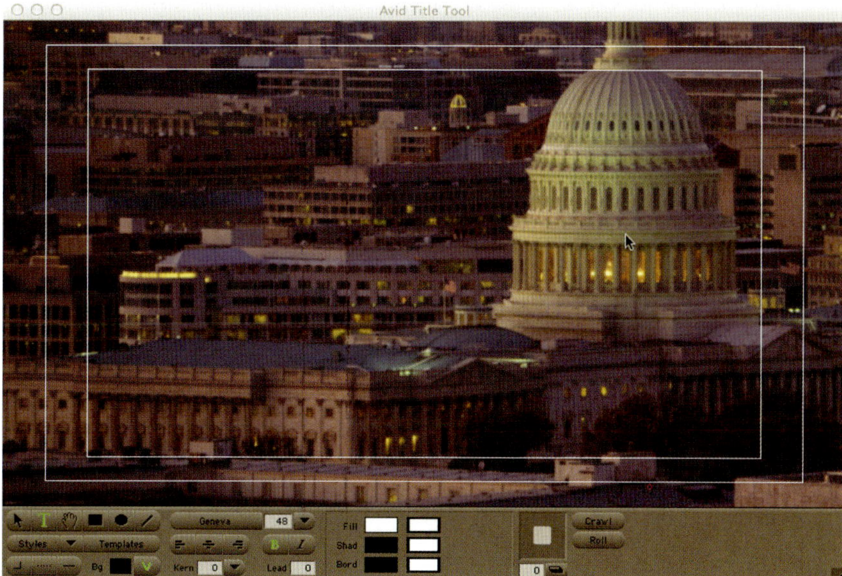

Fig. 4.40 This is the Title Tool. The majority of the tool is the window that shows the video fed by the Timeline. In this case, the image of the Capitol building.

bold. No one would intentionally make an artistic choice to use Geneva, 48 points or bold, so I'm not sure why this is the default. The other default is that the title is keyed against the background instead of a full-screen, colored background. That's a pretty good default. And finally it defaults to the Text Tool, assuming you'll want to start by typing something. Again, a good choice. The Text Tool cursor is a basic blinking vertical line, familiar to anyone who uses most text editors, such as Microsoft Word. You can click anywhere on the screen to begin typing at that spot. There used to be a way to force a temporary default font and characteristics on the Title Tool, but this capability went away when Avid broke the Title Tool off to create an independent app.

The Text Tool has two slightly different functionalities depending on where you click. If you click anywhere in the main window where there is no other text box, you can immediately begin typing and create a new text object on the page. If you use the Text Tool to click within an existing text box (text boxes are invisible when in Text Tool mode), then you are able to edit the text inside that text box. You can overlap text boxes in the Title Tool, but you can't click within one to create an additional box.

To change the font and font attributes while in the Text Tool, select them from the options at the bottom of the tool (see Figure 4.41). There is a pulldown menu for fonts (sadly, it is not WYSIWYG). To the right of that are a text box and a pulldown menu from which to select the font size. Entries can be made numerically in the text box. Below those two are options for justifying the font left, center, or right. Next to the justification options are buttons for toggling the bold and italics off and on. Standard keyboard commands for bold and italic don't work in the Title Tool.

To the right is a series of color swatches with the words: Fill, Shad, and Bord. These color swatches let you adjust the fill color of selected text, the shadow color or, if you've added one, the border color. I'll explain adding shadows and borders shortly.

The leftmost column of color swatches indicates and adjusts the color. To adjust the color of the fill of selected text (or other Title elements), click on the top left color swatch. This opens a dialog box with options for changing the hue, saturation, and luminance of your font (see Figure 4.42). An eyedropper lets you pick colors from the source video in the background. And for more specific color control, select the little "bullseye" icon at the bottom of the box to launch a more full-featured color editor from the computer's OS. You can select RGB, CMYK, or HSL sliders. There are also large color wheels and palettes of colors that can be selected from numerous tabs in the Color Editor. These colors can be saved and accessed from other locations, such as in the Effect Editor.

The shadow color and border color are the next two color swatches down and are editable in the same way.

The color swatches to the right are actually to adjust opacity, not color. The "whiter" the swatch, the more completely opaque the image is. The darker the chip, the more transparent the image is (Figure 4.43).

To the right of the transparency swatches are two swatches that allow you to edit a gradient of either color or transparency. With an object selected in the Title Tool, click on the color chip for Fill and set the color to red. Notice that both of the two far-right color chips change to red. Now you can click on one of those two gradient color chips to call up the color picker again. Change it to yellow (see Figure 4.44). This adds another box to the UI directly below the gradient color chips. This box allows you to click on any of its edges or corners to control the direction of the gradient.

Controlling gradients of opacity works similarly. With an opacity chip selected, click on one of the two gradient color swatches to adjust the amount of gradient and then down to the larger box to control the direction of the gradient.

To the right of the color and gradient chips are the controls for setting the direction and size of drop shadows. The "box within a box" can be dragged in any direction to create a shadow. Avid has added a slight "magnetic" feel to this if you want to constrain your shadow to a specific

TIP

Since the most-often-used tools are the Text Tool and the Selection Tool, Avid conveniently created a keyboard shortcut to switch between them without having to mouse over to the individual tools. Simply hold down the Option key (Mac) or the Alt key (PC) while clicking on a text object in the Title Tool to switch between the Selection Tool and the Text Tool.

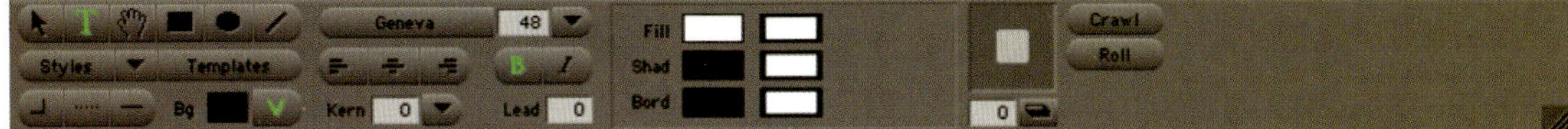

Fig. 4.41 Detail of the tools at the bottom of the Title Tool.

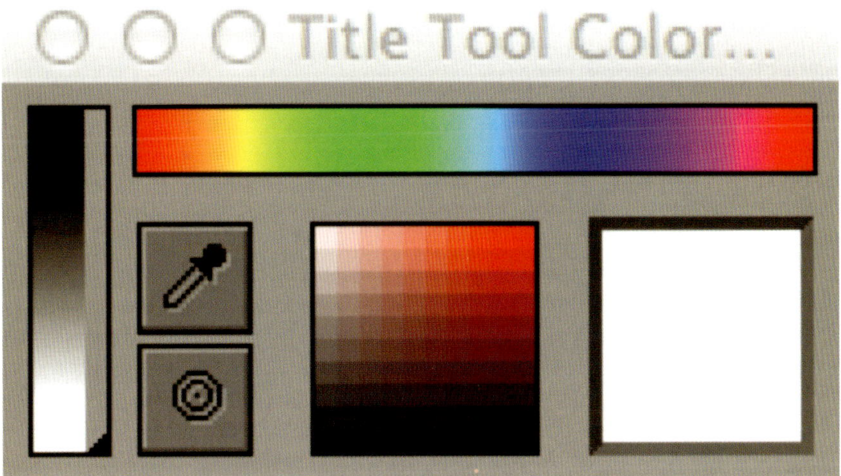

Fig. 4.42 This is the color adjustment UI launched from the swatches to the right of Fill, Shad and Bord.

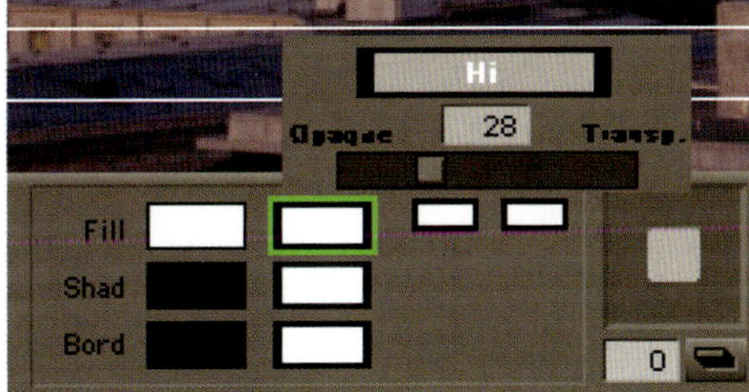

Fig. 4.43 This is the Opacity control in the Title Tool.

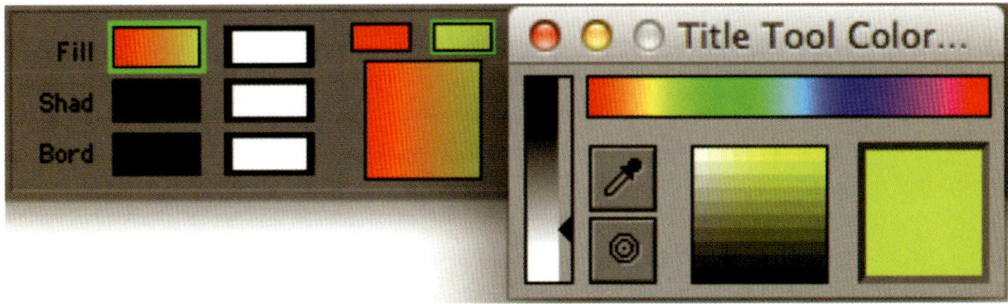

Fig. 4.44 This is the color gradation UI of the Title Tool.

direction. The shadow size can also be adjusted by dragging. There is a text box below the shadow box for entering in specific numeric values (see Figure 4.41).

Along the far left bottom of the Title Tool are more options. The Hand tool (formally, the Video Placement Tool) is pointless now. However, if your screen real estate is severely limited and you have shrunk your Title Tool to a small box so that you can't see all of the text or the background image, you can pan around the image with the Hand tool. It doesn't actually move elements around the Title Tool, it changes your view. The three shape tools to the right of the Hand tool allow you to create squares, rectangles, circles, ovals, lines, and arrows. As with many applications, if you Shift-drag out these shapes and lines, they are constricted in shape to perfect squares and circles or to purely vertical or horizontal lines.

At the far left corner are three mini pulldown menus that allow you to control the shapes of the corners of boxes (see Figure 4.45), the thickness of borders and lines (see Figure 4.46), and the placement of arrows onto lines (see Figure 4.47). The last option in the Box Corner pulldown and the Border Width Tool is an ellipse (...). Choosing this option allows you to type a numeric entry. The last option in the Arrowhead tool allows you to custom edit the shape of the arrowhead (see Figure 4.48).

Note that the Border Width tool serves two purposes: on selected text it can add a border around the text, or, when on a selected line, it adjusts the thickness of the line or arrow. The color of this border is determined by the border color swatch at the bottom of the color swatches area.

To the right of the Arrowhead tool is a black box and a green letter *V* (see Figure 4.41). The green *V* can be toggled on and off to choose between seeing video behind the text (when the *V* is green) or seeing a solid color behind the text (when the *V* is not green). The color of the background can be edited by clicking on the black box. If you are using the background color as a way to more clearly see what you're typing while you're working on the title, make sure to change the background back to *V*.

Avid Title Tool Menu Bar

On a Mac, the menu bar for the Avid Title Tool is not part of the Avid Title Tool window but, rather, is at the top of the computer screen itself. On a PC, it is on top of the Title Tool itself. The menus are File, Edit, Object, Alignment, and Help. The File, Edit, and Help menus are all pretty standard for any application. The two menus with important options in them are the Object and Alignment menus.

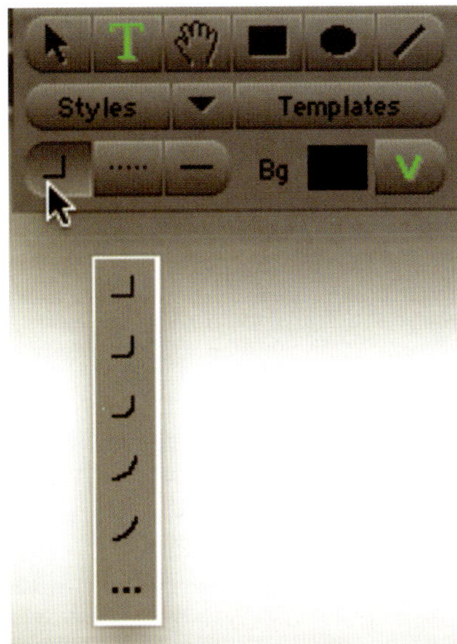

Fig. 4.45 This is the corner menu, allowing for rounded corners on boxes.

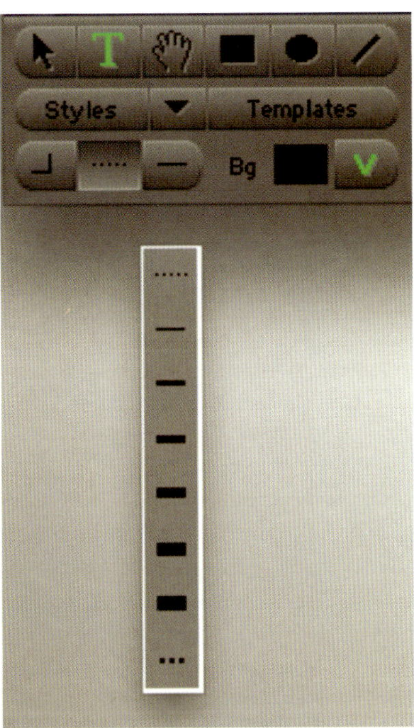

Fig. 4.46 This menu serves to select the thickness of both lines and borders.

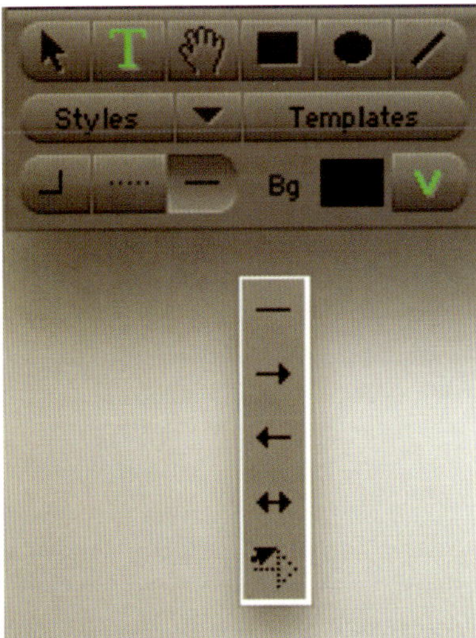

Fig. 4.47 This menu (the button with the solid line to the right of the button with five dots) allows you to create arrows.

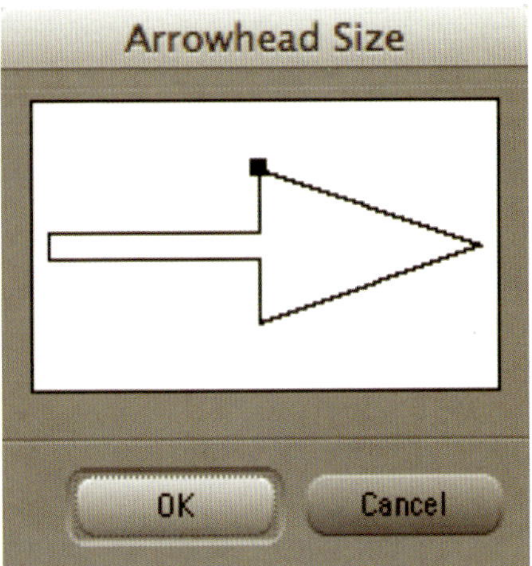

Fig. 4.48 The Arrowhead Size window allows you to create a custom shaped arrowhead and is accessible from the selection at the bottom of the Arrow menu. (Figure 4.47)

In the Object menu (see Figure 4.49), the Safe Title Area/Global Grid option is checked and operational by default, but if you want to see your Title without the Safe Title and Safe Action grid overlaid, this is where to turn it off.

The Preview option is unchecked and inactive by default. Turning Preview on allows you to have a better idea of what your final text will look like. Preview essentially just turns on anti-aliasing. It's off by default because it improves performance and responsiveness while you're working. Preview is sometimes good to turn on to check fine aspects of your composition, like thin borders or shadows, and also gives a better idea of the true kerning of letters.

If you have two title objects overlapping, for example, some text and a box, you can assign which object is "on top of" the other object using the Send to Front, Send to Back, Bring Forward, and Send Backward options.

One of the most useful options in the Object menu is Soften Shadow. If you have text or some other object that has a shadow, you can soften it with this option. Because there are no options for softening other objects, like boxes, for example, there's a great trick for creating a soft shadow behind text that just needs a little something subtle to help make it pop from a bright background. Instead of softening the box itself, you can soften its shadow. So create a box about the size that you want to have the soft box. Add a shadow to the box. Lower the opacity of the fill to 0. Lower the opacity of the shadow to a level that allows you to see the text but doesn't call attention to itself. Go to the Object menu and soften the shadow, possibly to the maximum level of 40. If you have some text on screen and the shadow is obscuring the text, use the Send To Back option in the Object menu to place the shadow box behind the text instead of in front of it.

In the Alignment menu (see Figure 4.50), you can choose to align, center, and distribute selected objects. To understand what these options do, create a series of words, with each word in a separate text box. Stack the words in a rough list from top to bottom and then, with the Select tool active, lasso or Command-A (Mac) or Control-A (PC) to select all of the words. Then use the Align menu choices to line them up neatly to the left or to the right. If they don't quite look aligned, it may be because the text inside each text box needs to also be justified left or right.

Drag out your row of text boxes so that they make a rough row from side to side. Then use the Align Objects Top or Align Objects Bottom options to make a neat row across. The Center in Frame Horiz and Center in Frame Vert options can be used to center the text boxes. As with Alignment, if the text doesn't look centered, remember that the text inside the box also has to be centered using the justification tools.

Object	Alignment	Help

Bring To Front	⇧⌘L
Send To Back	⇧⌘K
Bring Forward	
Send Backward	
Group	
UnGroup	
Lock	
Unlock	
Fast Save	
✓ Safe Title Area/Global Grid	
Preview	⇧⌘P
✓ Safe Colors	
Auto Size Mode	
Add Page	
Bold	⇧⌘B
Italic	⇧⌘I
Font Replacement	⇧⌘F
Make Crawl	
Move To Page	
Copy To Page	
Insert Page	
Remove Page	
Soften Shadow	⇧⌘H

Fig. 4.49 The Title Tool's Object Menu with the Soften Shadow option selected.

Alignment	Help

Align To Frame Left	
Align To Frame Right	
Align To Frame Top	
Align To Frame Bottom	⇧⌘Z
Center in Frame Horiz.	⇧⌘C
Center in Frame Vert.	
Align Objects Left	
Align Objects Right	
Align Objects Top	
Align Objects Bottom	
Distribute Left to Right	
Distribute Top to Bottom	
Distribute First to Last	
Show Alignment Grid	
Align to Grid	

Fig. 4.50 The Title Tool's Alignment menu with the Distribute Top and Bottom option selected.

The other useful Alignment option is distribute. This is good for creating *builds* of words. You can create a series of words or lines of text and use the distribute options to spread them out with neat, perfect spacing. I tend to set the top row where I want it and the bottom row where I want it and then use Distribute Top to Bottom to set the spacing between each row. Experiment with these options a bit to understand the behavior better.

GLOSSARY

Build: A text or title build is when a list of words "builds" onto the screen one line or word at a time, usually as the words are spoken in the audio.

Font Styles and Templates

Finally, also on the left side of the Title Tool are two pulldown menus for saving and selecting font styles and templates.

A quick tutorial: First create a custom look for some text. Maybe Trajan font, nonbold, yellow with a drop shadow. With your text looking the way you want, click the downward-facing arrow next to Styles. This calls up a pulldown menu (see Figure 4.51) that allows you to choose from previously saved styles or choose Save As to save a new style. Choosing Save As opens the Title Style Sheet dialog box.

To save this as a Style, you can simply click Done. But you can choose to rename the style in the text box that defaults to the simple name of the font. You can also choose to have the style apply only to certain parameters of the font. For example, by unclicking the check box for Font, you could apply this to any text and it would not change the font of the selected text, but it would change all of the other parameters, like justification and color or shadow. Though Avid does not allow you to have a default font style, you can save a style here and call it up quickly.

The Templates function is similar but includes positioning information and allows for multiple Title Tool elements to be saved and formatted instead of just a style of font. To try this out, create a quick lower third style with a name in one size and a descriptor line of text in another font and size below it. Perhaps add a rule or line to separate the two lines of text and maybe a box behind one of the text rows to act as an anchor. When you have it ready to go, click on the Templates pulldown menu and select the Save Template option (see Figure 4.52). This calls up a dialog box that allows you to save the template with a name. Click Save when done.

Now, delete all of the elements of the title so that you're starting fresh. Then, click on the Templates pulldown again and choose Include Template. This calls up a dialog box that allows you to select the template you just created. You can now edit the text of this template and save the resulting title.

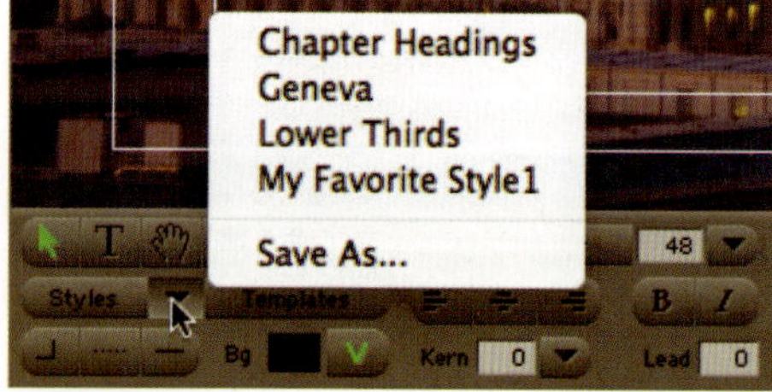

Fig. 4.51 The Styles menu in the Title Tool allows you to save Styles and recall previous Styles.

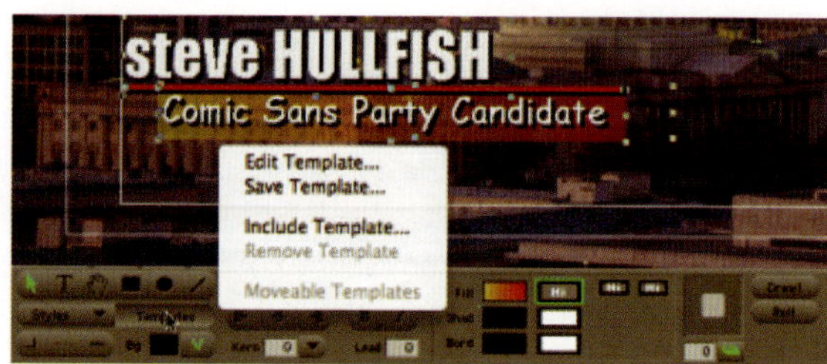

Fig. 4.52 The Templates menu in the Title Tool allows you to save Templates and recall previous Templates.

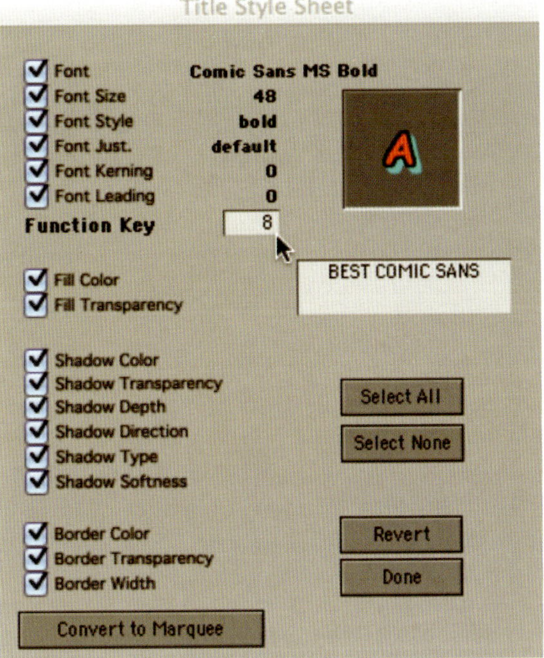

Fig. 4.53 The Title Tool allows you to save type styles using this window. Note the arrow (cursor) pointing to the number 8. This allows you to assign a function key to the style so that you can apply it to any text by simply pressing the corresponding function key.

Rolls and Crawls in the Title Tool

To the right of the shadow controls are two buttons that create rolls and crawls. Clicking on the Roll button adds a scroll bar to the Title Tool, allowing you to navigate up and down through your roll (like for credits). Clicking on Crawl provides a scroll bar for navigating horizontally within your crawl. While it is not possible to import text into the Title Tool, it is possible to copy text and paste it into the Title Tool.

Positioning and editing text and other elements inside of a roll or crawl works in a way that is similar to regular Titles.

Timing and animating a crawl is a little less than intuitive. Select a large body of text from a text document or even an e-mail, and paste it into the Title Tool. Click on the Roll button at the bottom right of the Title Tool and then Command-S (Mac) or Control-S (PC) to save the title. In the resulting dialog box, name it and (if you want to) choose a bin where the roll will be saved, a hard drive where the media will be rendered, and a resolution or media type from the pulldown menus. Click on the Save button. In the bin, you'll see a clip with an RT icon and the title "Rolling

TIP

Notice that in the Title Style Sheet there is a Function key option with a small numeric text box (see Figure 4.53). You can type a number into this box to instantly access your font styles by pressing on the Function keys while in the Title Tool. This does not affect any other Function key functions when using the Avid editing tools. If you ever operated an old Chyron text generator in the old days of TV, this is very similar. You can have some text selected and just press each of the function keys that has a mapped style on it to see what that text would look like with that text style. These styles could be very complete, with color and shadow parameters set, or you could use it simply to step through your favorite dozen or so fonts in rapid succession.

Title: (the name you gave it)." This should automatically be loaded into the Source window, but if it isn't drag the file to the Source window.

To see the text, scroll through the Source Timeline with the mouse. The rolling title needs to be rendered, so only scrolling or pausing on the rolling title will reveal it until it is rendered. The length of your roll is automatically set to five minutes. Determining a roll speed or a length for a roll is done by editing the entire duration of the rolling title in the Source into the desired length of time on the Sequence side. So if you know you need to do a thirty-second roll, simply mark a thirty-second span of time on the Sequence side and Overwrite the clip into the Sequence. If you want to key the rolling title over video on V1, make sure to patch the Source up to V2 on the Sequence Patch panel.

GLOSSARY

Leading: Leading is the vertical spacing between lines or rows. It is pronounced to rhyme with "heading" not "needing." The name comes from the thin lead metal spacing strips that printers used to place between lines of text.

If the roll is too fast or too slow, the only way to correct it is to change the length of the roll or to make the text in the roll smaller or reduce the leading (space between lines). If you'd like to have the first line of text in the roll to begin frozen on screen, then start to roll and then freeze on the last text, you need to use the Effect Editor. In the Effect Editor, position your text on or off screen and add a keyframe. For example, to start a text roll on screen, position it so that the starting text is in the position you want and add a keyframe. Add another keyframe at the point where you want the text to begin to roll. If you want the roll to "hold" on screen until it fades off at the end, add a keyframe several seconds before the end of the roll and position the roll so that the correct information is on screen.

Crawls work similarly to rolls.

To edit a title that is already in the sequence—perhaps to fix a misspelling—you can right-click the effect in the Timeline to go in to the Title Tool and select Edit Title from the contextual pulldown menu. You can also select a title in the sequence and click on the Effect Editor. In the Effect Editor, there is a button called Edit Title. Clicking on that button will launch Title Tool. Make your changes and save. This will load the new title into the Source monitor. It will also automatically make the change in the title that's already in the sequence, so there's no need to edit the revised title into the Timeline.

Marquee

Marquee is a hidden gem of the Avid universe that is rarely exploited by editors because they find it intimidating. Marquee is a very full-featured 3D animated titler (see Figure 4.54).

I was one of the first people to start playing with Marquee back when it was part of an early uncompressed on-line editing solution from Avid called Fusion/Illusion, which was developed by the SoftImage division of Avid. Marquee was also the character generator for the Avid|DS, which came from the SoftImage division.

The Marquee instruction manual is nearly 600 pages long. If you want a thorough understanding of how to use all of the power of Marquee, that

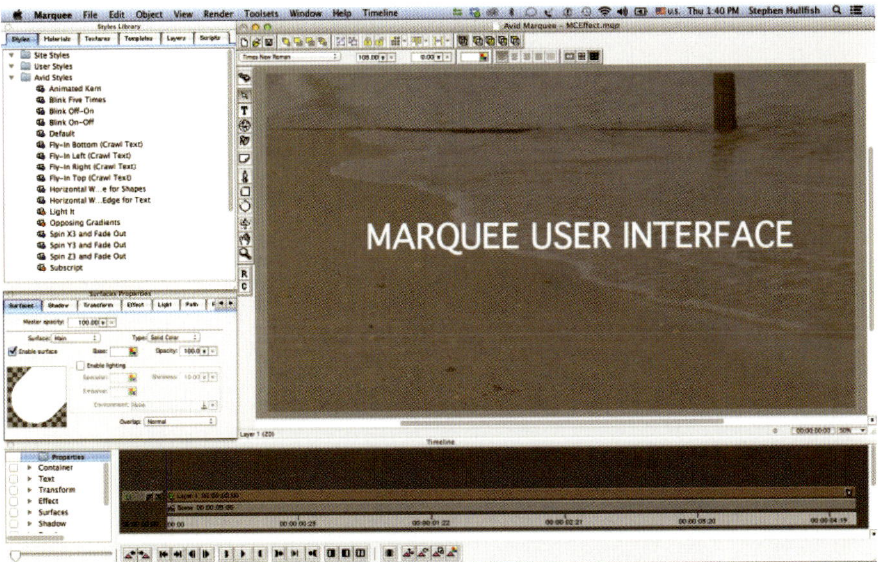

Fig. 4.54 This is the Marquee titler. It is a 3D titling tool with extensive animation capabilities and a learning curve to match. Basic unanimated titles, however, are quite simple to create. Your view may be slightly different depending on which tools or view are active.

is the place to turn. But to help you get started and begin to wrap your head around some of the value of Marquee, I'll cover the basics here.

Launch Marquee the same ways you launch the regular Avid Title Tool: Clip > New Title or Tools > Title Tool Application. When the dialog box comes up, choose Marquee instead of Title Tool. Checking the Persist box will let you bypass this intermediate step, but then you won't be able to access the basic Title Tool again until you relaunch Media Composer.

First off, Marquee is highly configurable, as is Media Composer itself. So to get us all working on the same page, we need to make sure our settings are the same so that we're looking at the same things.

The primary way to customize the interface is using the Toolsets menu in the menu bar. For the introduction to Marquee, let's switch over to Big Monitor to reduce some of the clutter. So, in the menu bar go to Toolsets > Big Monitor (see Figure 4.55). Then, so we can see all of our options, go to View > Toolbars and select all of the Toolbar options (place a check next to each one; see Figure 4.56); do the same in View > Toolbox > Full (see Figure 4.57). The Toolbars options add a series of icons along the top of the window, and the Toolbox > Full option adds a series of icons along the left side of the window.

Discussing these tools will get you familiar with about 80 percent of what you need to work confidently in Marquee.

Along the top of the toolbar, from left to right (see Figure 4.58): The first set of icons contains the typical file-based commands for New, Open, and Save. You could probably turn these off, since they all have keyboard shortcuts that are well known. The interesting thing about the Marquee

TIP

Anytime you pull down a menu option in Media Composer, Marquee or the Title Tool, note the keyboard shortcuts that are indicated next to many of the menu options. As you can see in Figure 4.55, the Big Monitor option is also accessible by pressing the F3 or Function 3 key.

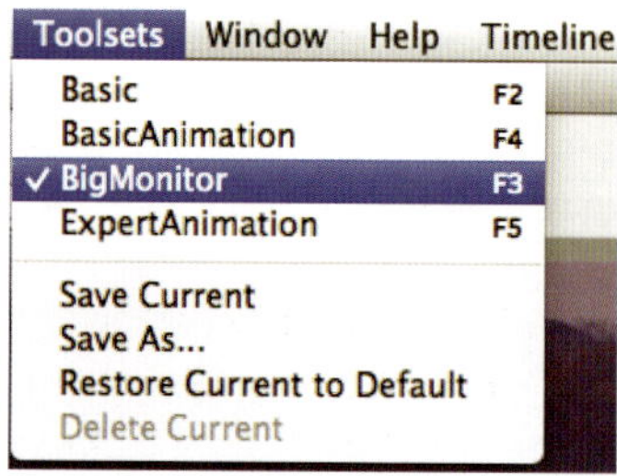

Fig. 4.55 This is the Toolsets menu of Marquee.

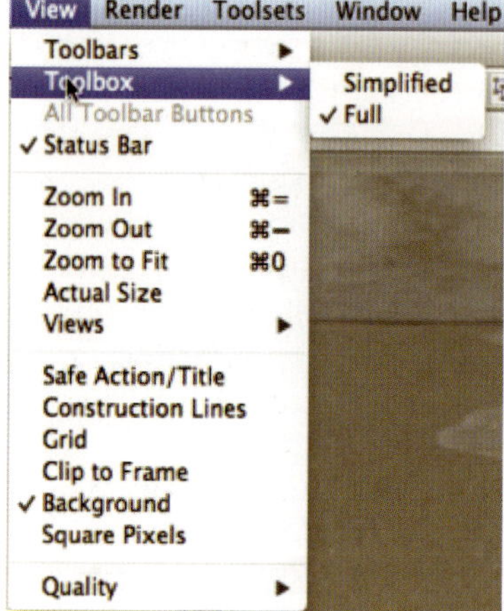

Fig. 4.56 The View menu of Marquee. Note that viewing a Simplified or Full set of Tools in the Toolbox can be chosen.

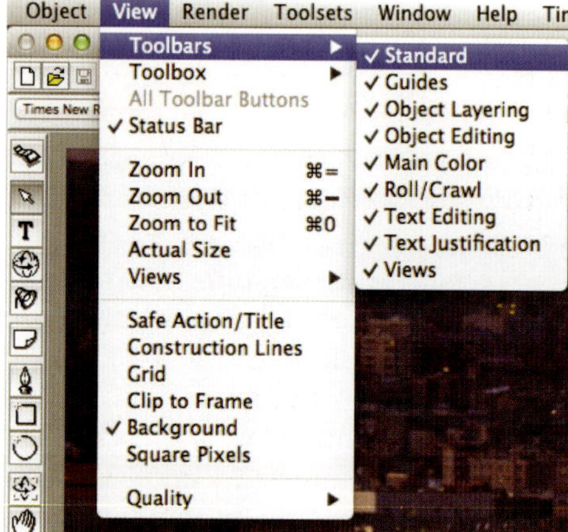

Fig. 4.57 The View Menu in Marquee. Note that the Toolbar can be configured with many options.

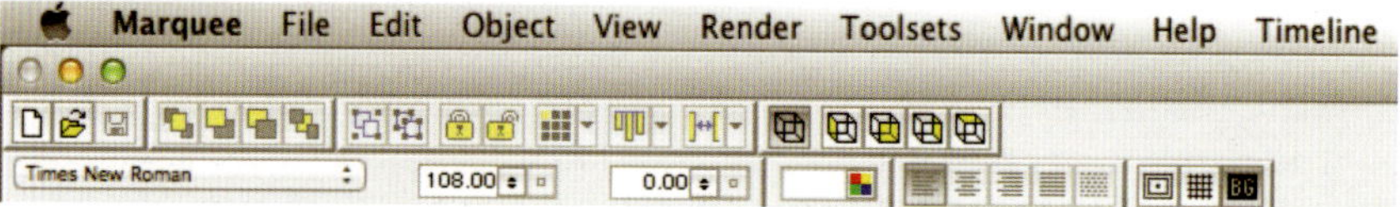

Fig. 4.58 Marquee titler toolbar.

toolbar is that the position of all of the tools is fixed, so if you hide a series of commands in the toolbar, the others don't slide to the left to fill the void. Instead each open or active set of commands is always loaded to the same spot in the toolbar. The next set of controls (overlapping gray and yellow boxes in the Toolbar), has to do with layering: Bring to Front, Bring Forward, Send to Back, Send Backward. If you want an object to appear in front of or behind another object, you can use these commands to do it. Another interesting thing about Marquee is that there isn't any kind of "object priority." Because it is a 3D application, the priority *should* be controlled by the position that the object is in 3D space, but Marquee actually uses layering *and* 3D special positioning to orient objects, which can sometimes cause some interesting "Escher-esque" anomalies.

The next two icons (looking like lassoed marquee boxes) allow you to group and ungroup objects. The following two icons (looking like open and closed padlocks), obviously, allow you to lock and unlock objects. The icon that looks like a grid of eight gray squares with a yellow square in the top left allows you to position (center or justify) objects. The pulldown menu to the right of it provides the options for positioning. The icon that looks like a bar graph has a pulldown menu that is used to align objects. The next icon with the arrows between two yellow bars is the Distribute icon. Using its pulldown menu, you can determine the numerous ways that objects can be evenly distributed horizontally, vertically, or even diagonally. The top row ends with five 3D boxes that allow you to look at your title (known as a "scene" in Marquee) from various 3D angles. The first box is the Scene View. The next cubes offer views from left, front, right, and top. I prefer to use one of the tools in the Toolbox, called the Tumble tool, to get an intuitive and interactive view of 3D space within the title, but we'll discuss that later.

The next row of tools in the toolbar (Figure 4.58) is for selecting the font and font size and for kerning the text. The size and kerning tools can be adjusted numerically or by clicking on the double arrow icon to the right and mousing up and down or left and right. The small square next to the numeric adjustment arrows is the reset button. (This is incorrectly identified in Avid's Marquee user manual currently.) To the right of the font section is a color swatch and adjustment button to change the color picker. To the right of the color picker are the alignment tools for justifying the text within text boxes: left, center, right, full, and equally spaced justification. Finally, there are the guide icons for setting up Safe Title/Safe Action and Grids. The final icon to the right (BG) allows you to change the Background from the default, which keys the title over the currently displayed video frame from Media Composer, or a colored background.

Since the space I can devote to Marquee is limited, I will assume that a simple description of these tools is enough to get most users moving along the path to actually using them, since they are very similar to Microsoft Word or other text editors with which most people are familiar.

The toolbox down the left side of the Monitor window (see Figure 4.59) has some tools that may require some additional explanation, however.

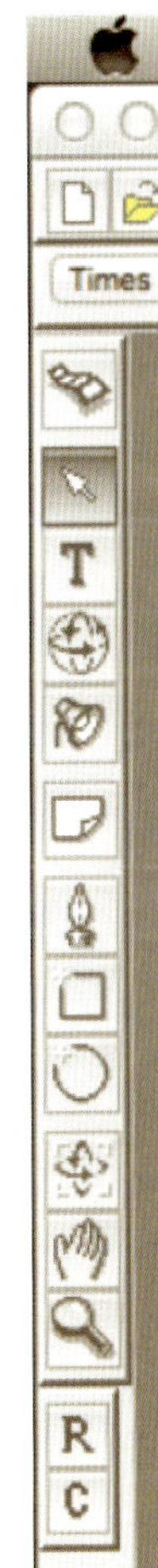

Fig. 4.59 Marquee titler toolbox.

Animation: The wavy film icon at the top of the toolbox is the Animation icon. If you don't see this, then go to the Toolset menu and switch to Basic Animation or Expert Animation. The Toolset you are using will have an effect on the icons you see in the toolbox. To follow along more easily with my descriptions below, switch over to the Basic Animation Toolset. If you are creating static titles, there is no need to enable the Animation button. If this button is red, it is active and you are creating an animated title. If it is gray, then you're creating a static title.

Selection Tool: The next tool (the arrow icon) is the Selection tool and is fairly common to anyone with any experience in most Adobe or Apple products. This tool lets you select, move, and manipulate objects.

Text Tool: Shown as a *T*,this icon allows you to type and edit text.

Rotate Tool: The globe-with-arrows icon, the Rotate tool, allows you to select objects for 3D rotation.

Light Tool: The lamp icon, the Light tool, lets you add and edit lights. Most of the manipulation of lights is done either from contextual menus that you can call up by right-clicking the lights in the "scene" or by selecting Windows > Properties > Light from the menu bar. Remember that Marquee is a 3D tool, and a lot of the "look" of 3D objects and animation is achieved through the lighting.

Page Tool: The next tool (looking like a rectangle with a dog-eared bottom right corner) is the Page Tool. A deep explanation of this will have to wait, but basically, you can create pages, almost like PowerPoint slides inside a "deck" of pages that can animate or transition between each other. If you are creating a single static page, you can ignore this tool for now, but to create a really cool series of animated pages, this is the thing to use.

Shape Tool: The pen icon allows you to draw various free-form shapes and includes the ability to create Bezier curved lines.

Rectangle Tool: The square icon with a plus sign allows you to make squares and rectangles. As with many computer programs, holding the shift key while using this tool will restrain the shape to a perfect square.

Ellipse Tool: The circle icon with a plus sign allows you to draw circles and ovals. As with the Rectangle tool, holding the shift key will constrain the shape to a perfect circle.

Tumble Tool: The Tumble Tool, which looks similar to the Rotate tool, does not manipulate the actual objects or the camera in 3D space, but it allows you to change your *3D view* of the 3D scene, so as to better understand the perspective and position objects within the 3D space.

Pan Tool: Like the Tumble tool, the Pan tool (hand icon) does not actually change the position of objects but merely allows you to move and view scenes when they are zoomed in on or if you need to see something that is off-screen.

Zoom Tool: The magnifying glass icon allows you to zoom in and out of a scene. When you are zoomed in, you may need to use the Pan tool to see various parts of the zoomed-in image.

Roll and Crawl Tools: Finally, the Roll and Crawl Tools (R and C icons) allow you create rolls and crawls.

For Mac and Windows users, there is a slight visual difference between the Marquee displays. On a Mac, the menu bar is at the top of the actual computer screen, like most menu bars for other apps. But for PCs, the menu bar is actually part of the Monitor window.

When you make a title, the next thing to understand is that Marquee acts a little bit like Photoshop in that it can save the files as an application-specific file—.psd for Photoshop and .mqp for Marquee—and as files that are more easily used by other applications, like JPEGs, PNGs, or TIFFs. It is a good idea to save each of your titles in Marquee's proprietary .mqp format so that you can edit them later if needed. However, it is also necessary to save them for use in Media Composer. To do that, go to the menu bar and choose File > Save to Bin. This also has a keyboard command that might be good to learn: Command-B (Mac) or Control-B (PC). Using the Save to Bin command jumps you from the Marquee application back to Media Composer. If you're going to do multiple titles, you might want to save each title in the .mqp format until you're ready to send them all over to the Media Composer. When the .mqp files are rendered to be sent to Media Composer, they create temporary TIFF files. Animated titles create multiple TIFFs depending on the length of the animation. Then Media Composer imports these TIFFs as title media, converting them into Avid's native .mxf files, complete with alpha. This title media includes metadata that allows Media Composer and Marquee to know which .mqp file needs to be called up to edit in Marquee. Rolling titles can be saved in two different ways. If you save them as Avid Rolling Titles, then you can adjust the length from within Media Composer, just as you can with rolling titles created from the Avid Title Tool. Or you can save them as an Animated Title, which takes longer to render and is of a fixed length. For Marquee to create an Avid Rolling Title, there are many requirements that the title must meet (see the Marquee user manual).

On a laptop, the Marquee interface takes up the entire screen. To switch back and forth between apps, you can use Command-Tab (Mac) or Alt-Tab (PC).

When Marquee Titles are cut into the Timeline, revising them is done with the same method as revising a standard Avid Title Tool title. Go into Effect mode on the title and choose the Edit Title check box in the resulting Effect Editor window. This will launch Marquee. Edit the title and choose File > Save to Bin. The edited title will be back in the Source window and will also be changed in the sequence.

If you save titles without sending them to the bin, the saved titles are included in a list that is saved in the Window menu. You can recall any of the saved titles from the list and save them to bins from there. You can also open previous titles from File > Open or File > Open Recent.

To get a sense of some of the other ways to modify and view titles, we need to look at the various windows. The options in the Window menu are Timeline, Layers, Properties, Library, Quick Titles, Scripts, and

Templates (see Figure 4.60). These windows provide a wide variety of options. Timeline allows you to navigate through animated titles. Layers provides a list of Layers so you can select, add, rearrange, delete, and copy layers. The Properties option in the Window menu offers a submenu of property windows (note the submenu in Figure 4.60): Info, Text, Transform, Effect, Surfaces, Shadow, Render, Light, Path, and DVE. These property windows provide the controls to adjust the parameters of objects in the title. The Library option has submenus that include Styles, Materials, and Textures. Calling up any of those Library options brings up a single window with a series of tabs, including Styles, Materials, Textures, and Templates, Layers and Scripts. Of those tabs, Templates has a number of choices that we should explore so that you can see the potential of what Marquee can do.

In the Templates tab is a series of three top-level folders that each contain multiple subfolders (see Figure 4.61). The Site Template is empty. It can contain templates that are saved on this specific editing computer by any user on the computer. The User folder is empty, too, but can contain templates saved by the specific User (you). Finally, there is an Avid Templates folder that contains these subfolders: Backdrops, Crawling Titles, Demos, Lines and Arrowheads, Lists, LowerThirds, RollingTitles, plus two other titles—finalcredits and Quote. Spend some time going through these templates, especially the Demos templates. The MarqueeSlogan, newsLogo,

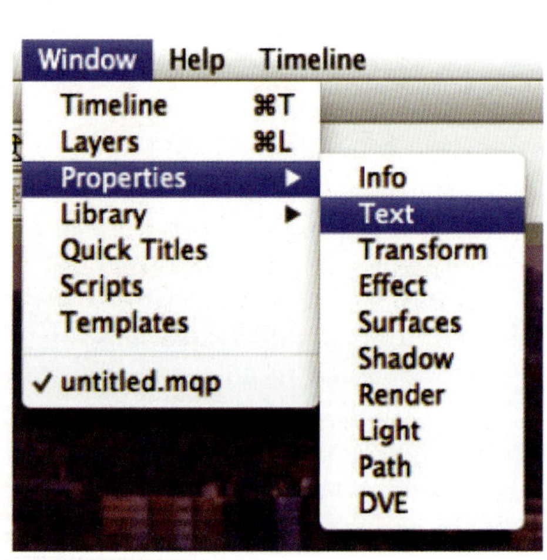

Fig. 4.60 Marquee Window menu.

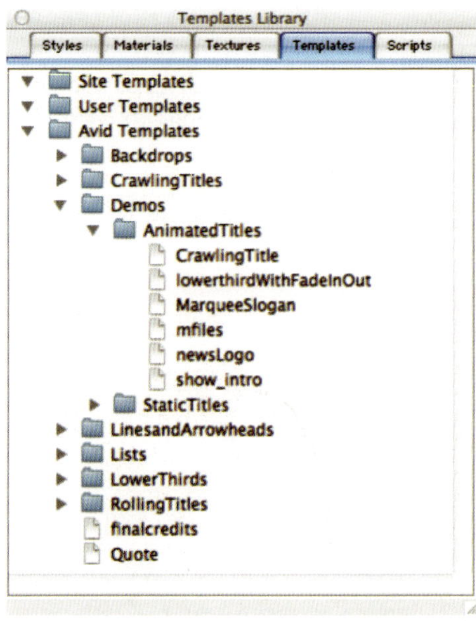

Fig. 4.61 The Templates tab allows access to Avid pre-made templates, which are a good way to see the possibilities of using Marquee for animated titles.

and show_intro templates will give you a sense of what's possible. It really would be nice for Avid, or someone, to create a set of templates that were really well designed. About a decade ago, Steve Bayes, who was a product designer at Avid and is now at Apple as product manager of FCP-X, created a training program for Marquee that included a bunch of professionally designed templates. It was released by Bayes' training company, called ArtFXual. The company's website is no longer active.

Third-Party Plug-Ins

NewBlue Titler Pro

If Marquee isn't your cup of tea and you aren't thrilled with the Title Tool either, you can check out the trial of NewBlue Titler Pro (see Figure 4.62). It's an AVX plug-in that you'll find in the Effect Palette. Drag the Titler Pro effect from the NewBlue Titler Pro column onto some filler in the Timeline and then go into Effect mode on the effect. In the Effect Editor, choose Launch User Interface. This calls up the NewBlue user interface. To me, this is a strange intermediary level of titler between Marquee and the Avid Title Tool. The main point of providing it for "free" seems to be in selling you a number of templates and

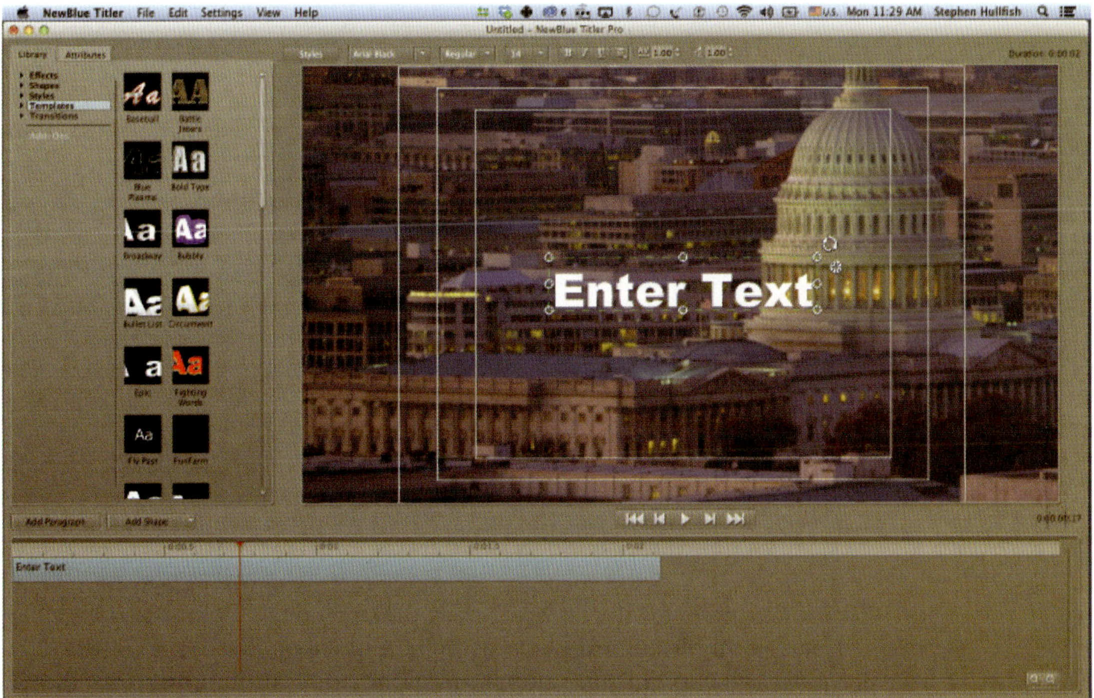

Fig. 4.62 NewBlue Titler Pro.

packs of effects. In the Library tab of the NewBlue UI, you can access a variety of freebie templates, but in each pulldown menu for Effects, Shapes, Styles, Templates, and Transitions, there is a Get More option that launches a web browser taking you to the NewBlue website allowing you to buy premade title templates. I wasn't impressed by any of them, but they may be just what you're looking for.

Boris

Boris has always been one of the premiere effects packages for Avid. Boris RED was essentially "Avid-ized" and provided for use as a serious effects package within the Media Composer UI. The other Boris package that is truly valuable is Boris Continuum, or BCC, which is basically a collection of filters to generate different looks.

Avid FX

Avid FX should install with the Media Composer program. Although it is an Avid-branded effects package, it is an AVX style effect. (See AVX effects explanation in paragraph three of this chapter.)

Avid FX launches from the Avid Effect Palette like all other effects and is usually at the top of the effects column. The three associated effects in the Avid FX category are Avid FX Filter, Avid FX Title-Matte Effect, and Avid FX Transition.

Avid FX Filter is the option you want to add a filter to a clip, but it does so much more. Basically it is for all of the other work that the other two Avid FX don't do. Avid FX Title-Matte is used for titles and basically anything with an alpha channel. Avid FX Transitions is, obviously, for transitions.

Avid FX Transitions

Let's start with transition effects (see Figure 4.63). The first thing to understand is that when you drop an Avid FX Transition effect on a transition and go into the Avid FX UI, the Timeline may be confusing without an understanding that the transition is displayed differently within Avid FX. The transition has been transformed into a short Timeline with the outgoing (first shot) on V1 and the incoming (second shot) on V2. The length of the transition is represented by the full length of Timeline.

First off, depending on your resolution, the monitor window in Avid FX may be zoomed in too much. Use the minus-magnifying-glass icon below the monitor to scale the view properly.

The various parameters of the two tracks can be had from the Controls window to the left of the main monitor window, but you can also

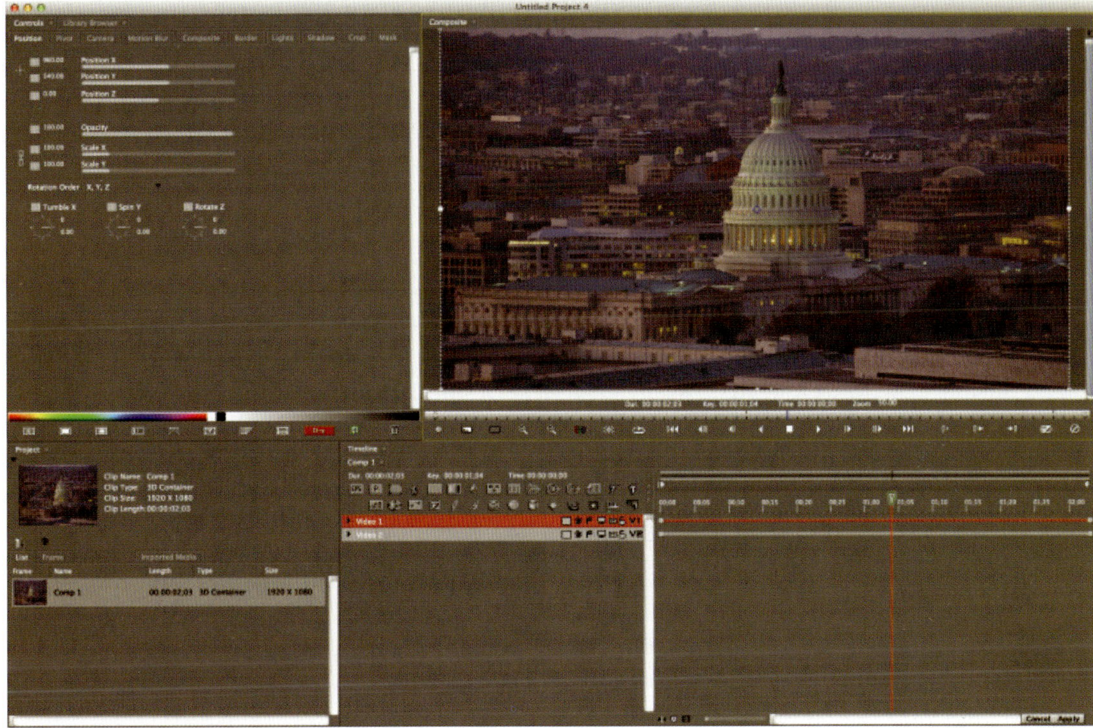

Fig. 4.63 Avid FX interface. This is essentially an "Avid-ized" version of Boris RED.

find almost all of these controls in the Timeline by spinning down each track by clicking the little black arrow to the far left of each track in the Timeline.

Before we build a transition from scratch, let's check out some of the preset RED transitions that come free with the system. There are also several companies that sell premade Avid FX effects sets. Some of these are in the Avid Marketplace.

For drag-and-drop ease, you'll want to go to the menu bar to Window > Library Browser. This is also accessible with the keyboard shortcut of Command-9 (Mac) or Control-9 (PC). Before you do this, though, you should realize that the Library Browser (or any other window) will get loaded as a tab in the currently selected pane (window) of Avid FX. So, if the Timeline is selected (thin yellow outline around the pane), then that's where the Library will be loaded. I like to load the Library to the same pane as the Controls pane.

Before we get to the drag-and-drop ease of these presets, we have some preliminary work to do, namely, adding thumbnails. In the Library window, there is a small icon that looks like a grid of four squares with a plus

TIP

The Avid Marketplace can be accessed from the Media Composer menu bar. On the right side, just left of the Help menu, is the Marketplace menu. It allows you to purchase stock footage, video and audio plugins, AMA plugins and even training. Avid FX presets are available in the Video Plugins section and include effects from EyeScream Factory, Kevin McAuliffe, Final Effects and others.

sign next to it. This is the Generate Thumbnails icon (see Figure 4.64). You can choose to create thumbnails of these effects on a per-category basis or, if you have some time to spare, load the entire library. These thumbnails will be created using whatever video image is loaded up from the Media Composer Timeline. It actually makes a bit more sense to load only thumbnails for a specific category or subcategory of effects that you want to look at. Some thumbnails, like blurs and particles, take a lot longer to generate. With the install of Avid FX that I had in 2013, there were two groups of presets: Avid FX presets and Boris RED presets. Just loading the Avid presets as thumbnails on a relatively fast Mac took over forty-five minutes. The Avid presets include one hundred categories of effects, with 10–60 effects in each category, and each effect is rendered with ten frames to show motion, so that's several thousand frames to render. Thumbnails for most categories can be seen in a minute or less. In the RED presets, there is a category of Dissolves with about twenty-four effects in it. Rendering out the thumbnails for that took about two and a half minutes.

With the thumbnails rendered, you can preview the effect in the thumbnail, which gives a vague indication of what it will look like (see Figure 4.65), or simply double-click the effect to load it into the Timeline using the media being fed from the Media Composer. A dialog will pop up asking if you want to save the current Timeline effect. If you are trying a series of transitions or filters, you don't have to save any of them. Really,

Fig. 4.64 Inside the Avid FX Library Browser is an icon that allows you to generate animated thumbnails to preview effects. The icon is indicated here by the black arrow icon/cursor.

the only time to respond to this dialog box with anything other than Don't Save is if you have modified an existing preset and you want to save it with a special name for later use.

Stunningly, in checking out these preset collections from Avid and Boris, I have found almost none that I'd consider using. The closest to something I'd actually use on a project is the Blur Dissolve in the RED collection of Transitions > Dissolves.

Let's make a transition effect from scratch and save it as a reuseable preset. The effect I'm thinking of is a kind of glowing, blurring dissolve. Which of these you decide to do first is really up to you. You could add the blurs and glows first and then make the transition between them with the dissolve at the end, but it makes more sense to understand the transition if we start with the dissolve. To do that, we're just going to adjust the opacity on the top track, video1, starting at opaque and ending at transparent. The Avid FX Timeline names its tracks the opposite of Media

TIP

I purchased a series of 100 Essential Transitions from EyeScream Factory from the BorisFX.com website and was impressed by the quality and usability of the transitions. They were $49.95 USD. I received a PDF with download and install instructions. The installer properly placed them in the correct folder, and I didn't even have to relaunch Media Composer to see them. I closed out of Avid FX, and when I relaunched Avid FX the EyeScream files were in a new folder in the Library Browser: Boris FX 10 Keyframe Library > Transitions > EyeScream Factory > 100 Essential Transitions.

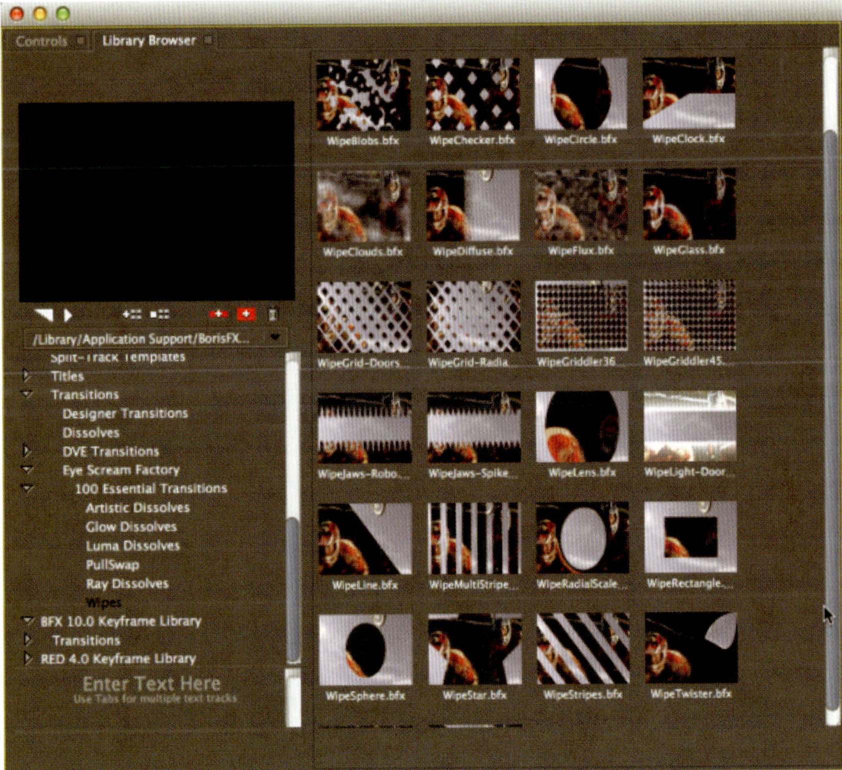

Fig. 4.65 This is part of the Avid FX UI. It shows the Keyframe library where various pre-made or purchased effects are stored. Note the Eye Scream Factory transitions on the left. These are effects I purchased. Thumbnails of the effects in the category are shown to the right.

Video Tutorial

This tutorial is a little difficult to show with static screenshots, so I created a video tutorial on this book's companion website, www.focalpress.com/cw/hullfish.

Composer, but the behavior is the same. So you look down through the top track to the track below it.

Click on the first keyframe (little gray circle on the far left of the top track) to select it, and it will turn red. There are two ways to adjust the opacity. One is by changing the opacity in the Controls window. Or you can spin-down the track with the black triangle on the left side of the track. Next, spin-down Transformations. Then, spin-down Position (having opacity in the Position category is a little unintuitive). Notice that the opacity is at 100. Then, select the last keyframe in the track and turn the opacity to 0. You can adjust the controls by clicking on the number in the parameter and dragging left/right or by entering a numeric value. With the opacity at the end frame set to 0, you can scrub through the Timeline or play the Timeline and see that a dissolve has been created from the first (top) clip to the second (bottom) clip.

Well, that's boring, and it can be done ten times faster in Media Composer. Now, let's add a ramped change in blur between the two clips. This could also be done in Media Composer, but not as easily as in Avid FX. Just as we selected the keyframes to create the dissolve, we'll do the same thing but in both tracks in opposite directions, so the top track will start "clean" and become blurry at the same time that the bottom track starts blurry and becomes clean. Select the top track, and then in the menu bar choose Filters > Effects > BCC Glow. This adds the filter to that track. Then select the lower track and add the filter to it as well. Instead of drilling all the way down through the Filters submenus, the top choice in the Filters pulldown menu should be the last effect that was selected.

On the first keyframe of the glow effect in the top track, move the Glow Threshold in the Control window (or in the Timeline Glow track) to 100. Then on the last keyframe move the threshold to 0. Because we added the opacity parameter first, it's hard to see the glow because we have dissolved to video track 2 completely by the last keyframe on video track 1. Then, on video track 2, apply the same keyframe parameters but in reverse, with the first keyframe having 0 Glow Threshold and the last keyframe having 100 Glow Threshold. Obviously you can also adjust the other parameters of glow, like radius, aspect ratio and intensity to tweak the effect to your liking. To get a better sense of the blurs and glow in the effect, you may want to change the dissolve effect so that—for a 30 frame transition—the dissolve starts 10 frames in and only lasts for ten frames.

It's also possible to execute a similar blur and glow without having to do the effect twice (once for each track) by deselecting both tracks and applying the Blur and Glow effects. With both tracks deselected, the Blur and Glow are applied above the two tracks, which applies the effect to both tracks. Because the effects are affecting both tracks though, your keyframes have to be applied differently. There must be three keyframes in this situation, with Blur and Glow parameters appled so that there's no effect on the first and last keyframes. Then set a keyframe in the middle and crank up the Glow and Blur on the middle keyframe. This creates a slightly different look. Which method you choose in the future is up to you.

You can save this effect as an Avid FX project. You can also save a copy directly to the Library from the File menu. With the effect saved, choose Apply in the bottom right corner of the Avid FX window to apply the changes and move back to Media Composer. Render the effect when you get back to Media Composer.

The easiest way to apply this custom effect to future transitions is to select the transition in the Media Composer Timeline and go into Effect mode. Drag the small purple effect icon to the left of the words *Avid FX Transition* to a bin. Give a memorable name to this effect in the bin.

Avid FX Title-Matte Effect

From the project files, import the Verascope Pictures logo from the project files on the book's companion website (www.focalpress.com/cw/hullfish). In the Import settings, make sure to check the Options, and in the resulting Import Settings dialog box, in the Image tab, in the Alpha Channel section, choose Invert on Import. This will bring the file in as a matte effect. Cut the Verascope logo in a sequence on video track 2 above some video on video track 1. If you prefer, you can just create a simple title using the Title Tool and cut that onto V2 instead of the matte effect.

Video Tutorial

This effect is also difficult to show with static screenshots, so I created a video tutorial to walk through this on the book's companion website, www.focalpress.com/cw/hullfish.

As I explained in the section "Promoting Effects to 3D," if you want to add 3D to a matte key or title, you don't drag a 3D effect onto the matte or title; rather, you promote the effect to 3D. However, with Avid FX, you do drag the effect onto the matte or title. If you want to use Avid FX on a matte or title, drag the Avid FX Title-Matte Effect onto the segment in the timeline. Go into Effect mode on the effect and select Launch User Interface.

On video track 1, add Filter > Effects > BCC Glow Alpha Edges. In the Controls window, change the color of the glow to a nice warm yellow. Next, change the Width amount of the Glow Alpha Edges on the first and last keyframe to 0. Then, add a keyframe in the middle and bring the width parameter up to 50. Add a keyframe on the Video track, about ten or fifteen frames from the beginning and another ten or fifteen frames from the end of the matte or title effect. Then, go to the first keyframe and change the Transformation > Position > Position X parameter to slide the title off-screen by entering a negative number, approximately −600.

Finally, on the last keyframe, do the same thing, but with a positive number. This seems to be nothing that you couldn't do more easily by promoting the title to 3D within Media Composer, but the difference is that Avid FX provides you with the opportunity to add motion blur to make the effect look much more organic and real. To add the motion blur, click on the Motion Blur tab of the Controls window. Check the Enable Motion Blur check box at the top of the Motion Blur tab and change the shutter angle from 180 to something higher, like 360. Also, since you want the motion blur to be the same throughout the effect, set the Interpolation to Constant. The Interpolation icon kind of looks like the pi symbol or like a Bezier curve. It's the icon just to the left of the shutter angle number in the Controls window.

Obviously, there's a lot more you can explore with the Title-Matte effect, but this tutorial should at least leave you feeling comfortable with the interface.

Avid FX Filter Effect

With titles, mattes, and transitions covered, everything else is covered by the Avid FX Filter effect. If you drop an Avid FX Filter effect onto a V1 clip in the Timeline, what may surprise you is that once you go into Effect mode and launch the UI, you will find that the Avid FX Timeline has two tracks. The top track is the clip on which you dropped the Avid FX, and the lower track is basically filler. In reality, this is the way Media Composer works too. Lower the opacity of an effect on V1, and the filler below it is revealed.

One of the main things that the Filter Effect can provide is the ability to composite footage with special composite modes like Screen, Additive, Multiply, Lighten, Darken, Soft Light, and others. Composite modes were an important part of FCP's Effects capabilities and are available in many other NLEs and effects apps. Avid FX is the only current way to access Composite modes unless you use another 3rd party AVX program, like GenArts' Sapphire (see Figure 4.66).

Boris RED (Avid FX) is an incredibly deep program. The Avid FX user manual is hundreds of pages long. One of the ways to explore the capabilities of Avid FX is to experiment with the many preset effects that can be applied from the Library Browser. Also, there are many presets from Boris and other third parties that can be purchased. These presets include charts, graphs, lower thirds, other titles, and looks. A simple web search of Avid FX or Boris RED presets will locate the possibilities. They can also be purchased from the Media Composer menu bar, under Marketplace > Video Plug-ins (Figure 4.67).

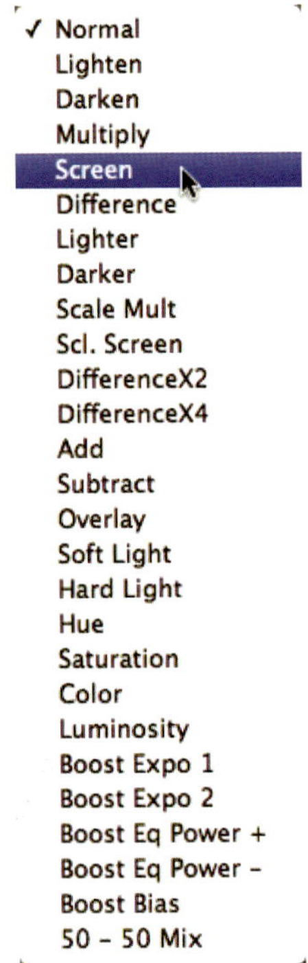

Fig. 4.66 One of the ways to get the FCP functionality of Composite modes is to use Avid FX.

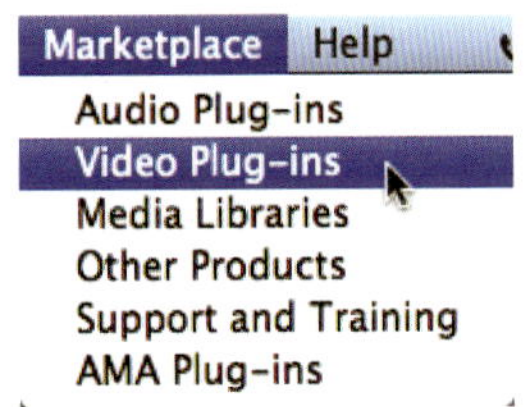

Fig. 4.67 Avid Marketplace.

GenArts Sapphire

Without a doubt, my favorite effects package is GenArts Sapphire. These are very high-end effects at a high-end price. I would highly recommend saving up or figuring out a project or two that can fund and justify the purchase. Most of the effects are beautiful and the default parameters are fairly well chosen to deliver a usable drag and drop solution without too much initial tweaking required.

Sapphire is very deep, and some of the effects are fairly complex. While there are plenty of easy-to-use effects in the package, there are others that definitely require the user to take a good long look at the user manual (horrors!) and spend some quality time figuring out how to exploit them.

Unlike some of these other third-party effects, there's not much to explain. Sapphire basically adopted the Avid effects UI, so it will seem quite familiar. Don't let it fool you, however, because there are a lot of secret powers hidden inside this package.

There are many more AVX effects, but those are some of the more mainstream. Two of the AVX Effects options are really color correction plug-ins, so I included them in the on-line chapter, so for information on the Baselight and Magic Bullet Looks AVX effects, look at the end of chapter 5.

For another series of AVX effects that gives real horsepower to Avid, check out AVX plug-ins from 3Prong.

CHAPTER 5

On-Line

As I addressed briefly at the beginning of the chapter 3, in many situations the line between off-line and on-line is blurring significantly. But in broadcast TV, feature films, and advertising, a definite line and workflow still separate the off-line "creative" editorial from the on-line finishing "technical" editorial.

One of the reasons I was convinced to write this book was because the workflows that helped bridge that transition have radically changed in the last several years. Many editors who were used to working in specific ways to hand an off-line edit off to the finishing folks are living in a whole new world. Most of this is because of the death of film and videotape as we transition into an almost entirely file-based world. Some of it is due to changes in the software we use, including new features and functionality built into the most recent software versions of Media Composer and the accessibility and power of DaVinci Resolve for media preparation and color correction. In film and broadcast TV, there's been a huge swing away from the traditional dailies workflow—being sent to a post house—and to almost entirely "on-set" dailies. All of these factors, plus the inexorable push for cost savings in an industry that seems to be fighting for its life, mean that old methods of off-line to on-line are being left behind and editors on both sides of the off-line/on-line divide need to adapt or die.

How a workflow is built clearly depends on the tools available for delivering the off-line and the on-line. Hopefully, one of the easiest workflows is when the off-line has been done in Media Composer and the on-line is done on Media Composer (hopefully with the Symphony option). But there are also times when an Avid off-line is completed in Smoke, Lustre, Quantel, or Resolve. Avid DS would also be a natural choice to on-line an Avid-edited off-line, but Avid discontinued DS in the summer of 2013. There will also be workflows for "round-tripping" out of Avid, into Resolve, and back to Avid.

High-end on-line has to start at the dailies stage and comprehensively provide for the linking of numerous generations of media. *Generation* isn't exactly the right term, but it's basically accurate when you think of the path that media takes through a production. The footage starts as the camera original file. Then the next "generation" is an off-line edit resolution. Then the next generation is either an up-rezed on-line version internal to the Avid or an up-rezed on-line resolution in an external application. The final generation is when that on-line application exports or lays the sequence to tape as a master. The trick is in following the metadata trail from one generation to the next.

On-Lining *Smash*

An example of following the metadata through the process is provided in the workflow for the popular primetime drama *Smash*. The on-line editor for the series was Mike Nuget of Techniolor–PostWorks New York. The colorist for the series was Sam Daly. The off-line was done on Avid Media Composer. The on-line was completed in Smoke, and the color grading on Lustre.

Nuget walked me through their workflow. Nuget's starting point is a little different from that for many on-line editors. Instead of having the off-line editor export an AAF file and send it to on-line, Nuget has Smoke and Avid running side by side, so instead of requesting an AAF, Nuget requests a bin containing only the final sequence. The off-line editors are tasked with preparing the sequence for on-line where they flatten everything down to one layer—unless multiple layers are required for picture-in-pictures or anything else that needs to be in multiple layers (especially comps, green screen, or audience comps). *Smash* features a lot of audience comps for which they "tile" people to create an a full-theater audience by shooting fifty people clustered in a section of the theater. They move and re-shoot these clusters of audience in different parts of the theatre and eventually put all of them together, generating the entire audience effect shot from as many as 15 separate layers. Nuget requested that the off-line editors not collapse or nest them. All titles are placed on a single layer. From there Nuget would take a more thorough look at the sequence to see if there was anything that could be done to flatten it down.

Nuget also shares another trick of the on-line: "If there were multiple layers, I would take that second layer, and let's say it was a picture-in-picture, I would take that second layer and copy that to the end of the sequence, so that when I did a decompose EDL it would all be on one layer. To me that was better than having to do multiple layers again going through Resolve and transcoding."

The core of the workflow was a column in the original bins that contained the original file name. Nuget explains, "In our workflow we used the file name as our master reference back to the original file. I would take that column and duplicate it into the clip name and into the sound roll column. We only used Sound Roll only as a reference. When I got into the Smoke I would relink to DPX files via the clip name and time code. So because we took the file name and put it into the clip name it was technically linking to the file name."

The workflow started with an EDL and an AAF. According to Nuget, "I would make an AAF, which would eventually go to Smoke, and I would also make an EDL of all the source material that I would need for the show. We use the EDL in DaVinci Resolve to do our transcode from ProRes to DPX files, which are needed for Lustre. So the AAF was for the Smoke to bring in a normal conform. Then the next step was to take the EDL, bring it into DaVinci Resolve, and tell the Resolve to look for all these original camera sources."

When the original camera footage came in during dailies, it was placed in folders indicating the date of the shoot. Everything was alphabetized in ascending order.

The Resolve was used to transcode the original sources to DPX. The resulting files were sent to one drive. Technicolor's engineering department had written scripts that would use the decomposed Avid sequence to grab all the shots that were necessary. The scripts would find each file in the main dailies folder based on the clip name, which was copied into and linked to the file name of each clip. It would put a shortcut clip into a folder of just the decomposed clips for that particular episode. Then in Resolve, Nuget would generate the media pool from the folder of shortcuts.

Then in Resolve, he would relink to the media. Nuget continues, "I would hit a transcode button and basically take everything with certain tails that I put in and change them all to DPX files on another SAN which is what I use for the conform of the episode in Smoke. But the EDL was also for Smoke because sometimes, when you are switching platforms, certain things don't come across or certain things come across wonky."

Nuget's other trick for prepping the on-line used Markers in Avid: "I would go in and make locators of any effects I see. *Smash* was a network show, so they only did blow-ups, dissolves; some timewarps here and there ... nothing out of the ordinary. I would make locators on the timeline and put notes on them so that when I went to Smoke if something didn't come across, I would go to those specific locators on the Avid and open up the Effect Editor and see that this is a push-in where it went from 100 percent to 110 percent and then I can manually enter them if they didn't come across in the Smoke."

The EDL also helped when the AAF would "lose" files. Nuget explains, "I could also use the EDL in Smoke when we would have what we called 'MIA EDLs' where I would bring in any clip that didn't come across right or [was] missing."

In Smoke, Nuget imported his AAF and told the AAF to look in the episode's folder on the SAN where the Resolve had written all of the DPX files and had it relink via clip name and time code. Nuget expounds on this, "So once I get into the Smoke, I have my AAF in and all my files linked up, I run through the episode and check everything, then re-create whatever kind of effects I needed to do. Then I would do what they call 'publish to Lustre.' We use the Lustre as the final color. Basically all I had to do was flatten the sequence down to one layer, copy my sequence to the server and e-mail the colorist the location. Then he would open that up and start working. Once the colorist was done, he reviewed it with the DP and rendered out new DPX files with color baked into it, which came back to me in Smoke, and then that would go out.

The entire on-line was done fairly quickly. Mike describes the schedule, "If I got the Avid bin Monday morning at ten o'clock, I can have a loose conform done—we have to output a loose conform version for promos for the network. This was basically just to get all the media in and get the show out to tape as quickly as possible. That whole process takes about six hours, which is really nice because there is no digitizing tapes. If you are doing a forty-four-minute network show if you had tapes, you would be doing close to eight hours of just changing tapes."

Mike expands on the six-hour schedule: "I would spend about an hour organizing the bin, changing the column names, and making markers to get the EDL and AAF made. Then I would go to Resolve, spend about twenty minutes getting the relinks going and then it was about a forty-minute transcode to the DPX files. That would put me back on the Smoke within two hours and then it would be just a matter of doing the conform. An average episode takes maybe two hours, maybe two and a half, to check every clip making sure everything is in sync and adding any kind of temp graphics. Then it takes forty-five minutes or so to get to tape. So the first leg would be about six hours. Then we would spend a little bit of time doing a graphics pull for the graphics house. There were times when there were ten shots to pull and some times when there were eighty shots to pull. The whole process of pulling eighty clips and having them named perfectly is only about half an hour."

Getting the VFX shots cut back into the show was much more low-tech than I figured. Mike explains, "On a forty-eight-frame shot they would actually give me back forty-eight frames plus eleven frame handles. So I would know every time I would get a clip back from them I would go to the eleventh frame and I was at the beginning the clip. Now, in my episodic Timeline I had a layer that had all the effect shots that I sent them, so when I got them back in all I had to do was really jump to that shot, go to the clip name in the bin, go eleven frames forward and replace it."

Avid-to-Avid Tips from NatGeo's *Border Wars*

Nuget also works on Symphony conforms that have a very different workflow. One of the differences with the workflow for *Border Wars* is that the off-line editors and Nuget are in the same building and on the same

Avid shared storage solution, called ISIS. "The on-line basically starts with just an e-mail that says, 'It's on this drive.' And I hook up the proper partition on ISIS to my system. Everything is on that partition. Nothing gets rendered or imported or captured to any other drive."

The system involved sharing a single bin, called "Sequence and elements," between off-line and on-line. That bin had the final off-line sequence placed in it by the off-line editor plus any other materials that later needed to be provided to on-line. Nuget copied those files from that bin and placed them in his own bin. From there, the on-line bins and off-line bins—in the same project—were kept carefully apart. Nuget didn't touch the off-line bins, and the off-line guys didn't touch Nuget's bins. All of the "communication" between them was centered on the shared bin. The on-line prep assistants that were part of the off-line team were actually responsible for up-rezzing the media before it got to Nuget.

As an on-line editor, he's very careful in organizing the project. Nuget elaborates, "I always have what I call an Elements bin, which includes things like blur presets or bars and tones or [GenArts] Sapphire presets that I know they are going to use a lot like a swish pan for transitions or a dissolve glow or something like that; credit bed templates and the other types of normal element stuff that you use on every episode especially for certain clients. If I am on episode two and I'm about to start working on episode three, the first thing I do when I open up episode three is copy that Elements bin over to that project. That way if I update anything in episode two, that would also be updated in episode three and I just keep doing that. I have had those same bins for six years and I just reuse them. I also have a bin called "Border Wars Color Correction Template" so that I can reuse color templates, for example, the opening or an interview that looks the same in every episode. So, I save it with [the subject's] name, and I carry that across with me every episode too. I do that with every client."

One of the critical things for on-line editors is to understand the entire set of deliverable requirements for each and every broadcaster or cablecaster. "I'm outputting a new show on Monday to Bravo for the first time and in my room I have a physical folder—not a computer folder—with all the specs from different channels which I have worked with in the past because I have worked with just about all of them, I think. But I haven't worked with Bravo yet, so they e-mailed me a forty-five-page spec that I printed out. I highlighted stuff that I needed to remember because it's not in my brain yet, so I am probably going to refer to it a couple of times more before I finish the first episode."

Up-Rez from Tape-Based Media in Avid

I spent about twenty years conforming tape-based projects in Media Composer and Symphony, but I have to say that I haven't had to do this for the last decade. Everyone is experiencing changes in the media world at a different pace depending on where they live and what types of projects they're working on. Tape is still alive and well in the broadcast world among other places.

The basics of tape-based up-rez (off-line to on-line) are very well established with Avid. It's been a primary method since the time Avid had the capability of delivering on-line quality, which, depending on your definition of "on-line quality" probably means somewhere between about 1992 and 1997. In 1992 we had a resolution in Avid called AVR27, which was touted as broadcast quality. I certainly delivered final productions at AVR27, but it was still a pretty compressed standard definition resolution. I know that *The Oprah Winfrey Show* used that resolution for national broadcast. Some people didn't want to use Avid for broadcast until a few years later when they came out with a 2:1 standard definition compressed video quality, which essentially matched DigiBeta.

Now Avid is capable of delivering broadcast quality HD resolutions that meet the requirements of all broadcasters. For many productions, there is enough storage even at these broadcast HD resolutions that it is not necessary to capture at an off-line quality and then up-rez to broadcast quality at the end of the process. But, many, many productions are shooting so much footage and requiring so much media storage that it is impractical to edit at anything other than even 20:1 SD, as Robert Landau described earlier in this book (Chapter 1, page 25) when describing the workflow for the *Crimes of the Century* series for CNN.

Many broadcasters and production companies are heavily invested in tape-based technology or have requirements or uses for the archived tape-based footage. For them, the need to take an off-line, tape-based production to on-line broadcast quality is still a necessity.

Primary in the ability to execute this workflow is the need to properly label all tapes, whether the production shot them or they came from an independent shooter or a stock house. Each tape needs to be labeled, tracked, and stored so that it can be found again when the time comes for the on-line. Labeling at larger productions can involve bar-coded tape labels, which is a great help. One of the things I discussed with Avid data management experts Ian Krabacher and Mike Hale was that Avid needs to recognize, label, and follow the "uniqueness" of the media. They were specifically talking about the data management built into Avid as media is ingested and stored in the Avid MediaFiles folders, but it also applies to the tape media *before* it is digitized into Avid.

A system must be developed to make sure that each tape has a unique name. That seems pretty easy, but with tapes coming from outside shooters and stock houses, it is definitely something that needs to be managed, and a plan must be in place. The tape-naming convention needs to be matched and followed and be *allowable* by Avid as it's digitized. In other words, character limits and character types that are required inside Avid, and possibly even by various EDL exports, need to be followed. I've always been partial to using dates as a primary naming convention. Not exclusively dates, of course, but dates can be used to hone the options for what a tape could be named. Then you can add lettering for cameras on multicam shoots and/or more numbers for the order in which a specific camera shot the tapes. Tapes from stock houses should probably be renamed using the dates they were received plus other designations. Including the name of the stock company or provider in the tape name—even if it's just an acronym or specific code—can be very useful in identifying and delivering the "rights bible" for the show later on. When a tape is digitized into the Avid system, you want to be very careful that the name that the tape is given in the Capture dialog is very clear and obvious, not just on the tape case but the tape itself. Tapes mysteriously swap boxes sometimes. Make sure the labeling matches clearly on the case, on the tape, *and* in the Avid system.

Whenever a tape is inserted into a deck that is attached to an Avid system, the first thing that Avid does is "sniff" the timecode of the tape. Ninety percent of the time, this sniff is something you can ignore because the Avid will take care of things from that point on. But with the other 10 percent, this sniff discovers some kind of issue. The main issue is usually with broadcasters who provide stock footage. Often, they will reuse tapes, and the type of timecode that is sniffed at one point in a tape will not match the timecode that is sniffed at another point. Or, the stock house will put several shots onto a single tape, each with a different type of timecode. Occasionally, a shooter will change the timecode by accident—after lunch or sometimes, to save time, they will record bars and tone prior to asking the producer what type of timecode they want, so the bars will have one type of timecode while the rest of the shoot has another. Avid can cope with only one type of timecode per tape. So, if this happens to you, the answer is to go with the flow. Simply rename the tape with the exact same name but with an NDF or DF suffix. This will make the conform a little bit harder in the end, but by keeping most of the tape name the same, you can at least easily identify the tape.

Avid is very good at tracking that tape name through an off-line edit.

When it comes time to conform the off-line quality media to the on-line quality, I think that the easiest workflow for most purposes is to start by duplicating the sequence. Keep the off-line quality media linked to the copy of the sequence. Then take the other version of the sequence and place it in its own bin. Decompose that sequence.

Now, if you've never heard of *decompose*, it just *sounds* disgusting. Basically it means "break the sequence down to its component parts." This is done by selecting the sequence and, from the menu bar, selecting Clip > Decompose (see Figure 5.1). Because of the variety of media that is present in most productions today, you will need to decide if you just want to decompose the just the tape-based media. Also, decide on the amount of handles you want to capture. The default is a good place to start.

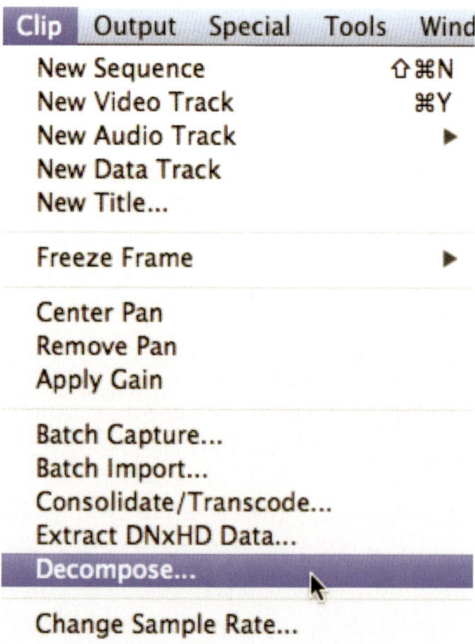

Fig. 5.1 Decomposing the sequence breaks the sequence down to its component parts.

But if you think you may need to make tweaks—adding transitions or trimming—then you might want to double the amount of the handles (see Figure 5.2).

Depending on how you're doing the audio, you may or may not want to strip the audio tracks off of the duplicated sequence before you do the decompose. Chances are that the audio you digitized during the off-line is not compressed at all, so there's no reason to recapture all of the audio. Recapturing all of that audio from tape can take a very long time, so simply deleting the audio tracks can save time and effort and head wear on the decks. Just double check that your duplicate sequence with the off-line media is still safe and sound. You can copy and paste the audio tracks from that sequence back over to the conformed on-line sequence at the end.

Decomposing will create a new clip for every single edit in the sequence. Theoretically, if you edited a documentary from a single tape that was captured as a single clip and then made a sequence from it that had one hundred edits, the decomposed sequence would be made up of one hundred new clips. The reason for Decompose is to make a bin with JUST the media that you need for your sequence. It allows you, for example, to not have to digitize an entire thirty-minute interview just to get the one five-second on-camera sound bite that made it to the final cut.

At the end of the decompose, all of your media, or at least all of your captured media, will be off-line. The next step is to organize all of those pieces

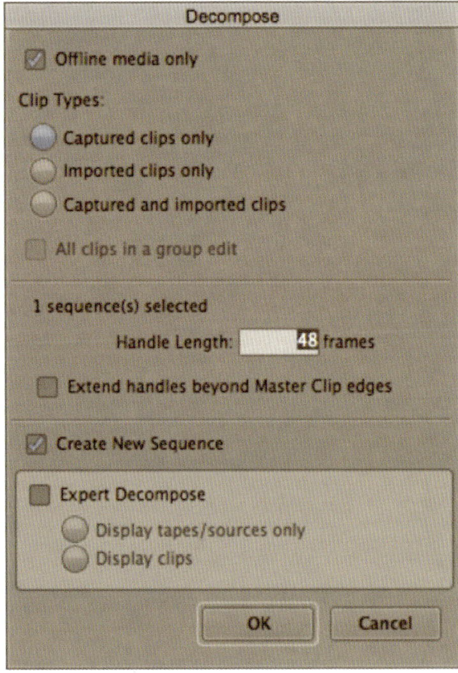

Fig. 5.2 The Decompose dialog has options on how the decompose will be performed.

so that the digitizing or capturing process can happen as quickly as possible. To do that, I create a special bin view that helps me prepare to batch capture.

There are a few things that are important to see at a glance when Batch Capturing (up-rezzing). With a bin selected, choose Bin > Choose Columns from the menu bar. Double-click the All/None button at the bottom of the Bin Column Selection window to clear all of the choices (so none are selected in blue). Then click the Camroll, CFPS, Disk Label, Drive, End, Format, FPS, Off-line, Project, Start, Tape, Tape ID, Tracks, Video, and Video File Format columns. Some of these are rarely needed, but they could help troubleshoot. When you're done selecting, click OK.

Drag the columns into a specific order. This is the order that I use:

1. Name
2. Tape
3. Start
4. End
5. Off-line
6. Project
7. Drive
8. Format,
9. Video
10. Video File Format
11. Tracks
12. FPS
13. CFPS
14. Disk Label
15. Camroll
16. Tape ID

You may find that a slightly different column order works best for you, but beginning with my order will get you to a good starting point from which to make your own decisions about what you like best for your project and the way you work.

The most important columns are to the left. Basically the really critical ones are Tape, Start, and Off-line.

Sort the bin by Tape and Start time. You do this by clicking or double-clicking on the column name at the top of the bin to toggle by Ascending or Descending order. To sort by tape name and then timecodes within each of those tapes, you have to organize the columns from left to right in order of importance in the bin. In this case, tape name would be first (to the left) and timecode would be next (to the right of the tape column). Use Command-E (Mac) or Control-E (PC) to sort in ascending order or Option-Command-E (Mac) or Alt-Control-E (PC) to sort in descending order.

You've put in a good deal of work to create and order these columns, so you'll want to save this as a Bin View by selecting the pulldown menu at the bottom left of the Timeline. (It might say "Untitled" or "Statistics.") Choose the Save As option and give the Bin View a name, like UP-REZ or Batch Capture (see Figure 5.3).

The first priority is to sort the clips so that all of the clips from a specific tape are together. The specific order in which the tapes are digitized doesn't matter, really, but the goal is to cluster all of the clips from each tape together. To save time and effort and to get a jump-start on any problems (lost or missing tapes), you can print out the bin at this point (File > Print Bin) and make sure that you organize all of the tapes for the show in the same order as the bin, making note of any missing or confusing tape names. This makes it faster to change tapes quickly. Also, it allows you to try to find missing tapes and skip over them while you're capturing all of the tapes that you have first. Another great tool is revealed by right-clicking on the program monitor with the sequence loaded. Select Sequence Report (see Figure 5.4) and you get a lot of options on reports that will show what's off-line, on-line, effects used, sources, and so on. It's a fast way to create a list of what elements are needed for finishing (see Figure 5.5).

It is possible to completely skip the decompose process, but if you do that, you have no control over the Batch Capture process. Media Composer simply picks the first tape it wants and tells you what the name is. If you don't have that tape, or can't find it, then you waste a lot of time that could be devoted to capturing from the tapes that you know you have. With a decomposed sequence, you (the operator) determine the order of the tapes, and having the information in advance lets you find the missing tapes instead of being surprised by them.

With the bin sorted by tape and the tapes located and organized, you can start the capturing process (or "digitizing," as it used to be called). You don't want to use the Capture command because that is used only the first time a clip is captured. To recapture a clip—either because you want to up-rez it or because it has gone off-line for one reason or another—you

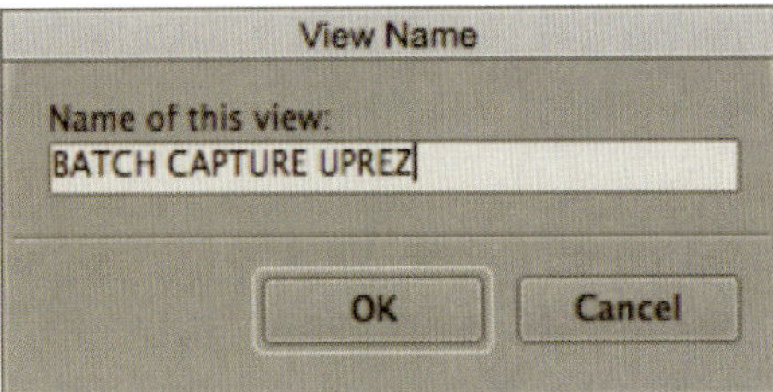

Fig. 5.3 This is the Bin View naming window. After creating a customized bin view, make sure to save it and name it something descriptive.

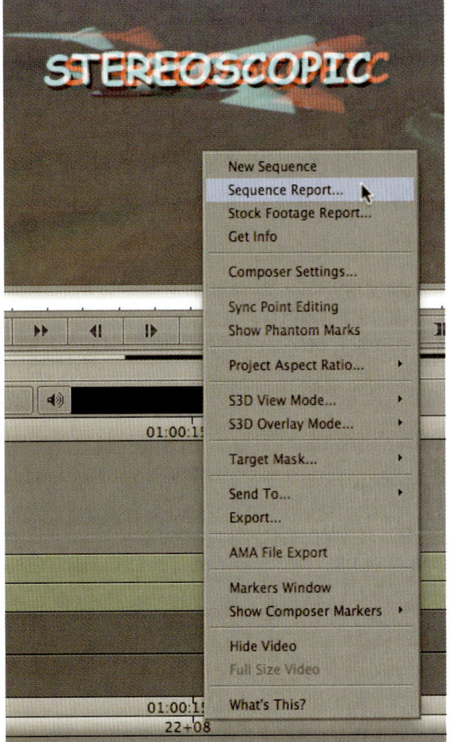

Fig. 5.4 A contextual menu (right-click) on the Sequence monitor reveals an option to create and print a Sequence Report.

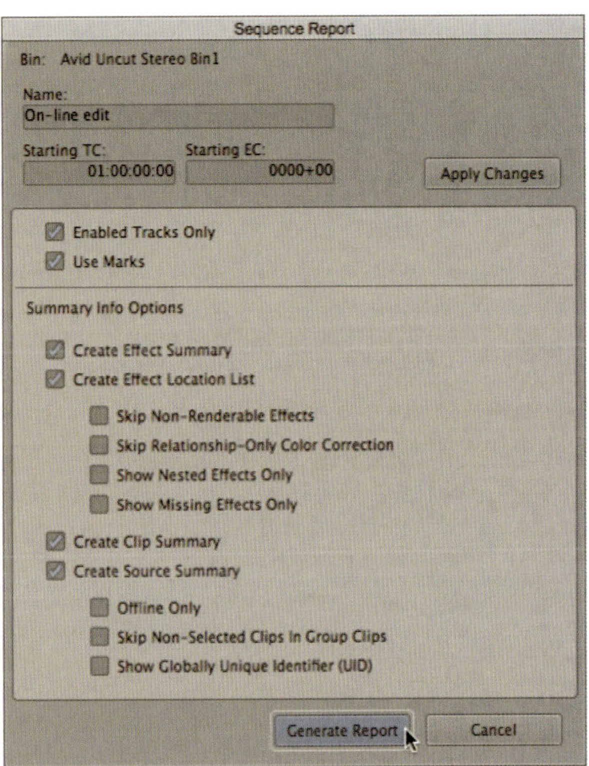

Fig. 5.5 This window, shows the options for creating the Sequence Report.

want to use the Batch Capture command. This is the same concept as Batch Capture in FCP. In the bin, select all of the clips that will be captured from the first tape. Either lasso them or click the first clip and Shift-click the last clip from the tape. Then from the menu bar, select Clip > Batch Capture (see Figure 5.6). This will call up the Capture Tool

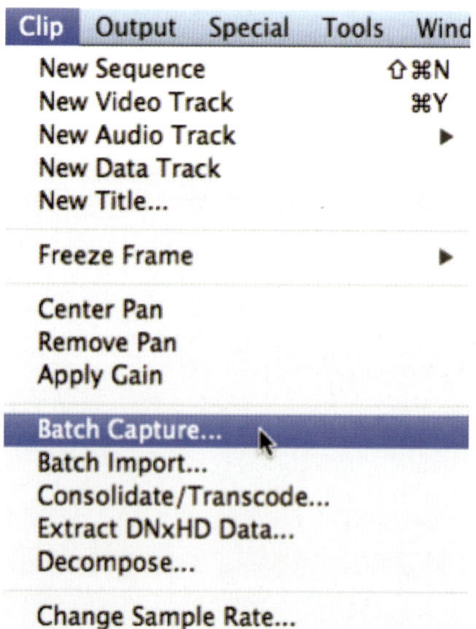

Fig. 5.6 This is the Batch Capture option in the Clip menu.

(see Figure 5.7). Make sure that you select the correct capture resolution. There's nothing worse than Batch Capturing for an hour at the original off-line quality! With the correct resolution chosen, the Batch Capture window will ask you to insert the tape into the deck that the clips you selected came from. Then simply follow the instructions from the Capture Tool. It will proceed through the clips, capturing each one, and when they're all captured, select the next batch of clips from a particular tape and continue the process. As you progress, the Off-line column and the Drive column will update showing the tracks and clips that have been captured. The Format, Video, and Video File Format columns will also update (see Figure 5.8).

If you have a second deck, you can save time by having the assistant (or you) make sure that the next tape is rewound to the head, and maybe even fast forwarded back to just before the first clip that is needed on the tape. For example if you have an hour-long tape and the first clip is in the first minute of the tape, but the tape was left parked at the end of the tape when it was first digitized, then you'll have to wait while the Avid system rewinds the tape back to the beginning. Or if the tape is at the beginning, but the only shot that was used in the sequence is at the very end, you'll have to wait while the Avid system fast-forwards to the end. With a second deck and some forethought, you can save a lot of time. When I was doing this a lot, I was able to Batch Capture an hour-long show from 130 tapes in about six hours.

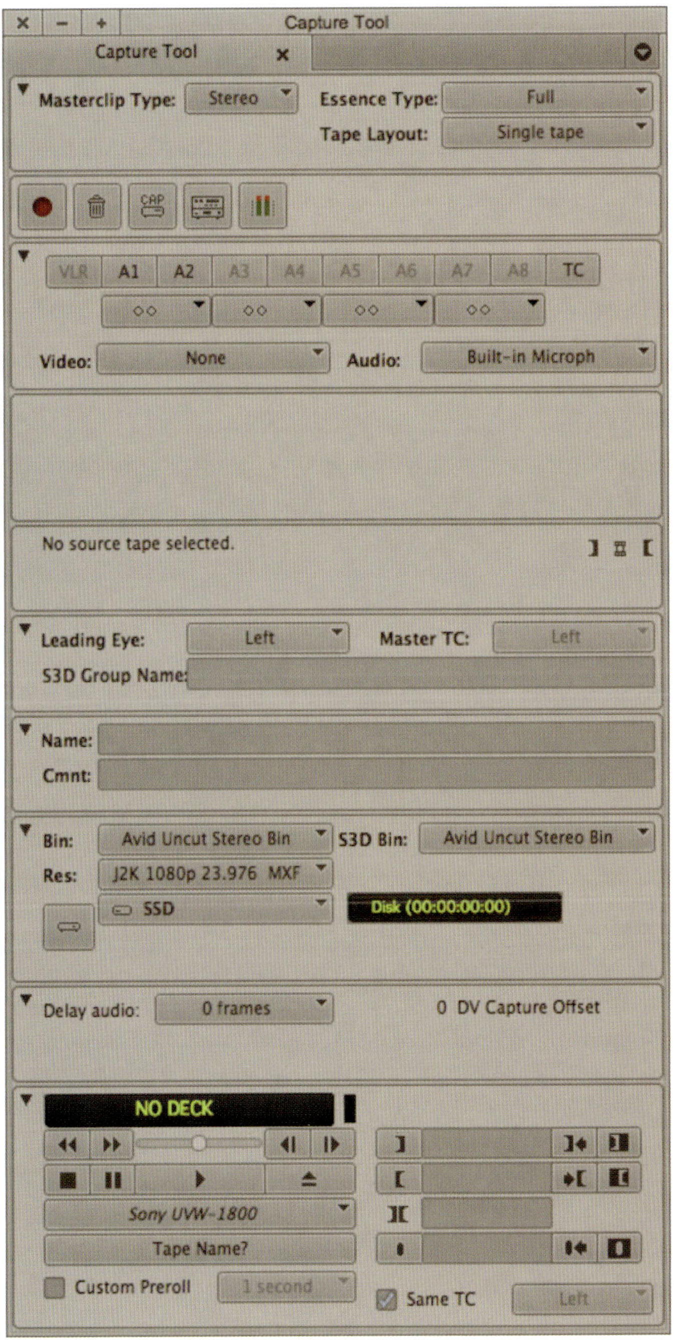

Fig. 5.7 The Capture and Batch Capture windows are similar, but the Batch Capture window has some minor differences and capabilities.

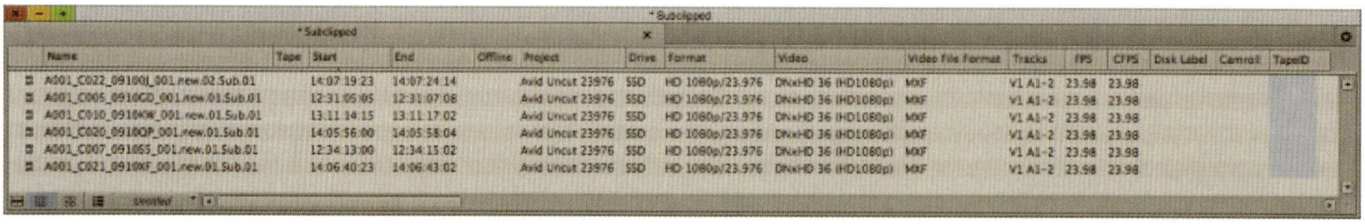

Fig. 5.8 The bin will update as the Batch Capture progresses. The Drive, Format, Video, and Video File Format columns will each indicate the Batch Capture clips.

Up-Rez from Inside Avid

If you used AMA to bring your footage in to Avid and then transcoded to an off-line resolution, like DNxHD36, to edit with, you can link back to the AMA clips and transcode just the material in the sequence to a higher resolution.

Place the sequence in a bin by itself and close all other bins. Open the bins with the original AMA-linked clips, select the sequence, press the Shift key, and select all of the AMA-linked clips in all of the open bins.

Next, right-click on the edited sequence and select Relink from the contextual menu (see Figure 5.9). In the resulting Relink window (see Figure 5.10), select Relink selected items to: Selected Items in ALL open bins, and check the Create new sequence box at the bottom. Select OK. This will create a new relinked sequence that is connected back to the AMA clips. You can test this in a number of ways. One way is to use Match Frame on a clip in the sequence to call that clip into the source window, on which you can then do Find Bin and see the clip in the bin, checking its resolution. At this point, you should be looking at AMA media. Another way to do this is to go to your Timeline Fast Menu settings and select Clip Text > Clip Resolution (see Figure 5.11). This shows the resolution of the media in the Timeline for each clip. If you look at the sequence you started with, you'll see DNxHD36 media in the sequence. In the Relinked sequence, you'll see AMA media. And, looking forward a step, when you transcode to an on-line resolution, your sequence will show the higher resolution media in the Timeline.

To transcode the clips to your final on-line resolution, right-click the sequence and transcode to your desired on-line resolution. Personally, I was surprised—because of the way that Avid works with tape-based media—that the Consolidate command wasn't the one to be used, but many types of AMA media can't be consolidated. So use the Transcode command to pick your desired on-line resolution. I chose DNxHD 175. This can be seen in the timeline. In the screenshot, I chose to display both Clip Names and Clip Resolutions. This screenshot was done before the transcode and after the relink to the original AMA clips (see Figure 5.12). Using this method allows you to double-check exactly which footage the current sequence is linked to—off-line, AMA, or on-line.

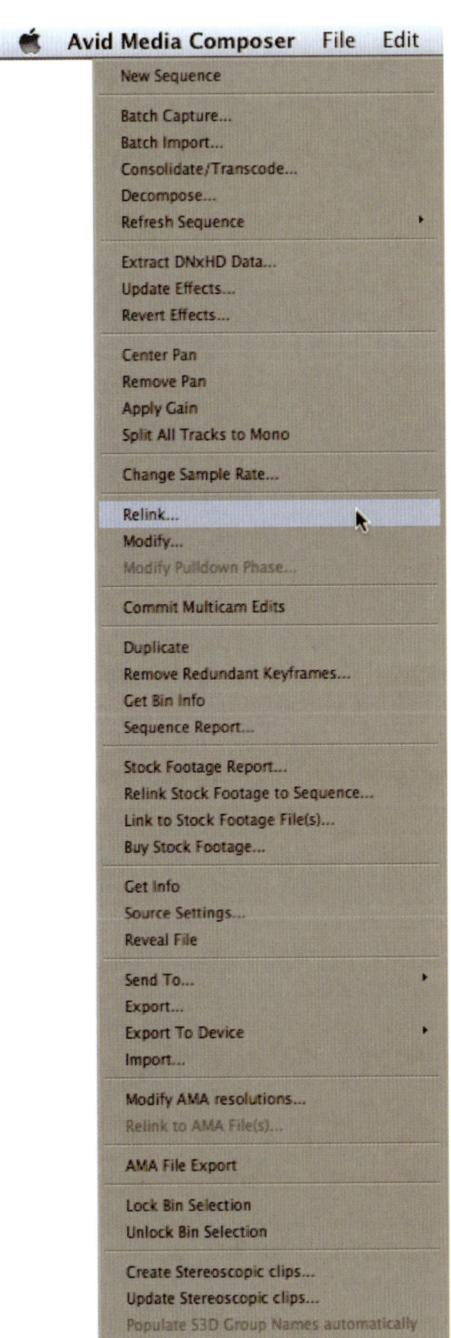

Fig. 5.9 The Relink option is available in the contextual menu for the edited sequence.

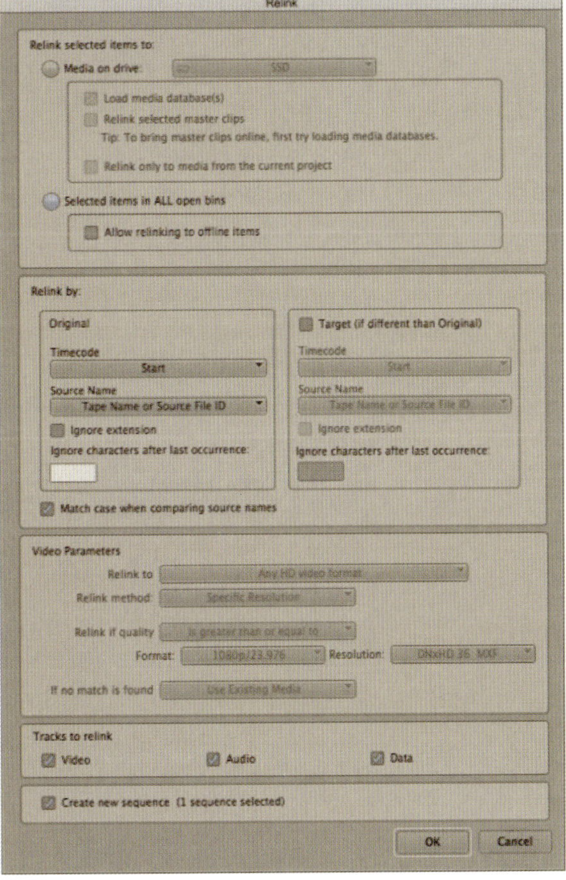

Fig. 5.10 The Relink window provides options for relinking the clips to the original AMA clips from the camera.

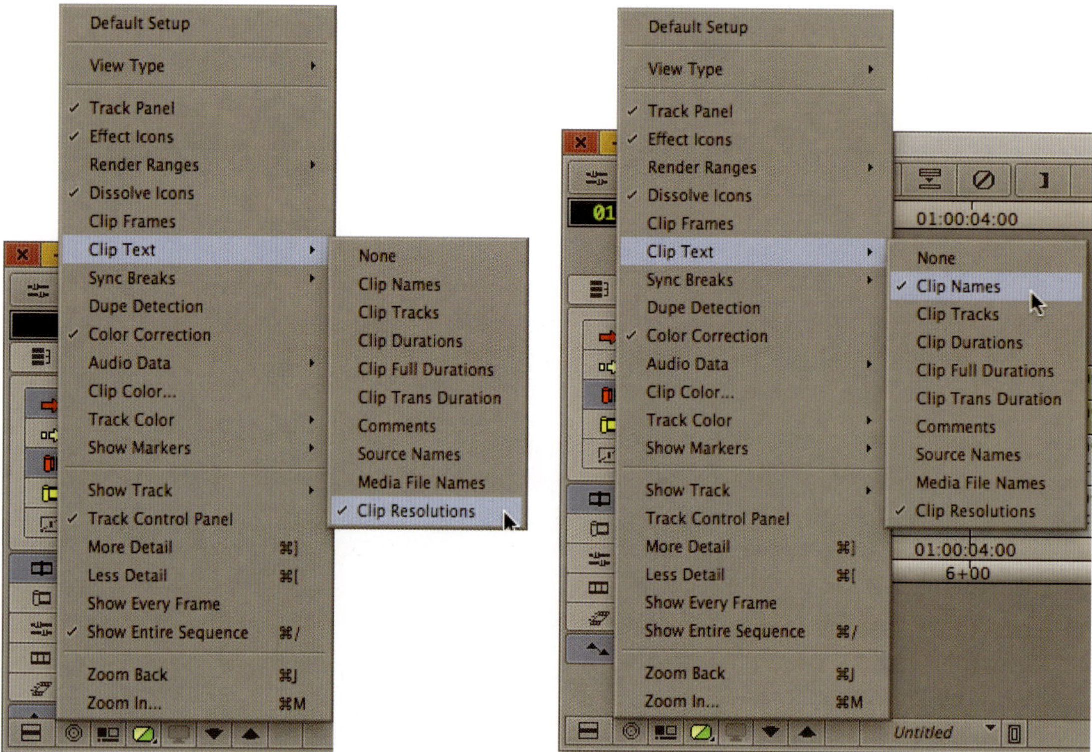

Fig. 5.11 It's possible to see whether clips are AMA-based in the Timeline from the Timeline Fast Menu with the Clip Text option.

Fig. 5.12 The Clip Text option allows you to display combinations of Clip Text options. Here, Clip Names and Clip Resolution are both selected as indicated by the checkmarks next to them.

Exporting for On-Line on Another System, or Round-Tripping

AAF is the latest in a string of file types used by Avid over the years to get editing and clip information from Avid to … well, anyone. The earliest Avid systems had no internal ability to create an on-line output. The only way that they could create a broadcast-quality project was to export an Edit Decision List (EDL), which was then opened in a linear, tape-based edit suite that used the original source tapes to conform the edit. As Avid became a major player in the feature film world, Avid FilmScribe gave it the ability to create a Cut List that could be conformed by a film negative cutter. Then came the OMF file, which Avid used until recently to act as a kind of "super EDL" to get Avid projects to other systems, including ProTools.

AAF is also used by some of the camera manufacturers or dailies software solutions to deliver some metadata to the Avid system before the edit even starts.

To complete the round trip from Avid to Resolve and back to Avid, there are a few things to remember. The frame size and frame rate in Resolve should match the Avid properties and if you're using DNxHD media, use RGB 709 color space.

There are dozens and dozens of potential workflows for round trips between Avid and Resolve. We can't possibly cover all of them in this book, so we'll set some parameters based on typical best practices and then walk through one method with the Alexa footage in the project media available on the companion website (www.focalpress.com/cw/hullfish). Then we'll walk through another workflow with the RED footage.

It's possible to edit with fast imported media or AMA-linked media, but those options add a lot of restrictions, and I wouldn't advise either of them for most projects. We could explain projects that are encoded outside of Avid, like in a dailies situation, but that adds a lot of complexity that we can't cover. So we'll deal only with internally transcoded media.

Because RED media is higher than HD, we will utilize a workflow that assumes that Resolve is on a separate system from Avid. We'll transcode to a lower-rez "off-line" DNxHD codec and then link back to the camera original RED media in Resolve.

For the Alexa footage, we'll transcode internally to a higher "on-line" quality DNxHD resolution and round-trip to Resolve running on the same system, using the DNxHD material linked to the DNxHD media in the Avid MediaFiles folder. Then we'll color correct in Resolve, render out to new media, and then link back to the newly rendered, color-corrected media.

Alexa Conform DNxHD Round Trip

Download the Alexa files from the companion website (www.focalpress.com/cw/hullfish) and AMA Link to the files. Then transcode them to whatever on-line quality DNxHD file you want to work with. Procedures for doing this are detailed in chapter 1 (starting on page 45), and it's possible that you have already prepared the files.

Video Tutorial

To follow along with this tutorial, download the Alexa files from the companion website (www.focalpress.com/cw/hullfish).

J. C. Bond and others suggest subclipping the transcoded footage and editing from subclips. This helps when the clips are decomposed or consolidated and if you are syncing double-system sound. For the purposes of this demonstration, it's not really necessary. Nevertheless, it's a good habit to get in to.

Edit together a short sequence and include some transitions and maybe a resize. Add several effects if you want to see how these effects move through the workflow. Or simply create a simple cuts-only sequence from the on-line quality media (see Figure 5.13).

To start the round trip, select the sequence and choose File > Export. You'll need to choose the Options button to open the File Export dialog.

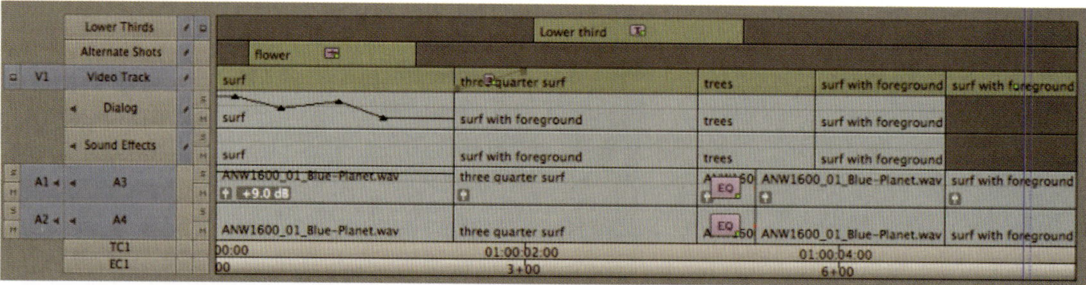

Fig. 5.13 This is a sample sequence of the Alexa footage cut together for the round trip.

In the File Export dialog, choose to export an AAF from the Export Settings drop-down menu. Because we have already transcoded to our final on-line resolution, and we are going to do a workflow where Resolve is on the same system as Media Composer, we are just going to link to the media (see Figure 5.14) as it sits in our Avid MediaFiles folders. Make sure this is checked in both audio and video tabs. Also, many projects have mixed media. Not all of it may be compatible with Resolve (though in this case it's all transcoded DNxHD), so you can select the Transcode Video To option to deliver usable media to Resolve.

It's always a smart idea to use the Save As button to save the name of the Export setting and the options. This way if you successfully make the round trip, you can have confidence that you can use this setting again. If you fail, you have a record of the settings and preferences you tried and can try a different combination the next time. Make sure to note where the exported AAF file is being saved. Executing the export will generate a new ".exported" sequence in your bin.

By using the Export (Linked) option, the export is very quick and doesn't generate any additional media that needs to be stored.

NOTE

If Resolve is not on the same system, it's possible to use the same workflow, provided that you manually move all of the native media that is linked to the sequence to the Resolve system or that it's accessible on a high-speed network drive. However, if Resolve is on another system, a better idea is to consolidate the media or copy all of the media in the Export Settings.

If you've never used Resolve, download the Lite version for free. The first time you run it, you will be asked to create a database. Click on the + sign in the bottom left corner of the database window and change the Driver from QPSQL to DISK to create the database on your local drive (see Figure 5.15). When that's completed, Resolve will ask you to set up a new user. Type in a username and password and maybe even add a photo (see Figure 5.16).

With the User set up, you'll move on to the main Resolve window (see Figure 5.17), where you'll need to create a project using the small + button in the bottom left corner. Set Resolve to properly read the timecode and reel number information from Media Composer. Click the gear icon button in the bottom left of the Resolve window to open the Project settings (see Figure 5.18). Choose Master Project Settings and then the Conform option. Set Use Timecode to Embedded in the source clip. In the option for Assist using Reel Names, choose "Embedding in source clip file." Check to

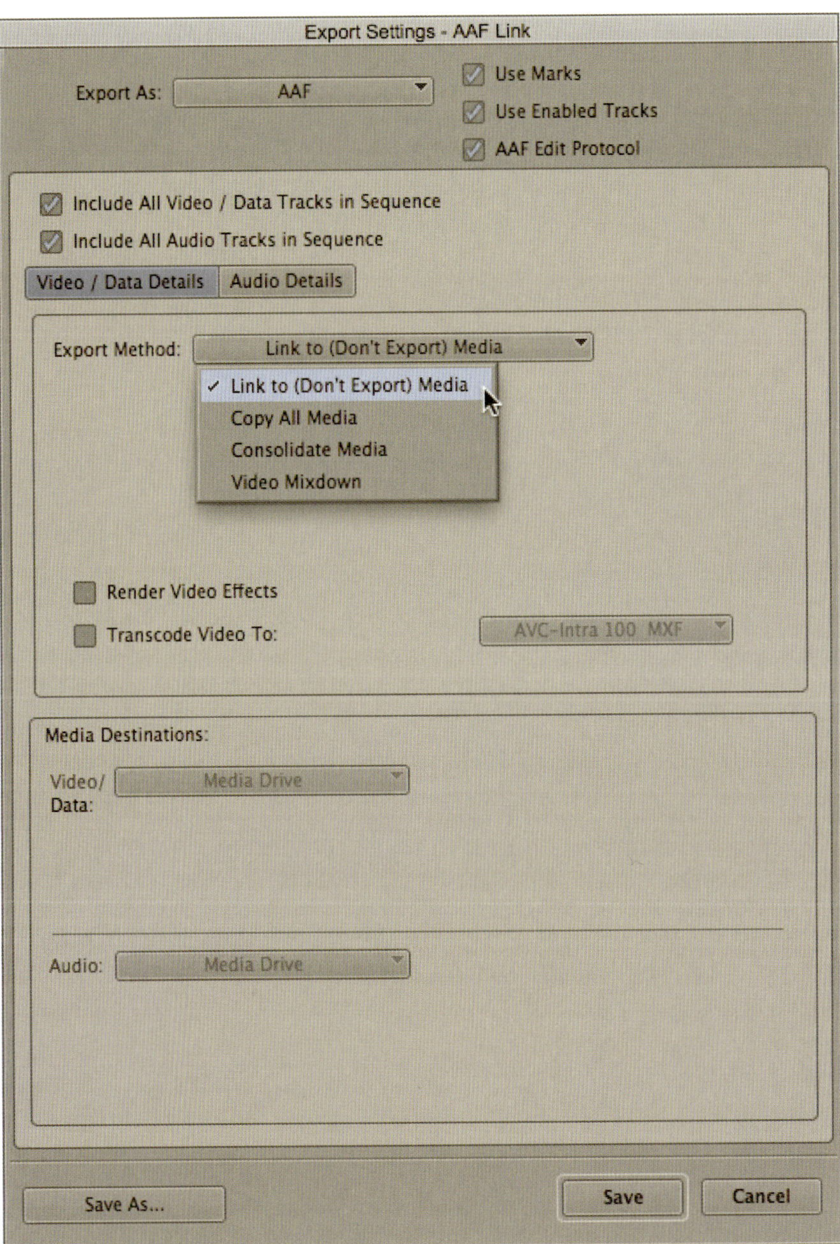

Fig. 5.14 This is the Export Settings window for AAF. For this tutorial, choose Link to from the Export Method menu.

make sure that the other project settings are a match for the media that's coming from Avid in regard to frame rate, frame size, color space, and so on. Then hit Save.

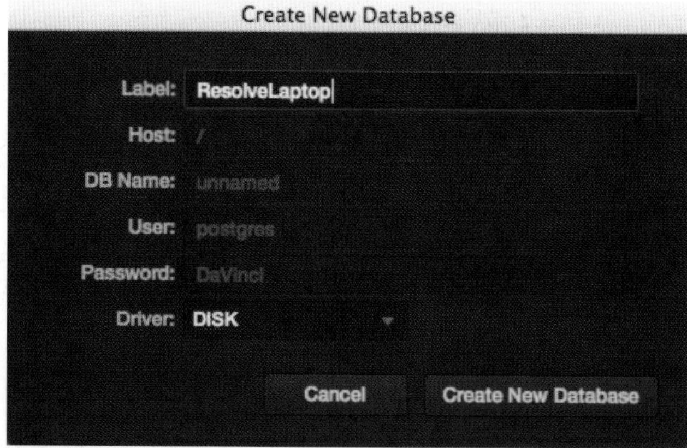

Fig. 5.15 If this is the first time you've used Resolve, you'll need to Create New Database before getting started. The following screencaptures are from Da Vinci Resolve.

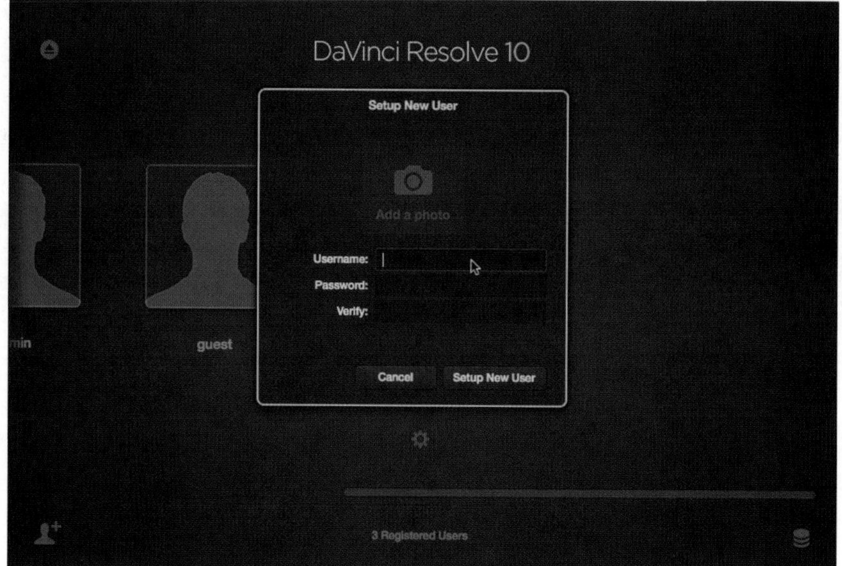

Fig. 5.16 If you've never used Resolve before, you'll need to set up a New User. This is the New User window from Resolve.

To open the AAF, choose File > Import AAF, EDL, XML. Navigate to the file you just exported from Media Composer and choose Open. In the resulting Import dialog, enable "Automatically import source clips into media pool" (see Figure 5.19). You can probably ignore the majority of the Resolve error messages that are displayed when the AAF is imported (see Figure 5.20).

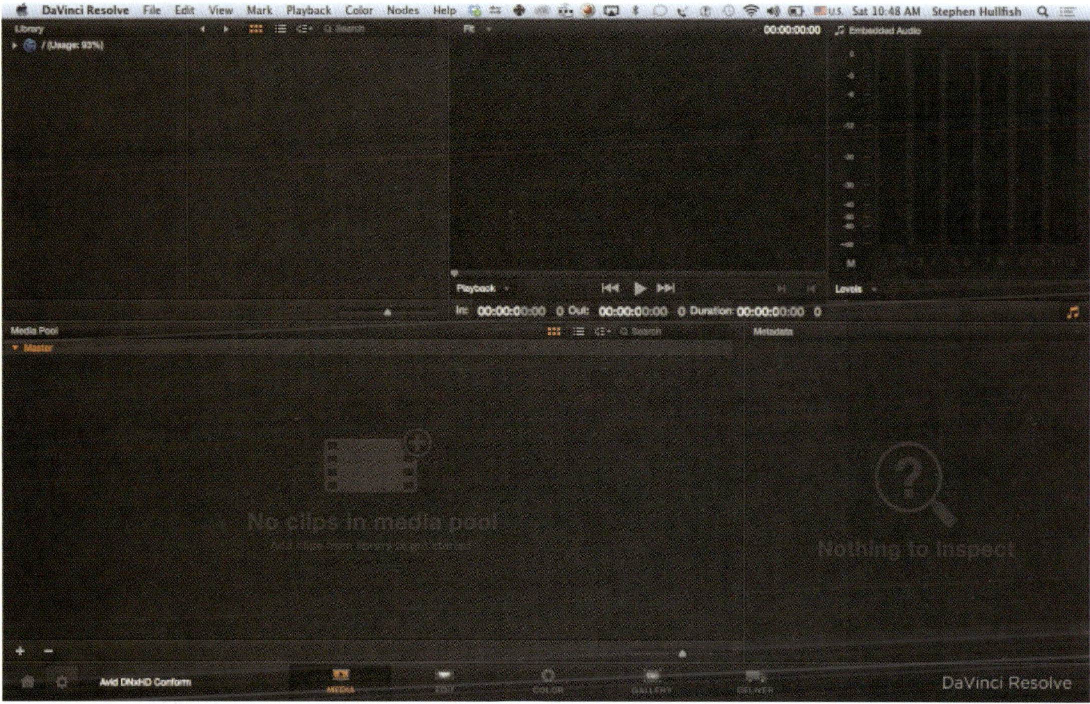

Fig. 5.17 This is the opening UI for Resolve. The choices along the bottom of the screen—Media, Edit, Color, Gallery and Deliver—swap between the main functions of Resolve.

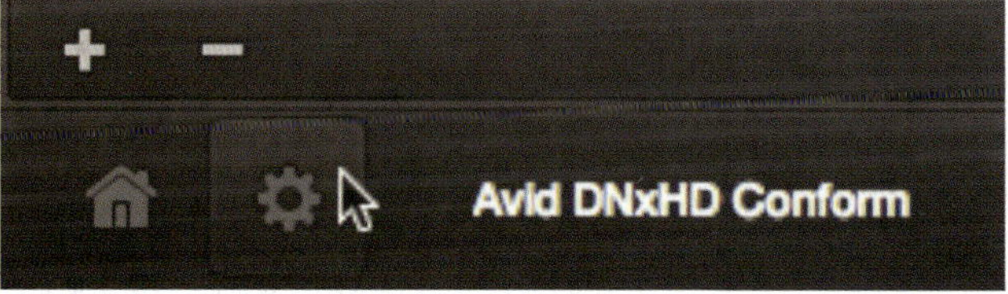

Fig. 5.18 The gear icon in the bottom of the screen launches the Project Settings.

This should populate the Resolve media pool with all of the clips in the imported AAF file. If not, you may need to add drives to the "drive pool" to allow them to be seen in Resolve. Note that in the example in the book, there is a title in the sequence, and it seems to make the file on the lower track appear to be not imported into Resolve when viewing it in the Color tab. By changing the track selection from V3 and V2 to V1, the clip appears properly in the timeline (see Figure 5.21).

You can then grade the project using the full power of Resolve. This is not the book to teach those skills. If you're trying this hard to get a round trip to Resolve, you should either already know how to grade using it or

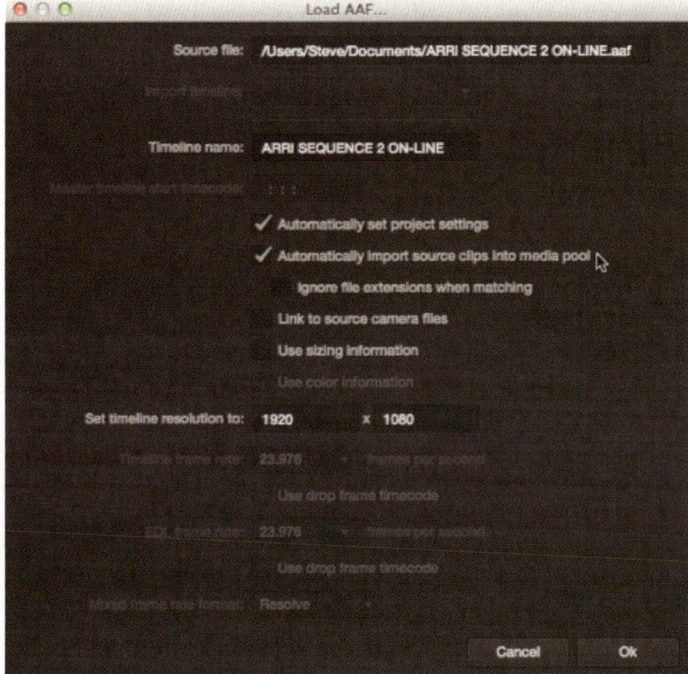

Fig. 5.19 This Window allows you to Import the AAF from Avid into Resolve.

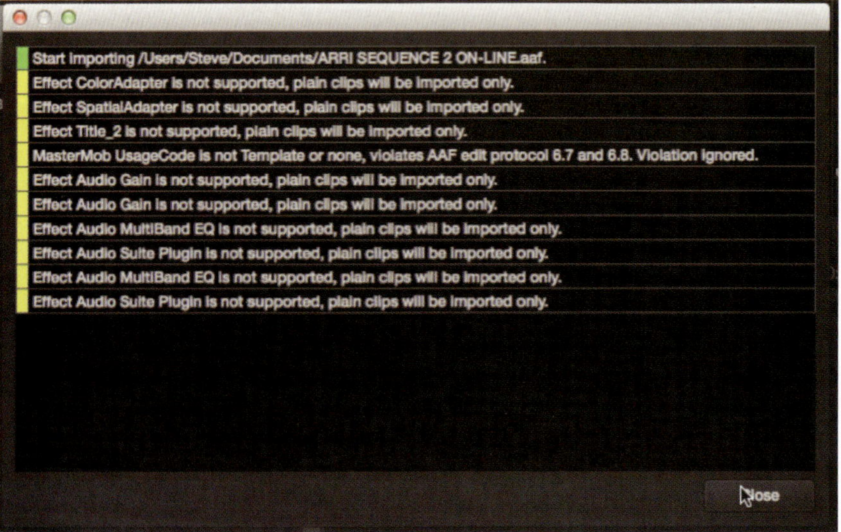

Fig. 5.20 When the AAF is imported, Resolve will display a series of errors. Most of these can be ignored.

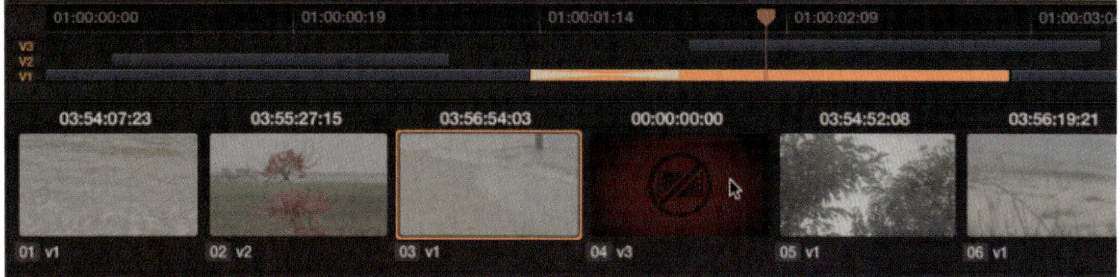

Fig. 5.21 After the AAF import—which also loaded the Avid media into Resolve's media pool—the timeline shows the sequence with thumbnails of the media.

know somebody who does. Otherwise, grade in Media Composer or Symphony. I explain some of the basics of that later in the chapter.

When you are done grading, you can export another AAF to go back to Media Composer to complete your on-line—possibly re-rendering graphics and effects, making small trims, and bringing in a mix from ProTools—or just using Avid to export the Resolve-corrected media. Exactly how you do this depends on whether all you did in Resolve was make color corrections or whether you made editing changes. Those two scenarios require two different export methods.

If you didn't edit anything in Resolve, using it only to do grading, then you can simply update the original AAF that you imported into Resolve. This is the easiest because it retains all of the effects and audio when you return to the Avid system. If you choose this option, don't move the original AAF file. It has to stay in the same place as when it was imported into Resolve. In my case, it was in my Documents folder.

If you edited inside Resolve, then you need to choose Generate New AAF when you export an AAF of the revised Timeline. The resulting AAF file will be missing audio and any unsupported effects. You'll need to manually re-edit the audio and effects from the original sequence back into the imported, color-corrected sequence.

To export the AAF back to Avid, go to the Deliver page and select the graded timeline. In the Deliver page in the Render Settings list, choose Avid AAF Round-Trip from the Easy Setup pulldown menu (see Figure 5.22). In the Output section of the Render settings, choose MXF and the codec you want to use for your output and eventual Avid on-line. Selecting handles is always a wise idea. (You will have to scroll down in the Render Settings list to see these options.) How much handles is a personal decision based on the potential of the project to need some revision when it is back in Avid.

You also have to choose the destination for the rendered files from Resolve. You'll have to scroll down a little further in the Render Settings list to see this. Since we're using a workflow where the Avid and Resolve are on the same system, it's wise to render right back into the Avid MediaFile folder. Inside the MXF folder in the Avid MediaFiles folder

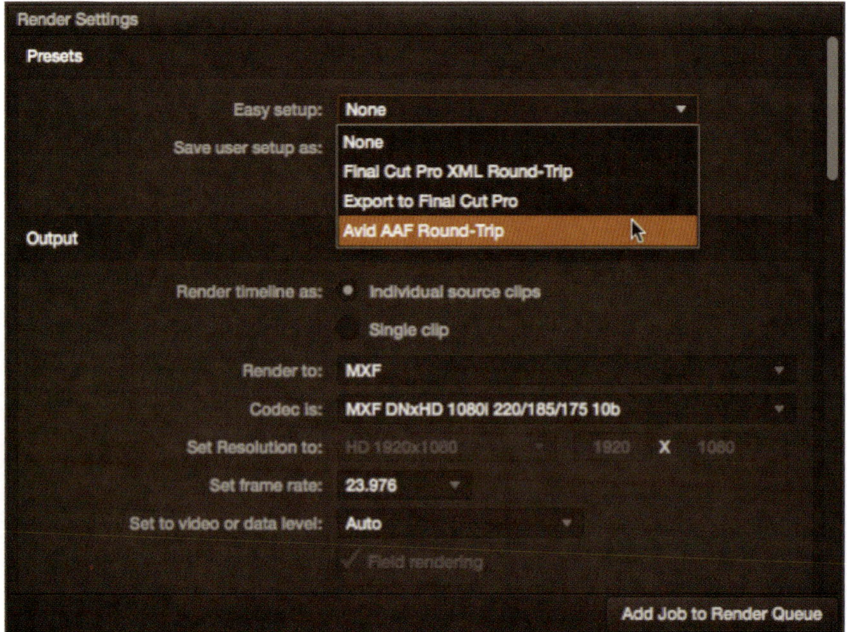

Fig. 5.22 This is the Render Settings dialog in Resolve, providing options for the rendering of the media which will be used for the Avid on-line roundtrip.

are a series of numbered folders. For some people, there is only a single folder named "1." You can use any numbered folder inside the MXF folder. I would suggest creating a folder with a name one greater than the biggest numbered folder currently in the MXF folder. For many of you, that means creating a folder called "2" and directing all of your Resolve rendered MXFs to that folder (see Figure 5.23).

Once the location for the rendered files has been chosen, the Add Job to Render Queue button will work. Click it to add the job to the Render Queue below and hit the Start Render button. For the tutorial material provided from the companion website, this render will take less than a minute.

If you didn't make any editorial changes to the sequence, go back to the Edit page and choose File > Save XML/AAF from the menu bar and choose AAF from the resulting window. Don't change the location from where you sent the original AAF from Avid; then hit Save.

Back in Avid, create a new bin to keep things separate and clear, and import the AAF file you just exported from Resolve (see Figure 5.24). This will automatically link back to the Resolve versions, which are graded. In my example, the last shot had gone through Color Management and FrameFlex before it went to Resolve. The resize that I did in FrameFlex worked perfectly, but since I had applied a LUT to the shot to make it look good before I sent it to Resolve and then in Resolve I applied the same correction again, that shot looks too bright and "contrast-y." Currently,

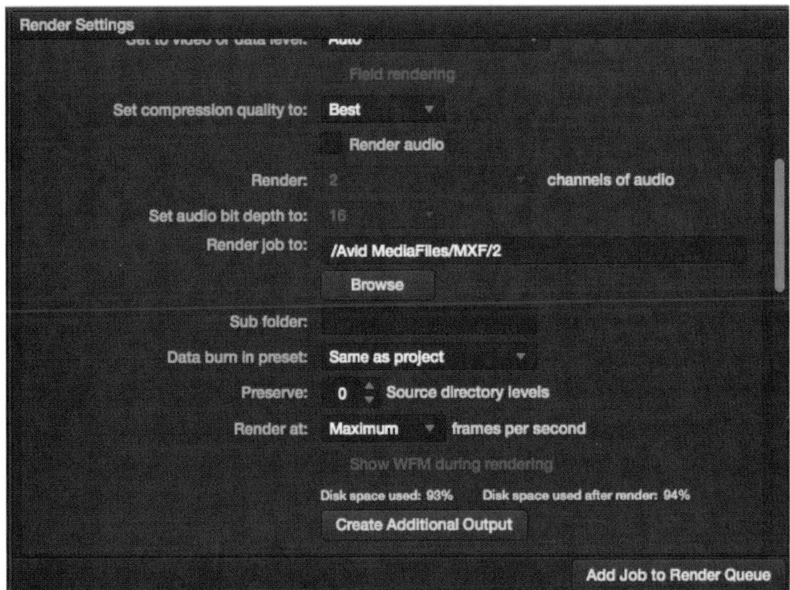

Fig. 5.23 Further down in the Render Settings dialog, you have the option for where to render the color corrections (Render job to). Note that the path selected is to the Avid MediaFiles folder, then the MXF folder and finally to a folder numbered 2. You will want this number to be one number higher than the current folder in the Avid MXF folder.

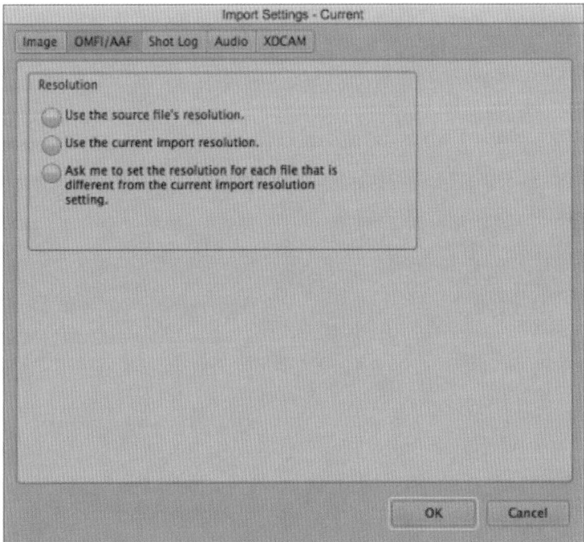

Fig. 5.24 Here we're back in Avid where the Resolve AAF that was just exported is imported back into Avid to complete the round-trip. In the Import options, choose the AAF/OMF option and elect to import the footage at the source settings (to maintain the quality and resolution created in Resolve).

Avid has no way to bypass all color management on export. For now, users would have to multi select the clips in the bin, open the source settings, and click on "Bypass all color transformations."

To have the change reflected on the Timeline, in the menu bar choose CLIP > Refresh Sequence > Color Adapters.

RED Conform Round Trip Using Camera Originals

Video Tutorial

To follow along with this tutorial, download the RED files from the companion website (www.focalpress.com/cw/hullfish).

We could have done these two round trips (Alexa and RED) the other way around, with the RED going DNxHD back and forth and the Alexa going back to the camera originals. It's not the camera itself that meant we had to choose a specific workflow. We just could only really afford to show two different file-based round-trip workflows, so we picked RED for doing it with the camera originals. If you've got Alexa footage and you want to round-trip through Resolve using camera originals, you can follow along with the RED workflow because it is basically the same. Or if you want to do a RED workflow and not have to go back to the camera originals, you can follow along with the Alexa workflow that we just demonstrated, and it will work the same. However, if you are using Alexa footage that was shot ProRes or DNxHD, then matching back to the original camera media instead of to the Avid footage won't improve the quality.

First, use AMA get the RED footage into Media Composer. This is covered in chapter 1 (starting at page 45), but here's a quick recap.

From the menu bar, choose File > AMA Link. Point to the .RDM folder level. Pretty much instantaneously the RED files will be brought in as AMA-linked files. Select them and choose Clip > Consolidate/Transcode. Select a drive to transcode to and an off-line resolution. The most typical would be DNxHD 36 (see Figure 5.25), but it is possible to choose a very highly compressed SD file on a project on which you are expecting a lot of material and you don't have enough hard drive space. Earlier in the book (starting at page 25), Robert Landau, of *Crimes of the Century* talks about

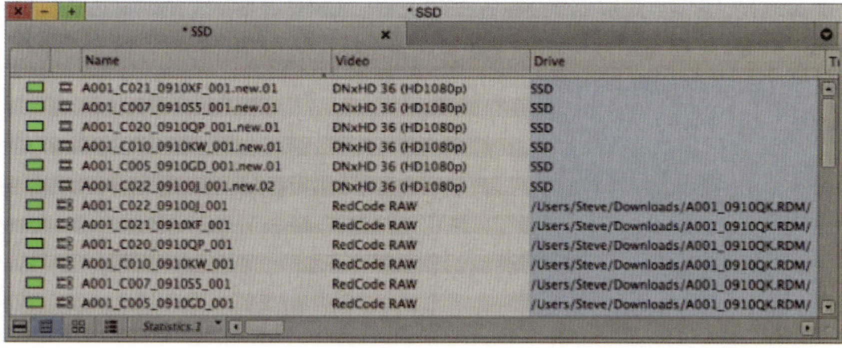

Fig. 5.25 Note that in this Avid bin there are AMA-linked clips at the bottom and transcoded DNxHD36 versions of the same clips at the top.

transcoding to 20:1 SD. With your options chosen, hit the Transcode button. Then with the transcoded media in the bin, you may want to move it to another bin, away from the AMA footage. From there, subclip most of each clip to another bin (see Figure 5.26). This would be common in a scripted piece, limiting the subclip to between the slate and the director yelling, "Cut!" or between "Action" and "Cut." In the case of the very short clips that we have from the companion website, just mark in and out points a few frames from each end and create subclips. Put those in yet another bin and edit only from the subclips.

Cut together a simple sequence, like the one made with the ARRI footage. Add some transitions, effects, and anything else for which you want to see how well the translation works or what problems might arise in the round trip (see Figure 5.27).

Select the sequence and choose File > Export. You will need to choose the Options button to open the File Export dialog. In the File Export dialog, choose to export an AAF from the Export Settings drop-down menu. Since this workflow will not use the DNxHD media and will utilize the original camera media in Resolve, we don't actually want or need any of the media from the Media Composer. In the Export options, there's really no option for "I don't need the media at all," so the best choice is "Link to (Don't Export) media." Make sure this is checked in both Audio and Video tabs. Actually with this workflow—and even in the ARRI workflow that we just discussed—you could leave the audio off entirely. Take note of where the AAF for this sequence is being saved. In this workflow, we're going to a Resolve system that is *not* on the same system as the Media Composer, so you'll need to transfer the AAF file to the Resolve system via e-mail, FTP, or even sneakernet. (If your Resolve is on the same system as your Media Composer, the workflow is virtually identical, you just

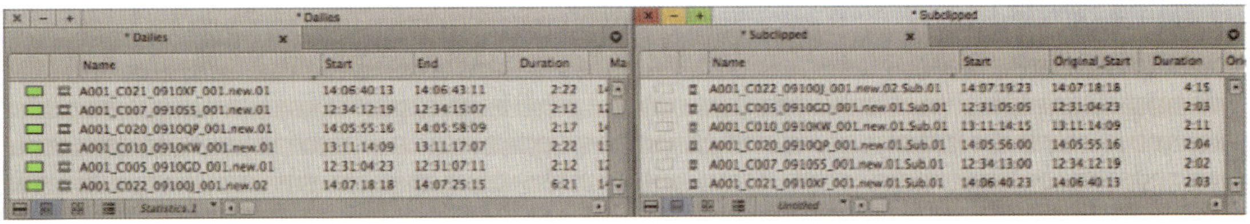

Fig. 5.26 The transcoded clips were placed in their own bin called Dailies. Then each clip was subclipped into a bin called Subclipped.

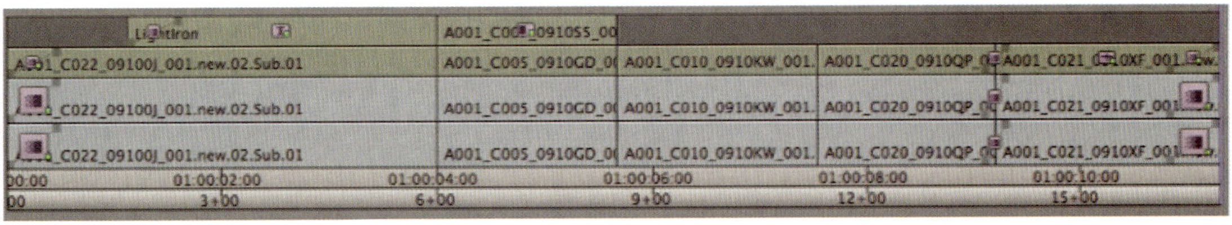

Fig. 5.27 This is a simple sequence cut together with the RED footage.

don't have to move the AAF file or copy the Resolve-rendered DNxHD files back to the Media Composer system.)

Launch Resolve, and create a new project (see Figure 5.28; see Alexa workflow for more in depth direction). Import the AAF into Resolve from the menu bar: File > Import AAF, EDL, XML. In the Load AAF options, choose Automatically set project settings, Automatically import source clips into media pool, and Link to source camera files (see Figure 5.29).

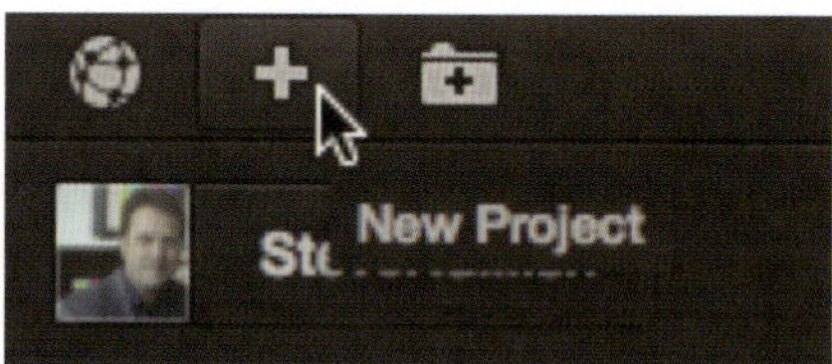

Fig. 5.28 In Resolve, clicking the + icon in the bottom left corner of the main UI will create a New Project.

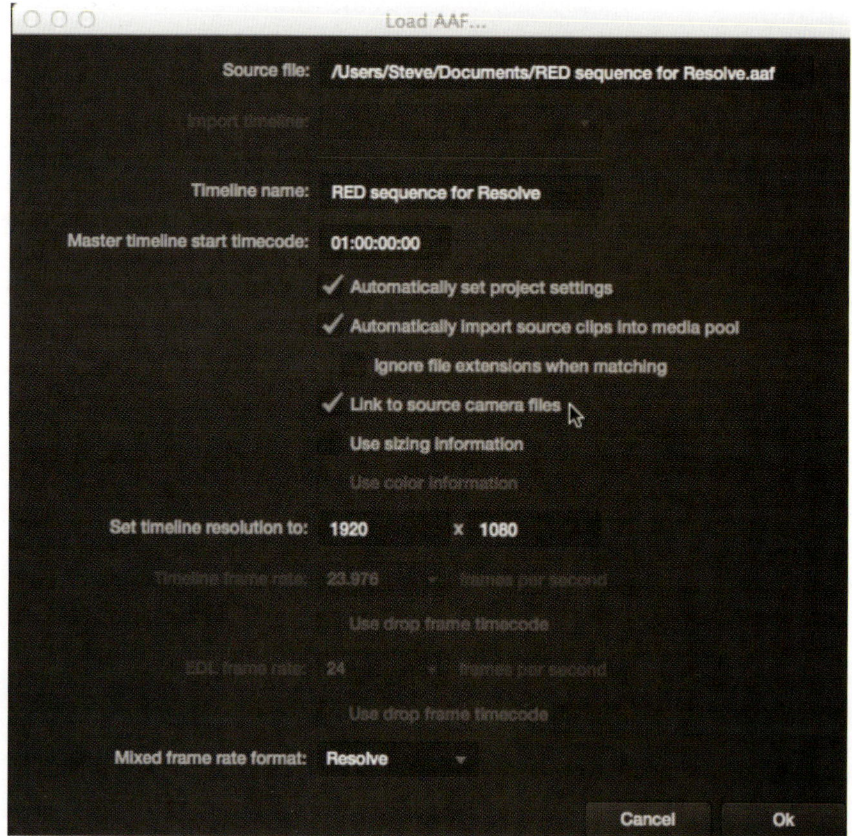

Fig. 5.29 This is the Load AAF (Import AAF) dialog window. Note the checkmarks next to the options you should select.

The real difference between the Alexa workflow that we just discussed and this workflow is choosing to Link to Source Camera Files.

To know that you are definitely linked to the source camera files instead of the off-line-quality DNxHD media, right-click on any of the clips for a contextual menu (see Figure 5.30) that allows you to reveal where the clip is from. In this case, you can see that the clip is an R3D file in the original RED RMD folder.

Color-correct the clips.

If you're going to render to another drive, make sure the drive is hooked up before you launch Resolve.

On the Deliver page, in the Render Settings (top right; see Figure 5.31) are the options for rendering out your rendered files. One of the simplest ways to get started is to use the Easy setup pulldown menu at the top of the Render Settings pane. The Avid AAF Round-trip setting is a good place to start. With that selected, scroll down in the pane to the Output section and, under Render To, click on the Browse button to select a drive or folder to render to. If you aren't seeing the connected drive that you want to render to, right-click on the File Browser window and choose Refresh from the contextual menu.

Sometimes it's easier to simply restart to get Resolve to see the drive. Or, you can create a new folder on your root drive and render to that (see Figure 5.32). Render the files using the source file names (see Figure 5.33) so the Avid system has a reference back through the AAFs to the original AMA import. Then send the renders to the render cue and hit Start Render.

With the render completed, you have to return to the Edit page and select File > Export AAF/XML... from the menu bar (see Figure 5.34). In the resulting window, name the AAF and choose where you want to save it. If you have an external drive where you rendered the media in the previous step, you can send it to that drive, since the AAF will accompany that media back to the Avid system.

If you rendered the media internally and exported the AAF to an internal location, you'll need to copy the media and the AAF to an external drive or network drive to get it over to the Avid system.

Move the drive to the Avid system, and copy the Resolve-rendered MXF files into one of the numbered folders in the MXF folder in the Avid MediaFiles folder. Personally, I like to have some control over this footage, so create a new numbered folder inside the MXF folder and move the Resolve files into that. When Media Composer is launched, it will scan the folder and generate the database files that it needs. Create a new project with the settings of the rendered Resolve media. Select the bin and Import the AAF file that you exported from Resolve, and, if necessary, relink the AAF-generated sequence in the bin to the Resolve-rendered media.

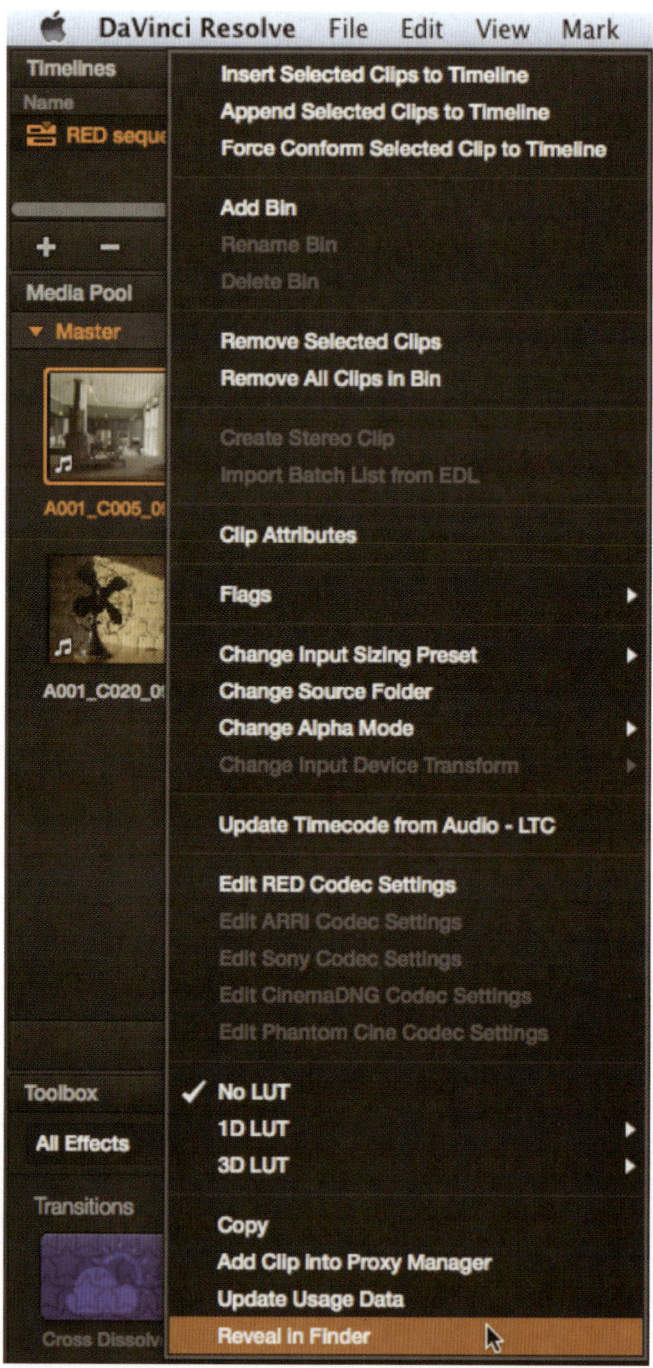

Fig. 5.30 This is a contextual menu in Resolve that allows you to Reveal the original media file. This allows you to insure that the media in the sequence is the correct version—not the Avid media, but the original camera media.

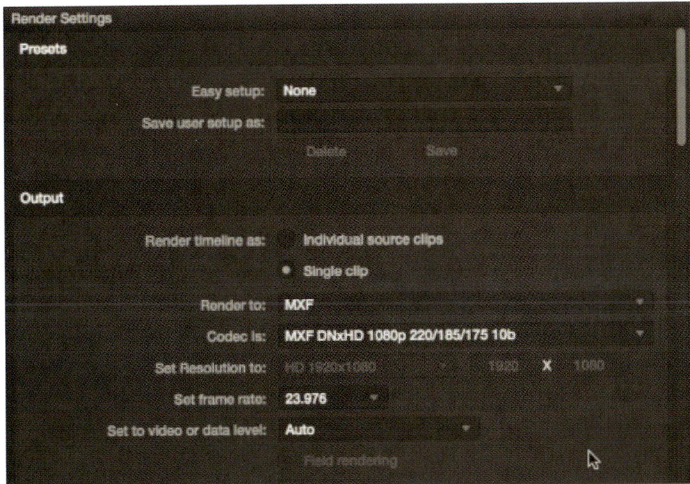

Fig. 5.31 On the Deliver page is the Render Settings Pane. Here you select how the color corrected clips will be rendered. The easiest way is to rely on the Easy Setup menu (at the top of this figure).

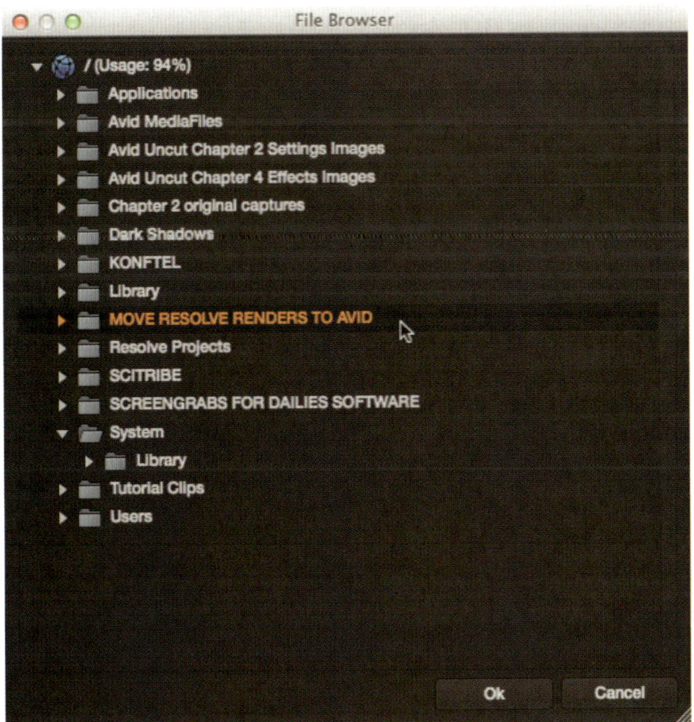

Fig. 5.32 Direct the renders to a folder that you'll use to move the files back to the Avid system. In this figure, I created a folder called "MOVE RESOLVE RENDERS TO AVID."

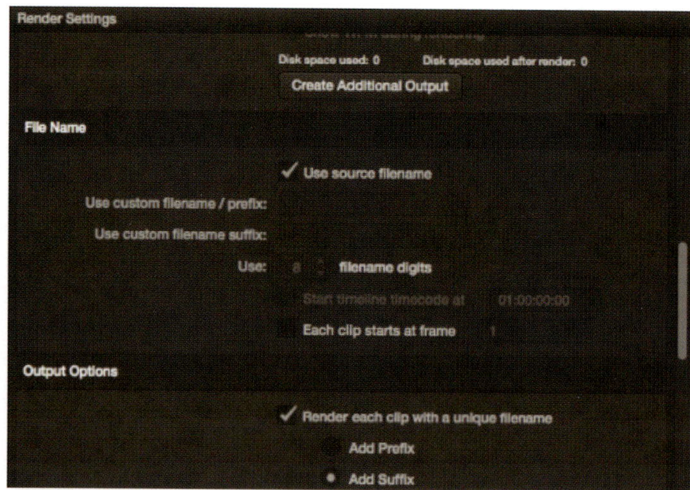

Fig. 5.33 In the Render Settings, note the checkmark next to Use Source filename. This is necessary for the round-trip to work.

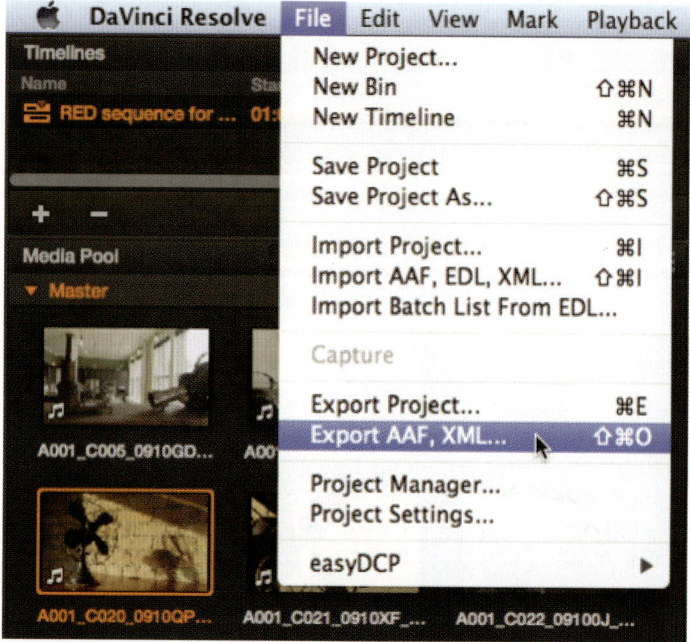

Fig. 5.34 After the renders are complete, export an AAF of the sequence from the File Menu > Export AAF, XML option. It would be wise to export the AAF to the same drive as the media was rendered to.

Color Correction

As an author, my most successful books are on the topic of color correction, so I feel pretty confident in my ability to guide you through this process. If you are interested in reading more than this half chapter on color correction, I would urge you to check out additional reading in the sidebar.

Color correction is important to delivering a good project. I like to tell the story of when I was first doing color correction on the beta version of Avid Symphony in the late 1990s. I am a pretty good editor, but none of my clients ever really oohed and aahed when I did a trim or juxtaposed the perfect shots or cleaned up a dialog edit that sounded perfectly natural. But when I started doing color correction, my clients really sat up and took notice. It locked in my clients because instead of appearing to be just a commodity, I was delivering something that they could see that they couldn't get from anyone else.

Of course, the color correction tools that make the most sense for the majority of Avid editors are the ones built in to Media Composer and the Symphony option. Symphony, specifically, has the ability to do relational color correction, which allows you to color-correct a shot and define one of a list of "relationships" to that shot and have all of the shots in that relationship color corrected instantly throughout the sequence without having to find them and apply the correction to each and every related shot. The relationships that can be defined, as of version 7, are Source Tape, Source Clip Name, Master Clip, Subclip, and Source Segment. Hopefully, in the near future, Avid will add some more relationships that will help with the new file-based workflows. We'll discuss the use of these relational color correction methods shortly.

One of the advantages of color correction in Symphony is that it's not treated like an effect that's applied to certain clips but rather as its own mode that allows for grading in context with the clips around it. The interface and its capabilities have scarcely changed from the late 1990s when the tools were originally implemented in Symphony. At the time they were created, they were pretty revolutionary for an NLE, but in nearly twenty years, they've grown a little stagnant. Though Symphony's toolset is definitely better than Media Composer's, it still needs updates, especially in the area of secondary color correction and with its interface with the Avid MC Color panel user interface (trackball controller; see Figure 5.36).

Let's take a look at the tools and workflow for doing color correction natively in Symphony. I'll describe the Symphony color correction tools and let you know when the Media Composer tools differ from Symphony. This should give you a good indication of what the differences are and whether the Symphony option is worthwhile for you.

The Art and Technique of Digital Color Correction, 2nd edition by Steve Hullfish. Publisher: Focal Press (May 11, 2012). ISBN-10: 024081715X ISBN-13: 978-0240817156

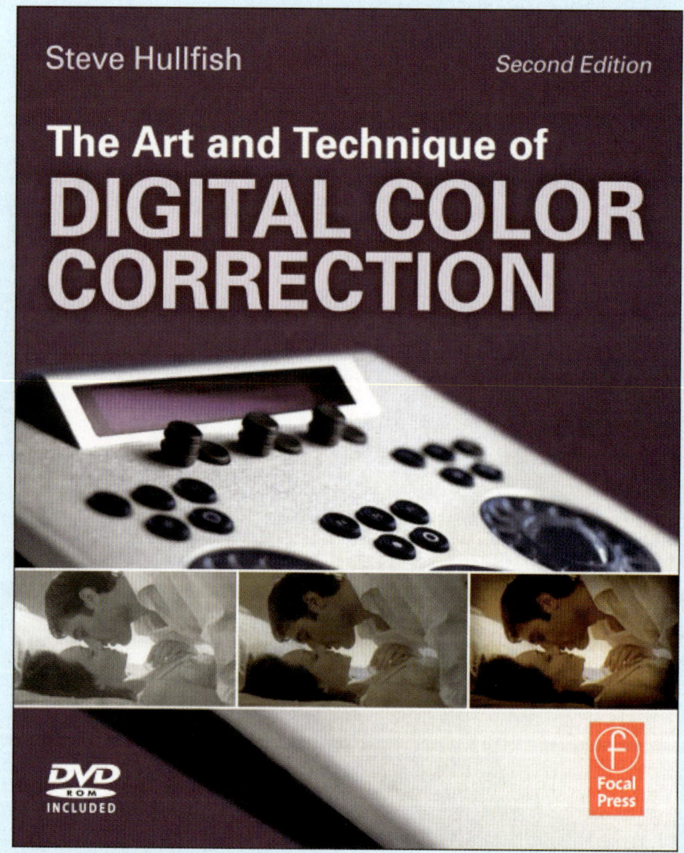

Fig. 5.35 The Art and Technique of Digital Color Correction, Second Edition.

Color Correction for Video: Using Desktop Tools to Perfect Your Image, 2nd edition by Steve Hullfish and Jaime Fowler. Publisher: Focal Press (October 27, 2008). ISBN-10: 0240810783 ISBN-13: 978-0240810782

Color Correction Handbook: Professional Techniques for Video and Cinema by Alexis Van Hurkman. Publisher: Peachpit Press (Nov 25, 2010). ISBN-10: 032171311 ISBN-13: 978-0321713117

Color Grading with Media Composer and Symphony 6 by Bryan Castle. Publisher: Cengage Learning (October 3, 2012). ISBN-10: 1285181417 ISBN-13: 978-1285181417

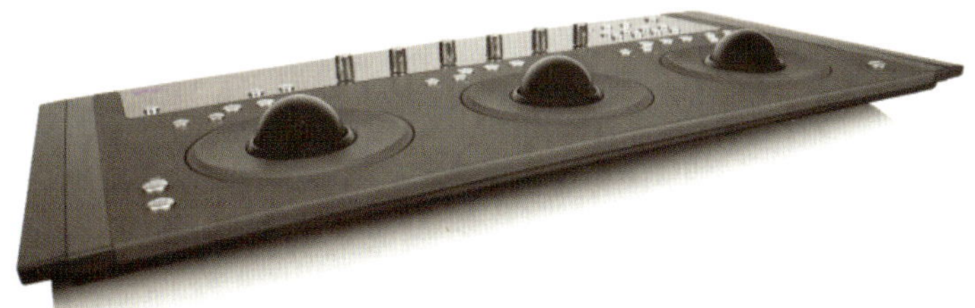

Fig. 5.36 The Avid MC Color panel can be used to control the Avid Media Composer, Symphony color interface, Baselight color correction plug-in, and DaVinci Resolve.

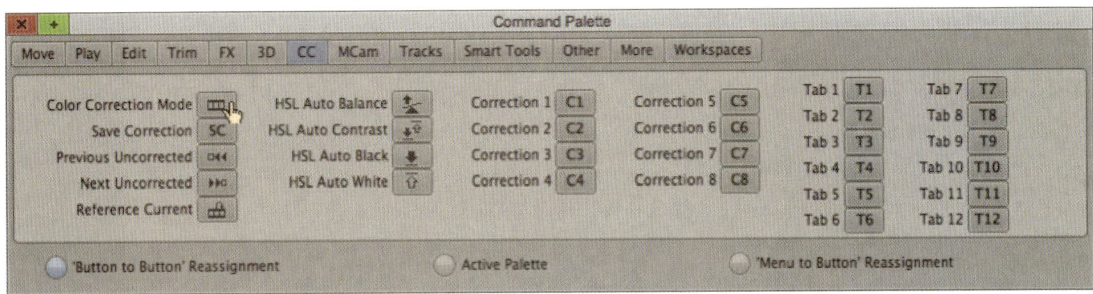

Fig. 5.37 In the CC tab of the Command Palette, find the Color Correction Mode button and map it to the user interface or keyboard.

Native Avid and Symphony Color Correction

The great thing about "workflow" in Avid color correction is that there really isn't a workflow. With a sequence of shots in the Timeline, all you have to do is go into Color Correction mode using the Color Correction Mode icon. The only place this icon is available on the default user interface (UI) is under the SmartTools to the bottom left of the timeline. Before we go too much further, it's probably a wise idea to map the Color Correction Mode button to your keyboard, or maybe above the Timeline (see Figure 5.37).

Instead of the Color Correction Mode button, you could also go to the menu bar and choose Windows > Workspaces > Color Correction to launch the Color Correction workspace. The value of using the workspace instead of just the mode is that you can customize the sizing and placement of all windows. The size and shape of the color correction tools window can't be adjusted at all, but the windows and settings around it are all adjustable. My Color Correction workspace has the Composer window (which, in Color Correction mode has three windows instead of just two) as big as possible, stretching from one side of the monitor to the other. Then I place the Color Correction Tools window centered in the monitor directly below the Composer window. The Timeline is then placed so that it goes directly below the Color Correction Tools window and all the way across the screen. It's good to have at least one bin open and accessible so that you can drag color corrections to it (see Figure 5.38). After making changes to the workspace, save them by using Windows > Workspaces > Save Current (see Figure 5.39). If you want to be able to trigger this workspace from the keyboard or the UI, it's mappable from the Command

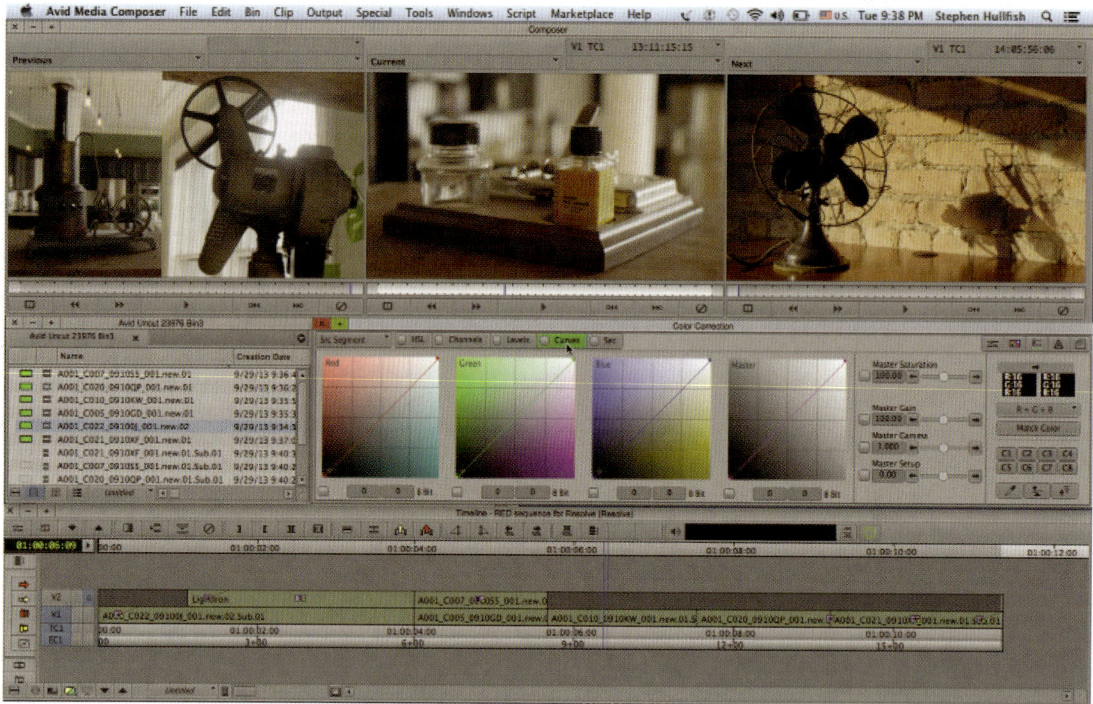

Fig. 5.38 This is my customized Color Correction Workspace for my laptop.

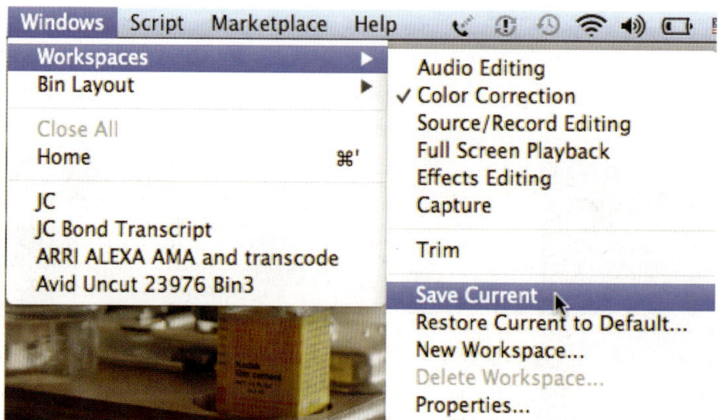

Fig. 5.39 With the Color Correction Workspace open and customized, go to the Windows menu in the menubar, then the Save Current option to save the customized Workspace.

Palette in the Workspaces tab; just make sure the bottom radio buttons are set to Button to Button Reassignment and drag the Color Correction workspace to a spot on your keyboard or UI.

The Color Correction Interface

The only windows that really matter are the Color Correction window, the Composer window, and, to a minor extent, the Timeline.

Composer Window

The Composer window has three monitors in Color Correction mode (see Figure 5.40). The default content of these monitors is (from left to right) Previous, Current, and Next, referring to which clip in the Timeline is displayed on the monitor. The center one, Current, displays the clip under the Timeline indicator. This is the monitor you watch to see the effect of the changes you make with the Color Correction window. The benefit of having the left and right monitors set to the default of Previous and Next is that it allows you to color-correct in the context of the scene, which provides a reference to the shots that come before and after.

However, if you want, you can change these monitors using the spin-down arrows above each monitor. The options for these monitors include Empty, Entire Sequence, Reference, Current, Previous, Next, Second Previous, Second Next, Quad Display, RGB Histogram, RGB Parade, Vectorscope, Y Waveform, YC Waveform, YCbCr Histogram, and YCbCr Waveform. Some thoughts on each of these choices:

Empty: This option is beneficial with effects-heavy sequences because it can save time, especially when the effects are third-party effects. It also helps reduce visual clutter.

Entire Sequence: If you want to be able to scroll through the entire Timeline in order to compare shots anywhere in the Timeline, Entire Sequence is a good choice.

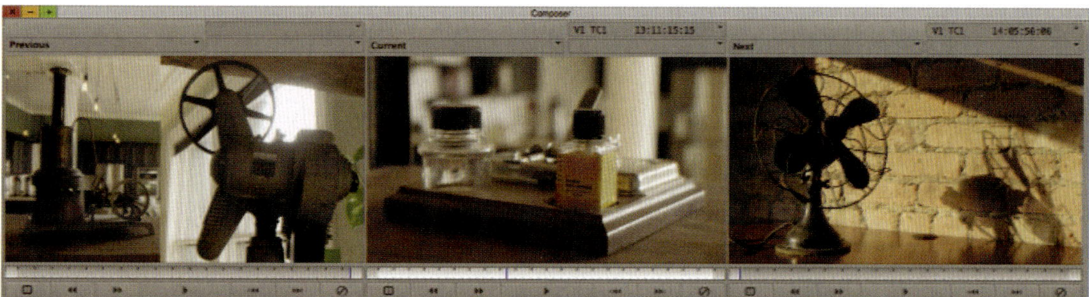

Fig. 5.40 The Composer window in Color Correction mode has three monitor instead of the two that are usually seen in the standard editing mode. The default for these monitors shows—from left to right—the Previous, Current and Next shot in the sequence.

Reference: With the Current monitor set to an image you want to use as a reference, choose Reference to keep the image as a reference or "freeze frame."

Current: This displays the footage under the Timeline indicator. It is necessary to have Current chosen on at least one monitor in order to monitor what you are color correcting.

Previous: This allows you to see an image of the clip before the Current clip.

Next: As expected, this allows you to see an image of the clip immediately after the Current clip.

Second Previous: This allows you to see the clip before the clip before the Current clip.

Second Next: This allows you to see the clip after the clip after the Current clip.

Quad Display: Choose this to see multiple "scopes" at the same time. The displays in the Quad are YC Waveform, Vectorscope, RGB Parade, and RGB Histogram.

RGB Histogram: An RGB Histogram allows you to see the distribution of red, green, and blue channels from black to white.

RGB Parade: RGB Parade shows a waveform monitor that lets you see the relative levels of the red, green, and blue channels. This is a critical display for doing good color correction.

Vectorscope: This shows the saturation and hue of the signal with hue indicated around the circular graticule of the vectorscope and saturation indicated by the distance from the center of the vectorscope.

Y Waveform: Y is an indicator of luminance, so this is like a low-pass waveform, showing only luminance and not chrominance.

YC Waveform: YC means that luminance and chrominance are shown, so this is similar to a flat-pass waveform display.

YCbCr Histogram: Y again is luminance. Pb is the difference between luminance and blue, and Pr is the difference between luminance and red. These three signals are shown in the form of a histogram. This is not an easy display to read.

YCbCr Waveform: Y again is luminance. Pb is the difference between luminance and blue and Pr is the difference between luminance and red. These three signals are shown in the form of a parade waveform monitor. This is not an easy display to read.

The other controls in the Composer window are the buttons under each of the windows (see Figure 5.40). The default buttons have always served me well, and so I leave them at the default.

From left to right the buttons are Dual Split, Rewind, Fast Forward, Play, Previous Uncorrected, Next Uncorrected, and Remove Effect.

Dual Split: This button allows you to see a split between corrected and uncorrected, unless the correction has been rendered.

Rewind: This is helpful for jumping backward from shot to shot, regardless of whether the shots have been corrected. Rewind jumps to the headframe of the previous clip in the Timeline.

Fast Forward: This is helpful for jumping forward from shot to shot, regardless of whether the shots have been corrected. Fast Forward jumps to the headframe of the next clip in the timeline. All three Rewind and Fast Forward buttons in the Composer window move through the Timeline the same way.

Play: Not very useful in Color Correction mode, this button plays only from the location of the current Timeline indicator position to the end of the current clip. Since most clips are pretty short, there's not a lot of value to this. Surprisingly, in Color Correction mode the buttons in the Composer window are not mappable. There are a number of better choices for playing video in Composer mode, but you can access them from the keyboard or the Timeline UI. So there are a number of buttons you'll want to map to the Timeline or to the keyboard. In Composer mode, Play and Play to Out operate virtually identically, so there's no need to map that button. The Play In to Out button can be useful if you want to specifically set in and out points before playing through the section of the sequence you want to watch. A very useful button is Edit Review. This button is essentially a macro. When you press Edit Review, the playhead jumps backward to the previous transition point (the beginning of the current clip), then it jumps back five seconds before that and goes into Play. In Color Correction mode, it will play through from that point all the way to the end of the sequence, or until it's stopped. The Play Loop button can also be used to bypass stopping at the end of the current clip.

Previous Uncorrected: I've never used this button because I tend to grade my sequences in order.

Next Uncorrected: I've not used this button much, but it would be useful if you had a lot of relational color corrections in a sequence that you knew were already corrected and you just wanted to move down in the sequence to the next uncorrected shot. Think of this like a Find button: "Find the next shot that I haven't color corrected."

Remove Effect: Park on a clip and use this button to delete color corrections. Oddly enough, the Remove Effect button doesn't really work in the Previous or Next Composer windows, but it does work in the Current window.

In the Composer window in Color Correction mode, with the Dual Split button active, the default position of Dual Split is a vertical wipe down the middle of the image. This can be customized using the small white triangles at the top and bottom of the wipe. While the wipe can't be diagonal, all four corners of the split can be dragged to place the "before" image so that it's on either side of the frame or even on the top or bottom or as a small rectangle in the middle of the image. If you drag the dual split out so that it fills the image, you can use the Dual Split button to toggle the entire frame between corrected and uncorrected.

TIP

It is possible to use Command-K and Command-L (Mac) and Control-K and Control-L (PC) to zoom in and out of each color correction monitor. Sometimes this is useful for selecting colors from the screen because it allows for more accurate placement of the cursor.

An important thing to note about Color Correction mode is that having the proper track selected is very important. Avid Symphony does not allow you to add color correction to filler, unless you have CC Effect selected in the relational color correction selection. But Media Composer does, and if you aren't paying attention, you can think that you are adding corrections to clips on, for example, V1, but instead you are adding the corrections to the filler of the track above it. Color corrections are added to the clip under the Timeline indicator in the Timeline on the selected or active *top* track.

Color Correction Window

The Color Correction window is organized into tabs. If you have the Symphony option added, the tabs are CC Effect (under Src Segment), HSL, Channels, Levels, Curves, and Sec (see Figure 5.41).

CC Effect Tab

The first tab lets you apply corrections to a clip; then Symphony automatically applies the same correction to any clip with a specified relationship to that clip. This option is only available in Symphony. This tab is not in the Media Composer interface. The relationships that you can use are Src Tape, Src Clip Name, Master Clip, Sub Clip, and Src Segment (see Figure 5.42). You can also add a correction to an entire Program track or Program segment, and CC Effect. This relational color correction is available only with the Symphony option.

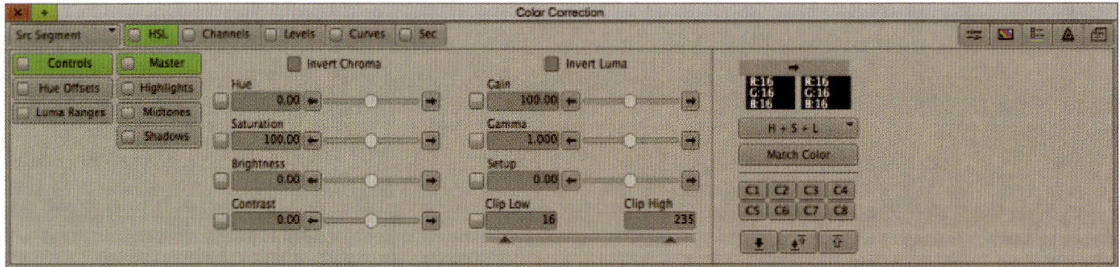

Fig. 5.41 This is the Color Correction Window. In this figure, the HSL tab is active.

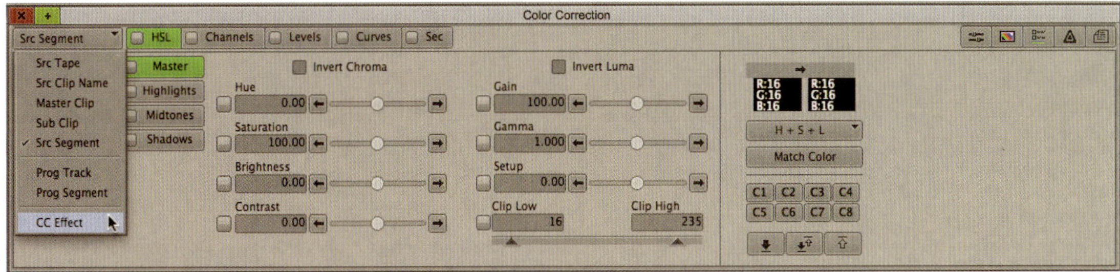

Fig. 5.42 This shows the CC Effect Tab, where you can select various relationships so that your current correction is applied to all of the other clips in the sequence that have that relationship.

Src Tape: This is becoming more and more useless with the overwhelming dominance of file-based workflows. And in some respects, it wasn't that useful to begin with, since it is a rare tape that has the same color correction needs from beginning to end—though for many interviews, an entire tape will have the same look and lighting throughout. However, if you capture or digitize a tape and need to apply the same correction to every instance of that tape in your sequence, this is the relationship to use.

Src Clip Name: This is one of the most important relationships that you can use. The reason for this is that usually color correction is applied during the on-line process, and to get to the on-line stage, master clips are often consolidated or decomposed when they are up-rezzed. This means that a longer clip that might need to have color correction applied relationally no longer has the master clip relationship because the up-rezzing and consolidating process breaks that longer clip into a series of smaller "master clips" labeled with the suffix ".new," so Avid sees each of these .new files as Master Clips. That means that trying to grade them relationally won't work, but since they all have the same name before the suffix, Src Clip Name will allow them to be graded based on the original Source Clip Name.

Master Clip: This works great for many clips unless they're decomposed or consolidated into new, smaller Master Clips.

Sub Clip: This is also an excellent relationship to use for many purposes. You can specifically choose the section of footage from a larger Master Clip that you want to relationally correct. For example, in the Master Clip, the camera may start indoors, but then the camera and subject walk through a door to the outdoors, and from that point on the color balance is wrong. In that case, you can subclip the outdoor section of the shot and then use the Sub Clip option to apply a relational color correction to grade

> **Warning! (or Tip)**
>
> When you use these relationships, it's critical to understand that once you've assigned a relationship, changing the grade on any segments later in the Timeline that have a relationship to an earlier segment will change *all* other clips with that relationship, including earlier segments that you *thought* you had already graded. So, you have to be very cognizant of the colored indicators that show that a segment has already been graded. If you want to change the grade on a shot that has been graded relationally, you have to remember to *first* change the relationship on the later segment to Src Segment so that you grade only that segment, not all related segments.

everything that happens after they go outdoors without affecting the indoor portion.

Src Segment: This means that you are grading the current segment with no other relationship. This is the default that doesn't affect any other clips.

HSL Tab

This tab is different depending on whether you have the straight Media Composer functionality or the Symphony option. Inside the HSL tab on Symphony there are sub-tabs for Controls, Hue Offsets, and Luma Ranges (see Figure 5.43).

In the Controls tab in Symphony, there are options to adjust the Master, Highlights, Midtones, and Shadows. In Media Composer these options are not there. The strange thing is that with Symphony, these four groups of settings provide access to parameters that include Hue, Saturation, Brightness, Contrast, Gain, Gamma, Setup, and White Point/ Black Point (Clip High and Clip Low). For Symphony it means you can adjust the highlights of the shadows or the shadows of the gammas or the midtones of the highlights. This doesn't seem very intuitive, but it essentially breaks the tonal range of your picture into nine controllable tonal ranges. Consider it to be really dark shadows, kind of dark shadows, lighter shadows, dark midtones, middle midtones, light midtones, darker highlights, middle highlights, and bright highlights. It also means that in Symphony you can adjust the saturation of highlights, midtones, and shadows separately.

While we're in this segment, let me provide my basic "rookie" color correction warning: Do not use the Brightness and Contrast controls. One of the most important things about color correction is having *control*, and these two parameters rob you of control. Brightness raises the entire level of shadows, midtones, and highlights equally. This is something you really never need to do. If your levels are too dark or too bright, you need to control these three tonal ranges separately, not all together. Using the

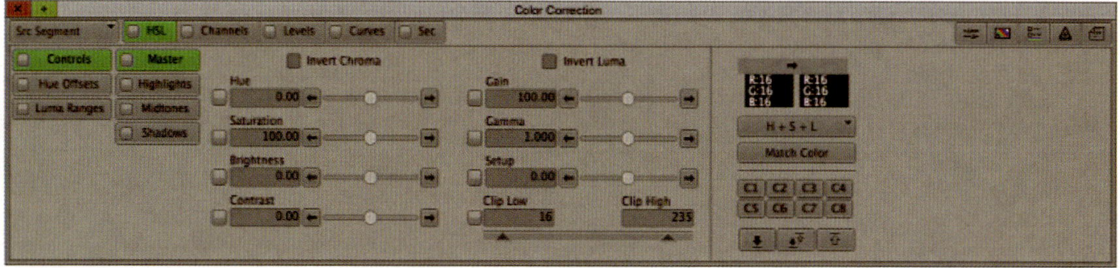

Fig. 5.43 This is the HSL Tab in Avid Symphony. Symphony provides three additional sub-tabs for Highlights, Midtones and Shadows. It also has sub-tab for Luma Ranges that Media Composer does not have. The rest of the interface between is similar.

contrast control is similar. It controls both the highlights and the shadows equally. When you lower contrast, it raises shadows and lowers highlights in equal amounts. When you raise contrast, it lowers shadows and raises highlights in equal amounts. If you want to increase the contrast of your images, you want to have separate control over each of those parameters (highlights and shadows).

Generally speaking you don't want to change the Clip Low and Clip High amounts, unless you're aiming for some kind of graphic look.

In these days of file-based capture, there's not much of a need for the Hue control unless you have a graphic background and you want to change the hue of the graphic. Otherwise, changing Hue is usually best done in the Hue Offsets tab, where you also have control of saturation and tonal ranges. Though Ra-ey Saleh, the on-line editor from *Mythbusters*, provides this little tip for using the hue slider: The color Green is by far the hardest to get out of skintones. Everything else in the grade could be perfect, but sometimes no matter what you try there is still a slight green cast to the skin. Move the Hue Slider 2 or 3 points to the right on the Avid. This will make the green turn more red, and the skintones will work.

Saturation is very useful, especially with the Symphony option where you can control the saturation in the shadows, midtones, highlights, or overall saturation. There are many instances where you want to lower saturation only in a specific tonal range, and this is definitely one of the benefits of Symphony. It is very common for colorists to want to limit saturation changes to just the highlights or just the shadows. With overall saturation or specific tonal range saturation changes, usually you do saturation changes near the end of your color correction because certain tonal range corrections will affect saturation, so you want to wait to see what the effects of those corrections are before you apply additional saturation or remove saturation.

The Hue Offsets controls in the HSL tab (see Figure 5.44) are the main tools that most "real" colorists use. The Hue Offsets in Avid are analogous to the three-way color corrector in Final Cut Pro and very similar

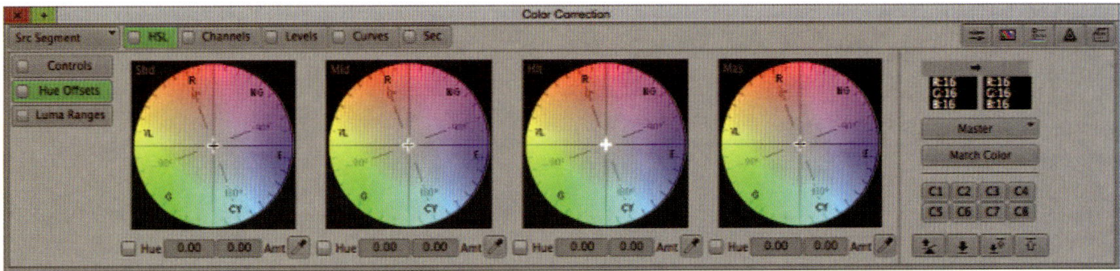

Fig. 5.44 This is the Hue Offsets tab in the HSL Tab. This is the Symphony version, which has an extra "Master" wheel. The Media Composer just has the three wheels to the left. This is similar to FCP's Three Way Color Corrector.

to the main UI controls on a DaVinci or other color correction system, allowing for hue and saturation control of each of the three tonal ranges. Symphony has an additional color wheel (hue offset) for the Master level, but other than that, the Hue Offsets tab works the same way in Media Composer and Symphony. Generally speaking, I don't use this master wheel because in most cases the hue and saturation adjustments that need to be done are all different in each tonal range. Sometimes you actually want the hue offset adjustments to go in different directions in different tonal ranges.

Manipulating the hue offset wheels is as simple as dragging the center crosshairs on any of the wheels in the direction that you want that tonal range to go. So, if you want warmer midtones (ideal for fleshtones, usually), drag the crosshairs up towards red, which is at about eleven o'clock on the wheel. Are your highlights too yellow? Drag the crosshairs in the opposite direction of the color cast that you want to eliminate. The opposite of blue is yellow (directly across the wheel). Opposite colors are always 180 degrees apart on the color wheel. So, the opposite of blue is yellow, the opposite of green is magenta, and the opposite of red is cyan.

The Hue Offsets controls are easiest to manipulate when monitoring a vectorscope, because the wheels mimic the shape and direction of the vectorscope.

Along the bottom of each color wheel is a button to the left of the word Hue. This is the bypass, or reset, button. If you click it so that it is gray, it is disabled. If you toggle it to blue, it is active. These enable/disable buttons are also available for each tab. Holding down Option (Mac) or Alt (PC) while clicking these enable/disable buttons resets the parameter or tab to its factory default. Also, along the bottom are numerical entry boxes in which you can enter or tweak specific hue offset numbers. You can also use the eyedropper under each wheel to balance specific tonal ranges. If you do this, remember that the eyedropper is expecting a color from the tonal range it is associated with, so when you use the shadow eyedropper (far left), you should try to select something that you think is pure black. When using the midtone eyedropper, select a color that should be pure gray. And when using the highlight eyedropper, try to sample something that is pure white.

TIP

When sampling pure white, make sure that you are not sampling a clipped white image. Clipped highlights can be deceptive because the three color channels can be clipped at different levels. For example, if your image is tinted blue and the highlights are clipped, the blue channel will be more clipped than the others. Using that clipped white as a reference will mean that the pure clipped white will be balanced, but as you get farther from that clipped tonal range, the blue will not be adjusted enough.

To the far right is a section for color matching at the top and for the "color buckets" in the middle and for auto-balancing and auto-levels along the bottom.

At the top there are two color chips with an arrow above them. The chip to the left is what you are trying to match, and the chip to the right is what you want it to look like at the end of the match. The cool thing with the color chips (they default to black chips with R, G and B at 16) is that you can Option-drag (Mac) or Alt-drag (PC) the chip to a bin to save it.

You can also drag a saved color from a bin to the right chip. That's useful if you have an approved client brand color or an approved skin tone that you're trying to match from one project to another. You can even load a specific color from the color picker into the right-hand color chip by double-clicking the right-hand chip and dialing in a color, such as pure black or pure white, or a sponsor's brand color.

The thing to know about matching colors is that the look of the match will be different depending on which tab you're in. For example, the only tab with NaturalMatch is the Curves Tab. There's also one in the Secondaries tab of Symphony. NaturalMatch is a trademarked (if not actually patented) technology that matches hue while allowing the luminance to be different, and then adjusts the saturation to have the correct relationship to luminance because the saturation of something is usually dependent on the luminance to some extent. Saturation is not possible in video at either end of the luminance spectrum. As you approach black or white, saturation automatically decreases. The effect of the color match is also dependent on the specific Match Type set in the pulldown menu directly under the color matching chips. The specific choices in the pulldown menu differ from one tab to the next. In the HSL tab, the choices are H+S+L, Hue, Saturation, Luminance, H+S, H+L, and S+L. You can determine which matching type you want to use for any given match. When you have selected a color to match in the left chip and a color to be matched in the right chip and a Match Type with the pulldown below them, use the Match Color button to execute the match.

Another consideration when using color matching is that it's dependent on the exact pixel that is chosen. As I pointed out earlier, it can help if you zoom in to the Composer window when trying to pinpoint a specific color. The other thing that can help is to customize the way that the eyedropper selects pixels. The default is that the eyedropper selects a single pixel that's directly under the tip of the eyedropper tool. However, this is a little dangerous because noise in the video or even minute variations, like dust, can mean that picking a pixel in the middle of what looks like a homogeneous patch of color can really deliver an unexpected result. However, in the Correction Setting you can adjust the number of pixels that are averaged together to create the color reading. The setting is in the Features tab of the Correction Mode setting. Just check the Eyedropper 3 × 3 averaging box to enable averaging. This is the setting that I use. This tab also lets you see the specific color data (which is the default) and a choice to let the color picker choose from anywhere in the application instead of just in the Composer window.

Under the Match Color button are eight buttons, labeled C1–C8. These are the "color buckets." They are temporary storage locations for color corrections. Especially when using a Media Composer without the

TIP

It is possible to map the color buckets to keyboard strokes. The value of this is that you can apply saved color bucket corrections to shots at the touch of a key. The color bucket buttons are mappable from the Command Palette in the CC tab. You can also Option-drag (Mac) or Alt-drag (PC) from the buckets to a bin. Then you can name them and even move them back to the buckets later.

Symphony option for relational color correction, these buckets are very useful for storing a correction for quick application to another clip. Once you have color-corrected a clip, you can Option-click (Mac) or Alt-click (PC) on one of the color buckets to load that color correction into the bucket for later recall to apply to another clip. When a color bucket has been loaded with a correction, the C# designation is replaced by the rainbow-colored color correction icon. You have to basically memorize which corrections you load into each bucket. It's one of the minor failings of the concept. There is a lot of wasted space in the design of the UI. As you can see from a screenshot, there is a lot of space to the right of the buckets. If this space was used to store thumbnails of the color corrections that are saved, it would be easier to remember which stored color corrections were in each bucket. Maybe by the time you read this, this brilliant idea will have been implemented. If not, it's a bit like playing the memory game in which you keep flipping cards over until you match up the pairs.

Below the color buckets are icons for Auto Black, Auto Contrast, and Auto White. The options for these automatic corrections change depending on which tab you are in. The Levels and Channels tabs don't have any Auto Correcting functions. The Curves tab has Remove Color Cast, Auto Balance, and Auto Contrast buttons.

The third control in the Symphony option of the HSL tab is Luma Ranges (see Figure 5.45). This control is unique to Avid. In most color correctors, the definition of shadows, midtones, and highlights is preprogrammed into the application. In Media Composer/Symphony, the Luma Ranges tab allows you to alter the definition of shadows, midtones, and highlights. These definitions affect the way that any controls that are defined by shadows, midtones, and highlights work.

One way that Luma Ranges can be helpful is if there is a specific tonal range, such as a very bright window, that you want to control. Instead of using the default definition of highlights, you can tighten that definition by adjusting the highlights curve to be more specific to the luminance of the window instead of the broader definition of a

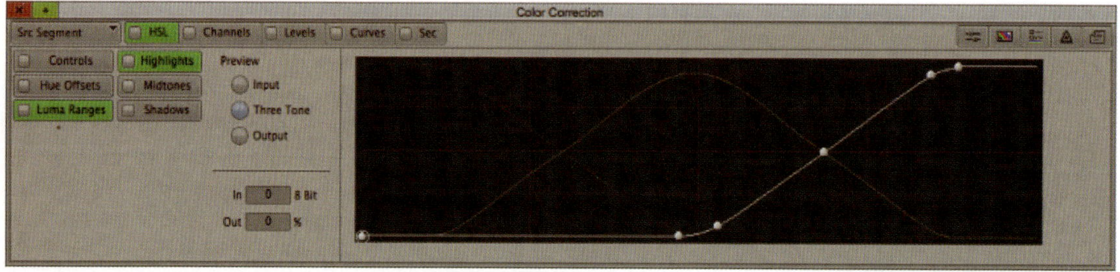

Fig. 5.45 The Luma Ranges tab in Avid Symphony (not available in Media Composer) allows you to control what range defines shadows, midtones and highlights.

highlight. The best way to see what you're doing as you make adjustments to the Luma Curves is to enable the Three Tone button under Preview. This turns the Current Composer window into a three-toned grayscale image. The shadow tonal range is depicted in black. The midtone tonal range is depicted in gray, and the highlight tonal range is depicted in white.

Channels Tab

The Channels tab is available only in the Symphony option (see Figure 5.46). This tab is not an easy one to use to do normal color correction. The way the Channels tab works is that it combines various color channels, including red, green, blue, luma, Cr, Cb, and Offset, by certain percentages to revise or even re-create a color channel. There are no controls in this tab to be able to adjust levels with the precisions and control of specific tonal ranges. One of the few places in which I've utilized this tool successfully is in the case of real damage or even the complete loss of a specific color channel. One example I salvaged was on the edit of a broadcast tennis match. The broadcast went perfectly, but it was discovered that one of the iso camera decks was connected improperly and was missing a color channel. I was able to use a combination of luma and one other similar channel to rebuild the missing channel. The colors were still off a little, but I used additional color correction tabs to bring the colors into closer alignment. Another place where this is sometimes useful is in the case of very noisy video. Oftentimes, this noise is present only in a single color channel. By changing the composition of that noisy channel (lowering the percentage of the noisy channel and rebuilding the missing level by adding a percentage of one of the other channels), you can actually lower the noise level appreciably. It's also good for the preliminary step when grading footage from Panasonic cameras that have been set up too red (this also happens to other cameras with certain overly strong color channels). You can take the red channel down to 90 or so and then continue grading from there.

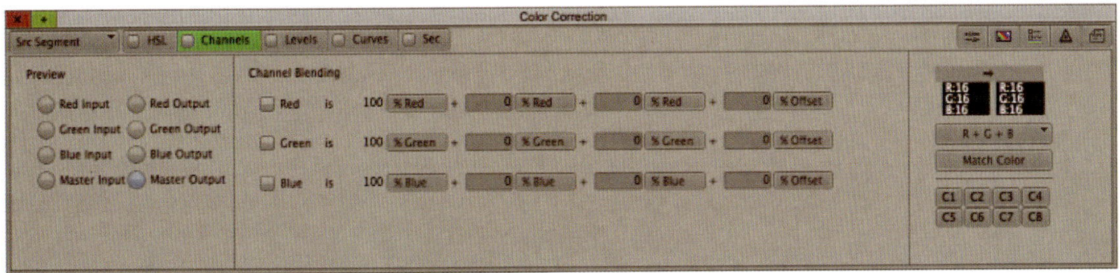

Fig. 5.46 The Channels tab is available on in Symphony. It is not very intuitive, but is very effective in rescuing difficult footage.

Levels Tab

The next tab in the Color Correction window, if you have the Symphony option, is the Levels tab (see Figure 5.47). Personally, this is not one of my favorites, but for many—especially those used to using the tools in Photoshop—this can be a very productive place to do color correction. Essentially, it provides the user with editable histograms. Numerous histograms are presented: Composite, Luma, Red, Green, Blue, and Master. For each of these options, the UI presents an incoming and an outgoing histogram on the left and right, respectively, with a curve in between. All three are able to be edited. But essentially, they all offer the same controls. The tricky thing is that the way the incoming and outgoing histogram controls work is backwards. The reason for this is that the incoming and outgoing histograms map their values to each other. What this means is that, depending on your point of view or experience in using histograms, the behavior of one of the histograms will be backwards from what you are expecting. For me, it is the incoming histogram. This is especially noticeable when adjusting the midtone point on the histograms. For me, when I move the midtone point left, towards black, I expect the midtones to become darker. That is the behavior you will get if you move the *outgoing* midtone point. However, if you move the *incoming* midtone point toward the left, the midtones will become brighter. The reason for this is because by moving the incoming midtone toward black, you are asking the system to remap darker midtones on the incoming histogram to the same middle-gray point of the outgoing histogram. This is something you'll want to play with a little to get the behaviors "under your fingers."

You can also control grades in the Levels tab using the "curve" in the middle, but this is nearly identical to the Curves tab, which is what we're about to discuss. The difference is that in the Curves tab, you can see and manipulate all of the different curves without switching between tabs, as you must do in Levels.

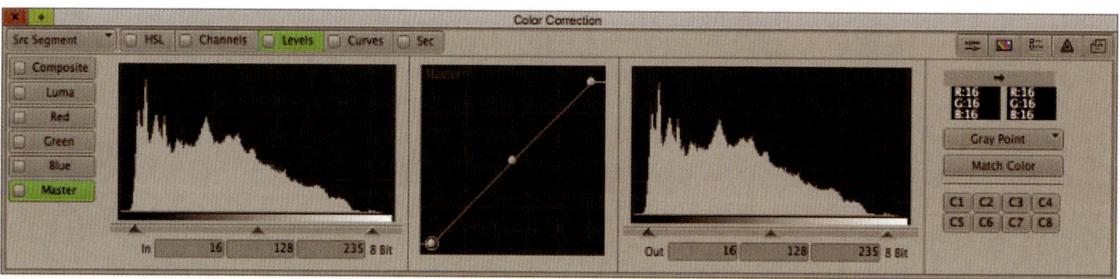

Fig. 5.47 The Levels tab of the Avid Symphony Color correction tool basically provides editable histograms.

Curves Tab

The Curves tab (see Figure 5.48) is not necessarily a place where a lot of experienced colorists on dedicated color correction applications spend a lot of time, but in Symphony and Media Composer, there are no numeric or slider-based controls to adjust the various tonal ranges of individual color channels, so Curves is the only place to do these critical adjustments. Most dedicated color correction applications have sliders or dials that allow for separate shadow, midtones, and highlight adjustments in each of the color channels: red, green, and blue. Media Composer doesn't have this, so color correction experts who use the Media Composer color correction interface actually spend quite a bit of time in Curves.

Curves also allows for very specific control, beyond basic shadows, midtones, and highlights. Curves will allow for as many points as you want at whatever point on the curve that you want. This level of control allows you to hold a nice, rich, legal black while stretching out some of the shadow detail. Or it will allow you to clip bright highlights to a legal level while still holding as much detail in the highlights as possible.

When adjusting Curves, the best monitoring solution you can use, for efficient and intuitive work, is the RGB Parade waveform monitor. The way to use the RGB Parade with Curves is to imagine that the bottom of each color channel curve is "connected" to the bottom of each color channel in the RGB Parade. The top of each color channel curve is "connected" to the top of each color channel in the RGB Parade and the middle point of each color channel curve is "connected" to the middle area of each color channel in the RGB Parade. So, if you see that the shadows should all be even but the trace (squiggly stuff representing the video signal) of each color channel are not even along the bottom, you can move the lowest left point on the curve or curves that are not even up to raise that channel in the RGB Parade or to the right to lower that channel in the RGB Parade. This is a little tricky to explain with text, so definitely check out the Curves Video Tutorial.

Video Tutorial

Curves are easier to explain when you see them at work, so there's a quick video tutorial on the companion website for this book, www.focalpress.com/cw/hullfish.

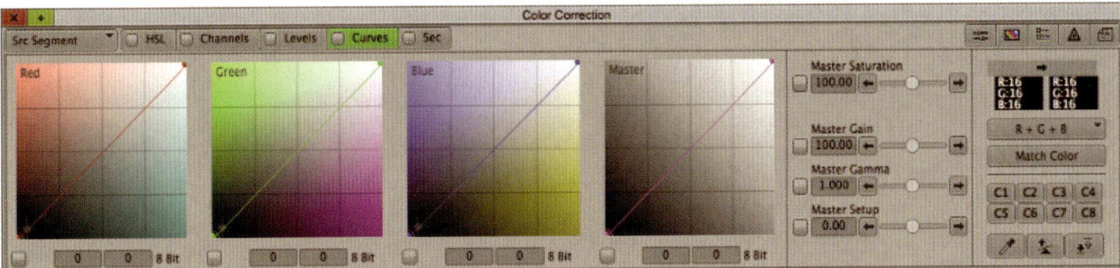

Fig. 5.48 The Curves Tab is a great place to do color correction in Media Composer or Symphony.

Video Tutorial

See the Secondary color correction video tutorial on the companion website, www.focalpress.com/cw/hullfish.

In addition to the individual color channel curves, there is also a master curve that is ideal for adjustments to overall luminance levels, gamma adjustments and contrast adjustments.

Secondary Tab

For Media Composers, the Curves tab is the last (furthest right) tab. But, the final tab in the Symphony option is the Secondary Tab (see Figure 5.49). Secondaries is a color correction term that is slightly antiquated with modern color correctors, but it refers to the fact that primary color corrections are done to the overall image using all of the tools that were just described in this chapter. And after those corrections are done there are secondary corrections that affect only specific geographic regions of the image or specific color values.

Secondary color correction is often demonstrated by selecting the color of a car or a piece of clothing and changing it to a different color. While this is certainly a good visual demonstration, the most common purposes for using Secondary color correction is to adjust skin tones, grass color, and sky and water tones without affecting the other colors in the image. Notice that these four adjustments fine-tune the natural colors that our eyes are very used to seeing; we have an innate sense of what they should ideally look like. But often times, making these colors look natural means making adjustments that make the rest of the image look bad. Secondary corrections, however, allow these colors to be isolated and corrected without affecting the rest of the image.

Secondary color correction is a two-part process. The first part of the process takes place on the left side of the Secondary tab. This process is known as "qualifying" the color. *Qualifying* basically means "selecting." So in the case of a skin tone, for example, you want to "qualify," or select, the skin tone first. There are two ways to do this. The first way is to adjust the colored wedges on the left vectorscope so that the skin

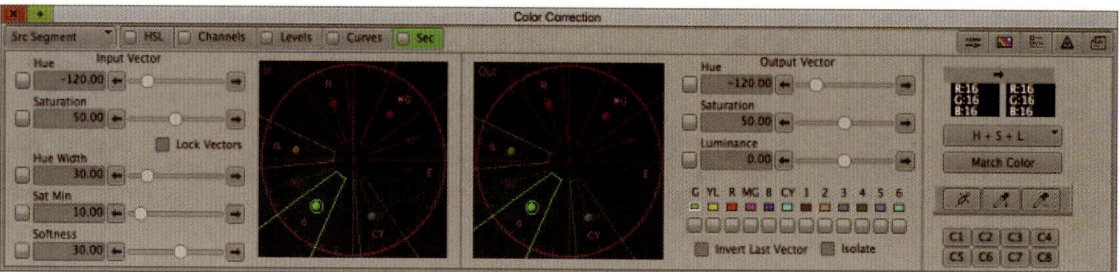

Fig. 5.49 The Secondary Tab of the Symphony color correction tool. This is not available in Media Composer without the Symphony option.

tones are selected as completely as possible while excluding any other similar tones. Skin tones are probably going to be closest to the red vector. To "edit" the vector, look at the central Composer window and note that red colors in the image are in full color while other tones are represented in shades of gray. Adjust the Hue, Saturation, Hue Width, Sat(uration) Min, and Softness sliders on the left side of the window (see Figure 5.49) to best "dial in" the skin tones so that all of the skin tones are in full color while all non-skin-tones are gray. When this is accomplished, move to the controls on the right side. The first thing you want to do is deselect the Isolate check box. This will bring the entire image to full color. When you do this, the skin tones are still selected as before, but now you'll be able to see the corrections you make in context. To adjust the skin tones, you can alter the hue, saturation, and luminance with the sliders on the right, or by manipulating the red vector shape on the right side.

The other way to qualify and adjust a secondary is with what Symphony calls a "custom vector." Custom vectors are created by using the Syringe and eyedroppers on the right side of the Secondary UI. Technically, these controls should be on the left side, since they're qualifying tools, but keeping them on the right side keeps the color chips and eyedropper location consistent with all of the other tabs. So, to select the skin tone, click on the syringe icon. Then, in the central Composer window, click and drag over representative areas of the skin tones. As you do this, you'll see more skin tones represented in color. If you accidentally select an area outside of the skin tone, you can use the "– eyedropper" to deselect the accidental color tones. You can add in additional tones with the "+ eyedropper." Notice that the shape that represents the skin tone on the vectorscope is not a wedge shape; rather it's a circle or oval. You can tweak this oval using the same parameter sliders we used before. To actually execute the adjustment of the skin tone, deselect the Isolate check box and then click and drag the circle or oval shape in the right-side vectorscope. If you want the skin tones to be more saturated, drag the oval away from the center of the vectorscope. If you want them more yellow, drag down and to the left, towards the yellow vector. To make skin tones less saturated or more blue, drag down and to the right. All of these adjustments should be done with a light touch.

The number of secondary colors that you can have in any single color correction is limited to twelve standard vectors (wedge shaped) and twelve custom vectors (oval). So if you want to qualify and alter a skin tone, then qualify and alter a grass color, and then qualify and alter a sky color, you can. You can even switch methods of qualifying colors between the standard vectors and the custom vectors.

Baselight Color Correction

There are definite advantages to staying within Avid's built-in color correction. There are also advantages to the Symphony relational color corrections, but even the Symphony tools are getting a little old. One of the top names in color correction is Baselight. DaVinci has definitely been the gold standard in color correction for nearly thirty years, but Baselight is critically regarded as a true contender. Luckily for Avid editors who want some serious color correction power inside of Avid, Baselight has developed a plug-in version of their stand-alone color correction application.

Baselight's color correction plug-in is far and away more powerful than the built-in Avid tools. Baselight is a real-time effect, so it can be previewed in the Timeline without having to render it. It can also be applied either directly to the clips or to the filler above them. Adding to filler allows you to grade a series of shots in a row by simply dragging or trimming out the filler to cover the shots that have the same grade. Just like regular Avid effects, AVX effects, and color correction, Baselight effects can be copied to a bin and applied from the bin to any other shots in the Timeline, so even though Baselight doesn't have the true relational color correction capabilities of Symphony, there are several ways to apply one grade to multiple clips throughout a sequence, saving significant amounts of time and effort and maintaining a consistent look.

Baselight installs like most other AVX effects Adding Baselight corrections to shots is similar to other AVX effects: open the Effect Palette with Command-8 (Mac) or Control-8 (PC) or from the Project > Effects tab. The Baselight effect is in the Filmlight effect category (see Figure 5.50).

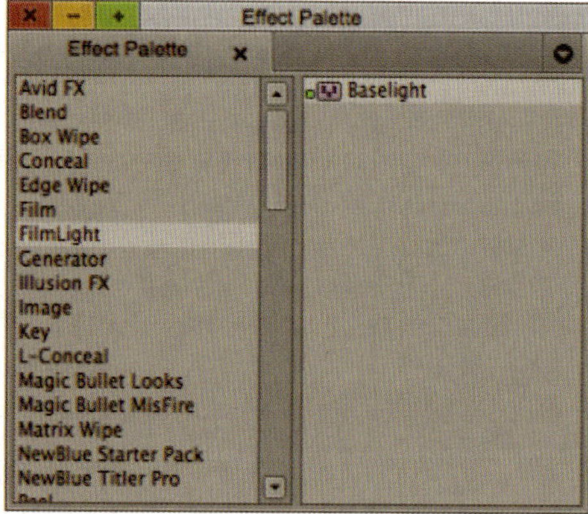

Fig. 5.50 In the Effects Palette, when Baselight is installed the FilmLight category has the Baselight AVX effect which can be applied to segments in the timeline just like any other effect.

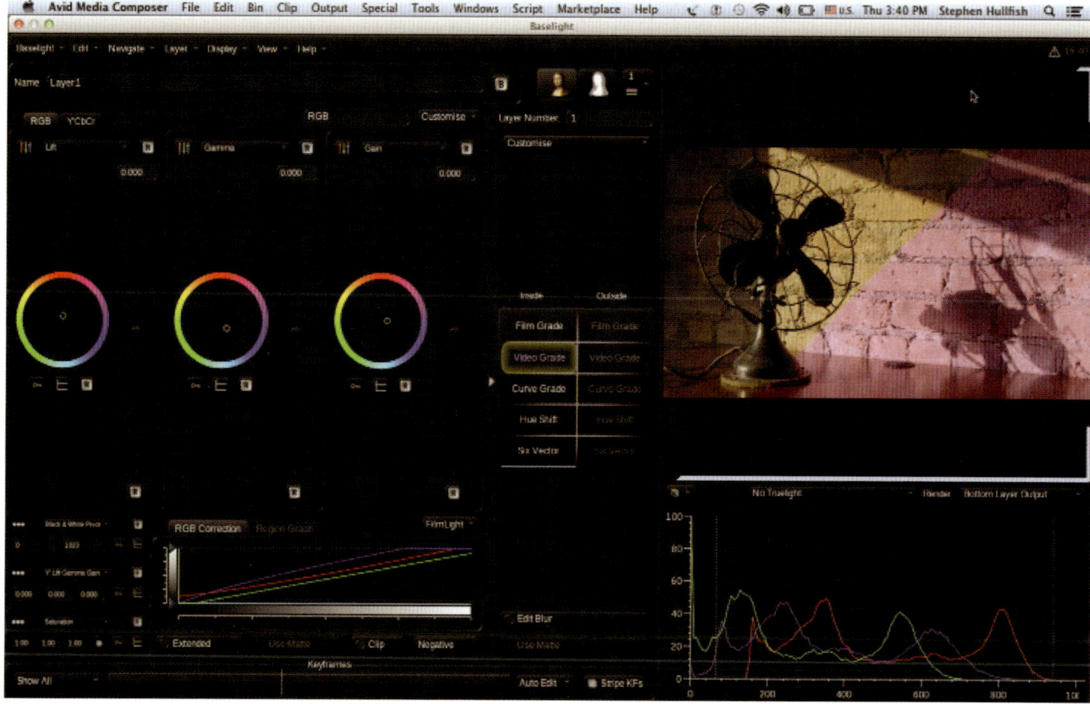

Fig. 5.51 Baselight user interface in Media Composer.

Simply drag it onto the clip or filler segment that you want to control and go into Effect mode. In the Effect Editor, click on the check box next to Baselight to launch the Baselight UI (see Figure 5.51). You can use the UI the way it launches or expand the window to full screen to get a bit more screen real estate.

Baselight works using layers. These layers can be stacked to form complex grades. The camera original is layer 0. New layers can be added with the P keyboard shortcut. The controls for these layers are in the Grade Operators Panel, which can be called up with the keyboard shortcut L, or, you can spin down a small left-facing triangle just to the left of the main image to reveal the Grade Operators Panel, which includes Film Grade, Video Grade, Curve Grade, Hue Shift, and Six Vector (see Figure 5.52). You can collapse these choices by clicking the little right-facing arrow to the left of the word Curve Grade, or you can use the keyboard shortcut L.

If your footage needs a LUT—for example, to convert a RAW image to a standard video color space, such as REC709—you can add it by adding an extra operator to the Grade Operators Panel. Right-clicking on one of the operators allows you to select from a longer list of additional

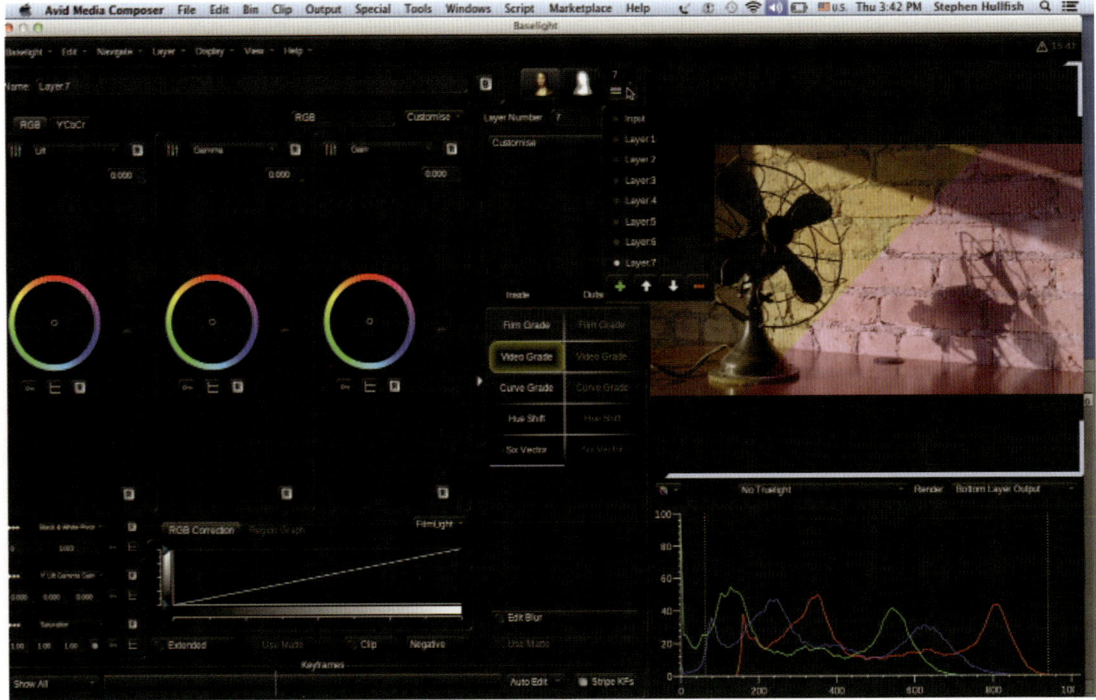

Fig. 5.52 Layers, or Grade Operators, Panel.

operators. You can either replace operators in your current list, or add more rows of operators in the panel. One of the additional choices in the operators list is TrueLight, which provides several factory default LUTs and the ability to add custom LUTs.

The Video Grade option is similar to the three-way color corrector in FCP or the Hue Offset tab in the HSL tab of the Avid built-in color corrector (see Figure 5.53). These color wheels work identically to the Avid built-in versions. If you want to raise and lower the setup (black level), midtones, and highlights, there is a very difficult-to-see slider that's to the right of each of the color wheels. You can adjust these levels by simply dragging up and down.

One big advantage of the Baselight color corrector is that it allows you to create vignettes (in Apple Color lingo), power windows (in DaVinci Resolve lingo), or, to put it as Baselight does, mattes. Near the top middle of the Baselight UI is an image of the Mona Lisa alongside a ghostly version of the Mona Lisa. That Baselight chose to use this specific image, which was painted by Leonardo da Vinci, is rather ironic. The full-color version of the Mona Lisa is the icon for the grade itself. The ghostly Mona Lisa is the icon

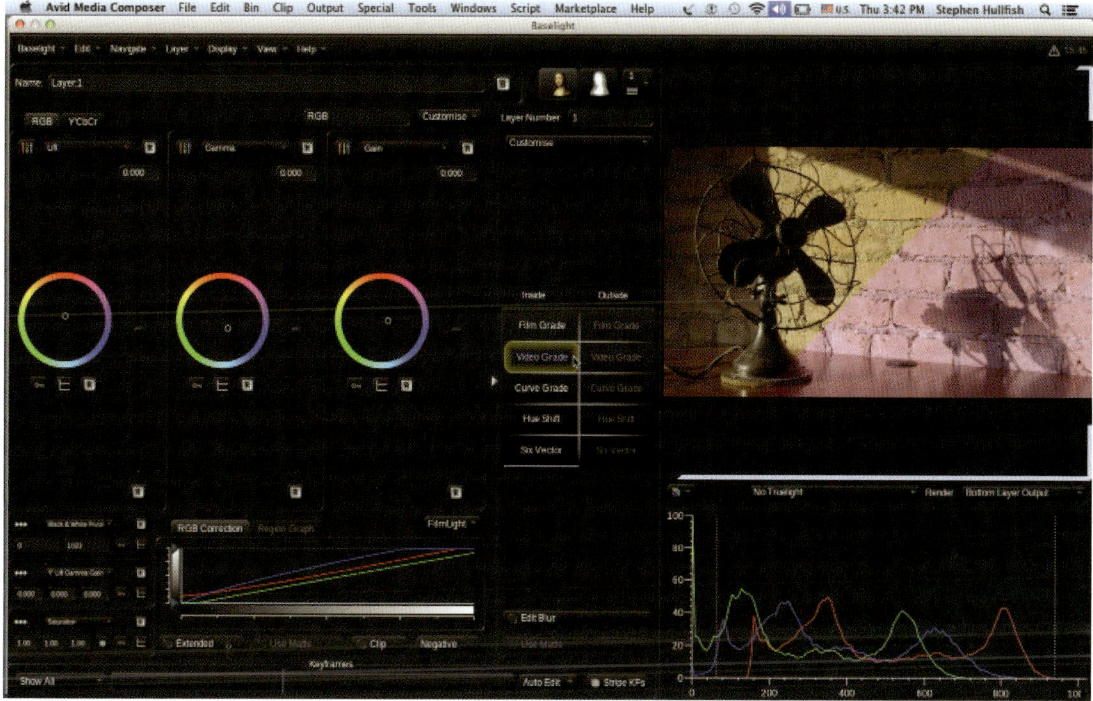

Fig. 5.53 Video Grade option.

for the matte or vignette for the grade (see Figure 5.54). There are a number of methods for generating the matte (power window or vignette), including drawing a freehand shape, rectangle, or oval or by defining a key using a luminance, hue or saturation value that you want to affect. Once these mattes are created, Baselight allows you to grade either "inside" or "outside" that matte selection. You can also create combinations of mattes, including having shapes that can be subtracted from other shapes. Baselight also has powerful tracking software that allows these shapes to follow the action in the video.

The Baselight users guide is 180 pages long, which I have to condense here to maybe three or four pages, enough to give you a taste of how to use the system. Baselight also has an excellent series of free tutorial videos on its website, which I strongly suggest that you check out.

Note that there are built-in scopes but that they are dependent on your GPU (graphics card), so you may not get the same options on a laptop as on system with more horsepower. The scopes are customizable from the View pulldown menu at the top of the Baselight UI.

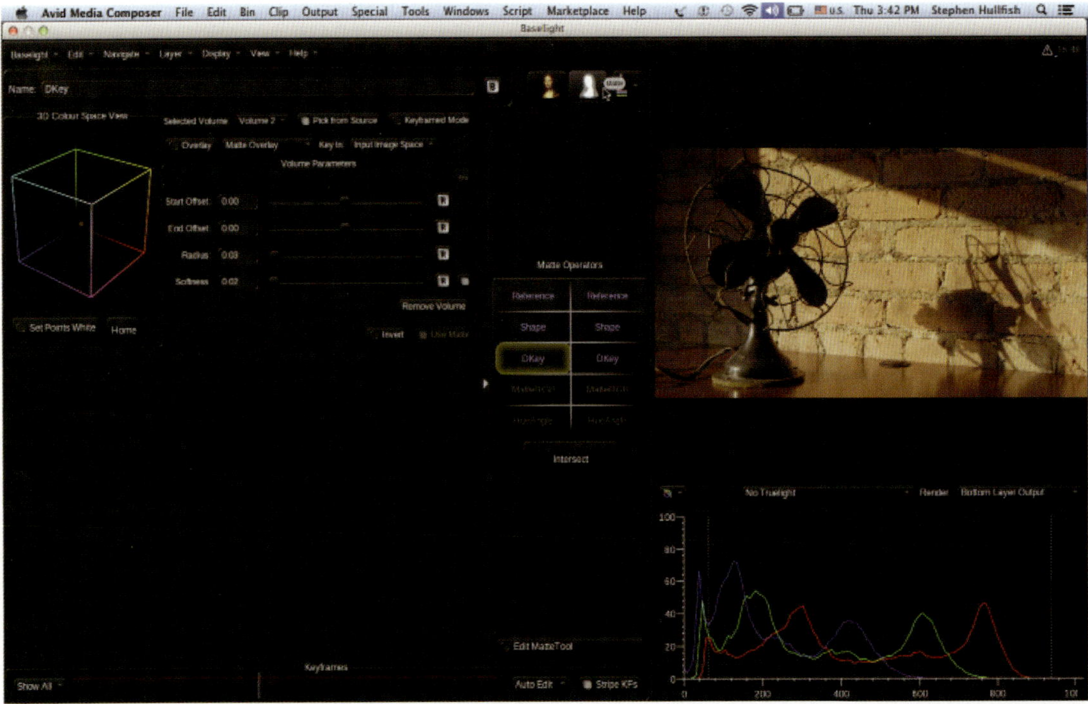

Fig. 5.54 Mona Lisa vignette/matte.

Red Giant Magic Bullet Looks

I fought against installing Looks for a long time, thinking that it was simply a sleazy drag-and-drop replacement for doing real color correction. As a guy who has written two color correction books, that's not a good thing. But I was wrong about Looks. Looks has some pretty amazing stuff that is far, far beyond basic color correction. Sure, there are drag-and-drop "looks" that don't require much thought, but drag-and-drop is not really a bad thing. When I interviewed famed colorist, Bob Festa, he was quite proud that he had developed a bunch of DaVinci Resolve presets that he could easily apply and show to clients. The difference with Festa was that he actually created those looks himself. But the point of having an easy "look" is not a bad thing.

But even at that, Looks is more than just having a preset look. It's the ability to apply many different tools for the user to integrate to create a unique look that goes far beyond a simple drag-and-drop preset.

If you don't own Looks, head over to the Red Giant Software website and download a free trial. You can play with the looks and tools all you want and follow along with a quick little tutorial while I get you up to speed. Without a paid version, though, you'll get a big ugly X watermark. The download and install was pretty basic, and when I rebooted the

Media Composer and went to the Effect Palette, there was Magic Bullet Looks and a Magic Bullet MisFire category waiting in alphabetical order in the Effect category list. Inside Magic Bullet Looks, there is only one effect called Looks. Drag that to a clip in your timeline, or select the clip or clips and double-click the Looks effect to apply it.

Go into Effect mode on the clip and check the Edit Look ... box. This launches the Magic Bullet Looks UI, which is pretty cool (see Figure 5.55). I really like how it keeps the workspace very clean by sliding the tools and libraries off to the sides as you work. My biggest problem with the interface was that I keep my Mac dock hidden along the right side of the screen, instead of the bottom, and the same mouse behavior that unhides my dock also reveals the Looks controls, which hide and reveal along the right side as well. To the left side of the Looks UI is the Looks Library, called the Looks Theater. It remains out of sight until you hover near the left side of the UI. Then it slides in, revealing a list of Look categories: Basic, Cinematic, Diffusion and Light, Monochromatic, Stylized, People, Classic Music Videos, Classic Popular TV, Classic Stock Emulation, Tints, Stock Emulation, Special Effects, Black and White Tints, Music Videos, Popular Film, Popular TV, Horror, Diffusion, and Custom. These are the

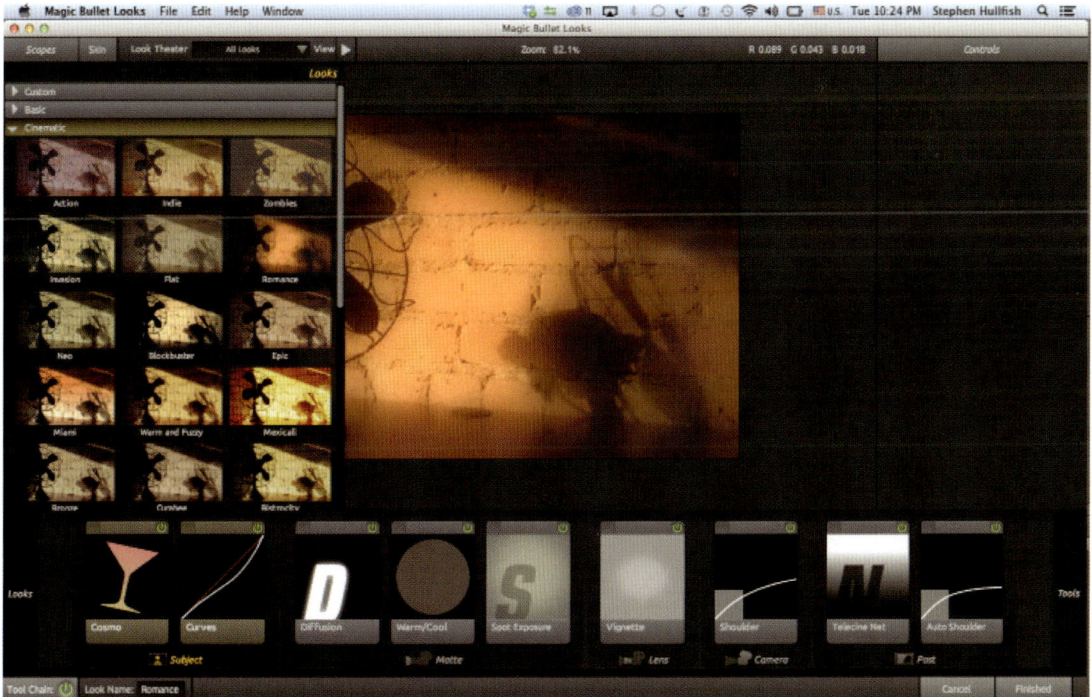

Fig. 5.55 Magic Bullet Looks user interface.

basic drag-and-drop premade Looks. When you double-click on one of the thumbnail images within the categories, it applies a set of tools along the bottom of the UI (see Figure 5.56). When you use your mouse to select any of the tools, the controls for the selected tool will appear to the right of the main image. Use these controls to manipulate and customize the look. To bypass the effect of individual tools in Look, use the little green on/off-power icon in the upper right corner of each Tool. You can also create subtle changes by rearranging Tools, dragging them to one side or the other. Or, you can drag tools completely out of the bottom area to delete them. Finally, at the very bottom left of the UI, you can bypass the entire look by toggling off the Tool Chain's green power-button icon.

If you'd like to add more tools to a Look, hover along the right side and Tools will appear. Tools are organized in tabs, including Subject, Matte, Lens, Camera, and Post. Each tab has between a half-dozen and two dozen tools. I would suggest that you go to the main menu bar and choose Magic Bullet Looks > Preferences. In Preferences, change the default for Tool Tips to "on" and you will get quite detailed descriptions of what each tool is for. You can also name Looks that you have created in the text box at the bottom left of the UI. Hitting return after typing a name will save the new look under Custom Looks in Look Theater.

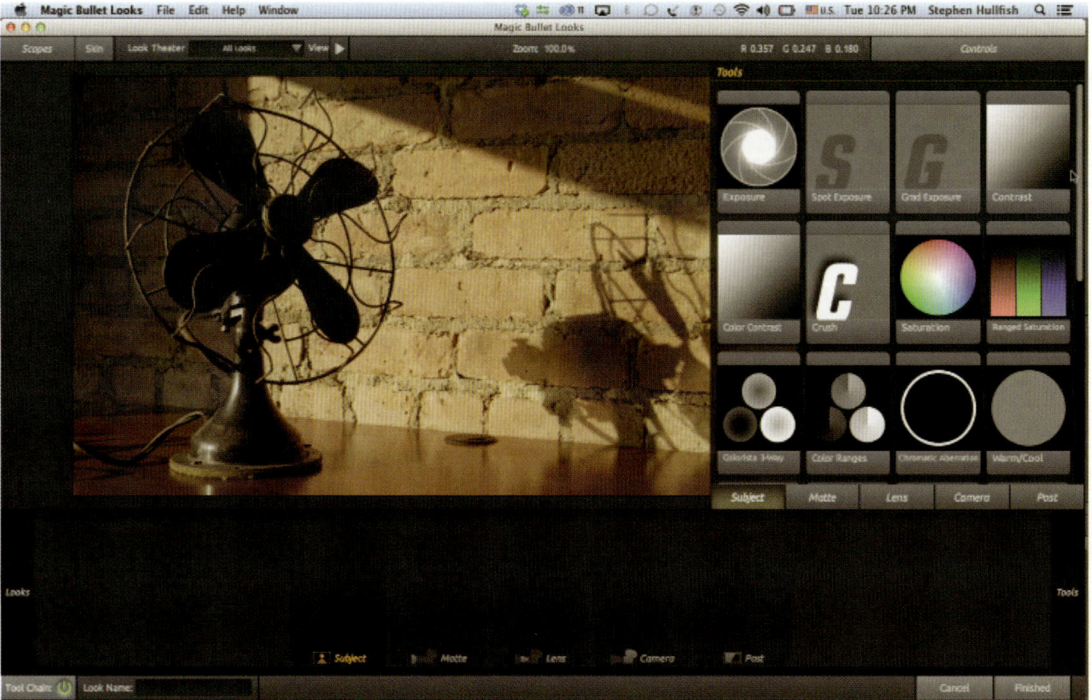

Fig. 5.56 The Looks Tools allow color correction beyond the preset drag and drop Looks.

When you're done with a specific look, click the Finished button in the bottom right corner.

Output

When the on-line has been conformed and color corrected, it's time for output. This is done from the Output menu in the menu bar. There are basically four different options here: Export, Export to Device, AMA File Export, or Digital Cut.

Digital Cut

A Digital Cut is a layoff to tape. Honestly, this is pretty much limited to broadcasters at this point. And even that, as we'll discuss in a moment, is about to move more toward file-based delivery with the advent and promotion of the AS-02 and AS-11 standards.

To attempt a Digital Cut, you must have Nitris or Mojo Avid hardware, or one of the third party-hardware solutions that include deck control. To have the option for a Digital Cut, the Avid has to detect a connected tape deck. On my MacPro, I have a BlackMagic card with deck control. On my PC, I have an AJA Kona 3 card with deck control, and on my MacBook-Pro, I use the AJA IOXT, which uses Thunderbolt to connect to a small outboard box that includes deck control.

Personally, I always used an assemble edit to do Digital Cuts. This basically meant that I only needed to "black" the first fifteen seconds or so of the tape. Make sure that when you black the tape that the timecode that you use to black it is identical to the type used in your sequence.

There are a couple of settings that are critical to the check if you're having any problems controlling a deck. Check the Deck Configuration setting (see Figure 5.57). It scans a folder to see all of the possible deck configurations. (If you want to speed this process up, you can just delete all of the decks that you're sure you'll never own or rent from the list in the folder. I keep only about four decks in my list.)

The "channel" is the pairing of what you use to "talk" to the deck—which should almost always be the same—and a deck. On the left is the communications section. Double-click it to launch a little Channel dialog window (see Figure 5.57). Select the hardware that you use to plug in the RS-422 cable. For me, it's an IoXT Serial connection. This is an AJA IoXT, which connects to my MacBookPro via Thunderbolt. Then the RS-422 control cable to the deck comes out of the IoXT. Because they were connected when I launched the setting, Media Composer correctly identified the deck on the end of the RS-422 cable. If yours doesn't "read" correctly, double-click on the box to the right and the Deck Settings dialog (see Figure 5.58) comes up. Choose a

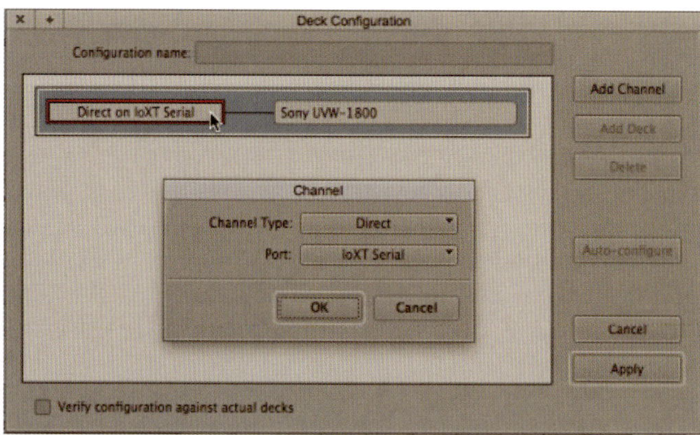

Fig. 5.57 The Deck Configuration window lets you pair a particular control (like RS-422) or a particular video card with a particular deck. The "channel" is the way the deck is controlled.

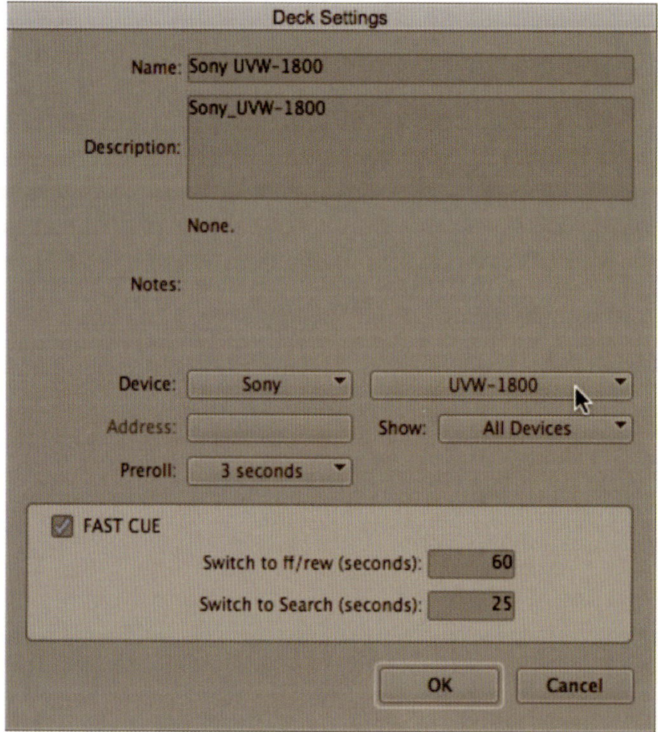

Fig. 5.58 The Deck Settings window allows you to set the deck ballistics by choosing from a list of decks, categorized by manufacturer. You can rename these decks to nicknames instead of the formal deck name, like "Beta Deck" or "Deck 6" instead of "Sony UVW-1800."

manufacturer from the first Device pulldown menu and a model from the second Device pulldown menu. Then save this combination for later use.

Before we get to the Digital Cut tool, there's another important setting to know about if you want to be able to Assemble edit. Without it, the Digital Cut tool will show you only CRASH record or Insert Edit options in the Deck Control section. In the Deck Preferences setting in the Settings list, check the box for Allow assemble edit & crash record for digital cut (see Figure 5.59). You may also want to check the box for Relax coincidence point detection. That setting allows a little bit of leeway when searching the tape for the exact preroll point before the deck goes into record mode.

GLOSSARY

CRASH RECORD: Pressing the Record button on the deck and the play button on the Avid with no deck control. This does not allow for the timecode on the tape to match the timecode from the Avid sequence. But in an emergency, it can at least lay off the sequence to tape.

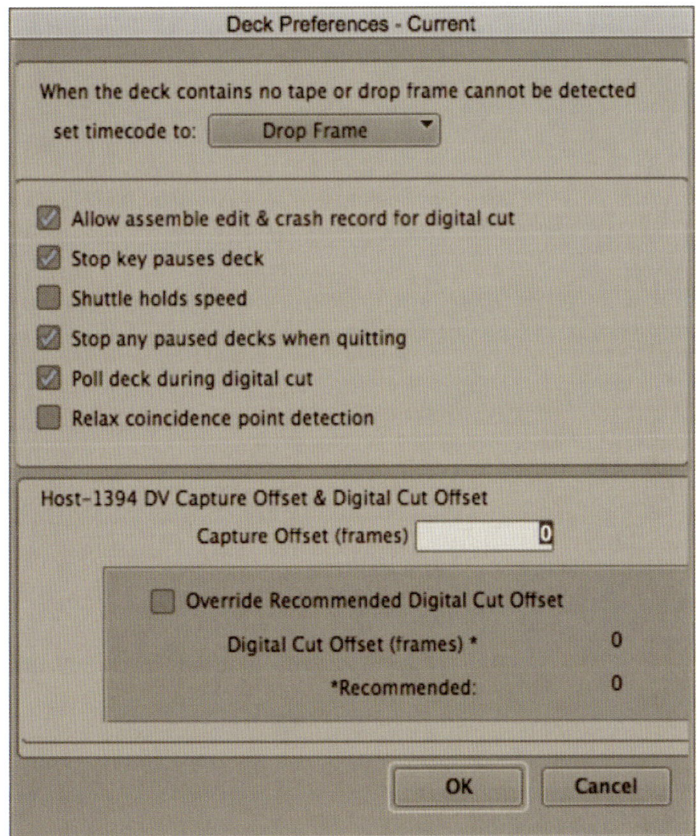

Fig. 5.59 The Deck Preferences window has the settings for the tape deck behaviors. The top check box allows you to do Assemble edits. This feature is turned off by default to insure that users are very intentional about using Assemble edits which can be very destructive.

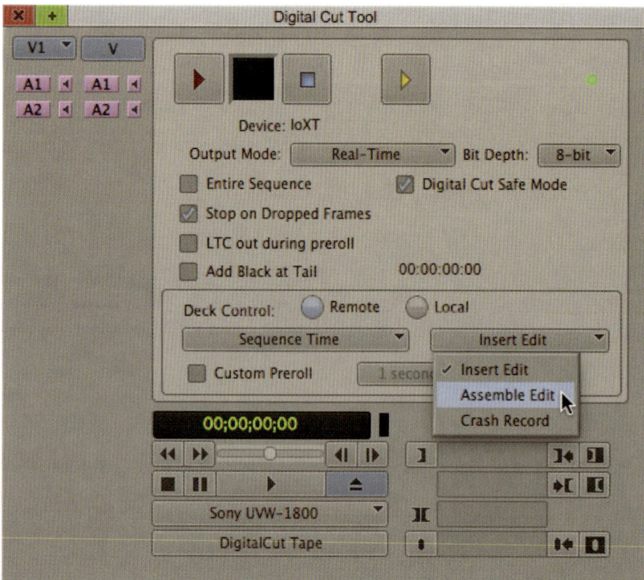

Fig. 5.60 The Digital Cut Tool allows for laying-off to tape. Here, it is ready to Assemble Edit at Sequence time, meaning that it will match the timecode in the Avid Timeline to the tape.

The default settings for the Digital Cut tool (see Figure 5.60), which depend largely on what and how you're laying off, are the following:

- Check the Entire Sequence box to lay off the entire sequence.
- Check Stop on Dropped Frames.
- Adding Black at Tail is good if you use an Assemble edit. Then add a minute or two of black using the numeric entry box to the right of the check mark.
- In the Deck Control section, the radio button for Remote should be on.
- Below that is a pulldown to determine which timecode to use for the layoff. I almost always use Sequence Time. The other options are Record Deck Time, Mark In Time, and Ignore Time.
- Next to the timecode pulldown is the pulldown to determine what kind of edit Media Composer will execute. I use Assemble edit. The other choices are Insert and Crash.

To the top left are icons that allow you to determine which audio and video tracks you want to include in your layoff and patch where they go on the tape. These should be set up correctly. Then just hit the red triangle button to start your Digital Cut.

Export

Exporting is pretty basic (see Figure 5.61). Clearly there are a lot of options for exporting, including AAF, AFE, Quicktime Reference, HDV, XDCAM MXF OP1a, Audio, Graphic, Avid Log Exchange, and Tab Delimited.

Let's quickly run through some of the exporting options (see Figure 5.62):

AAF: AAF is Advanced Authoring Format. This is essentially a supercharged EDL format that allows you to send the sequence information out to another application. This is similar to the old OMF option or to the XML option in Final Cut and some other apps. This is also the preferred file type for exporting audio to ProTools.

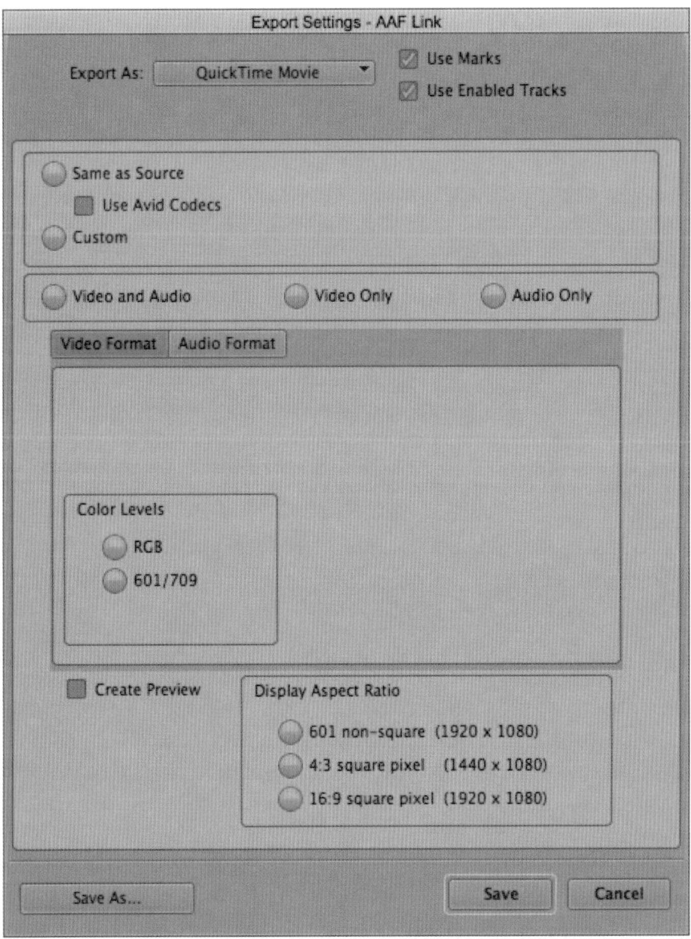

Fig. 5.61 The Export window is used to select the settings used to export a sequence or footage.

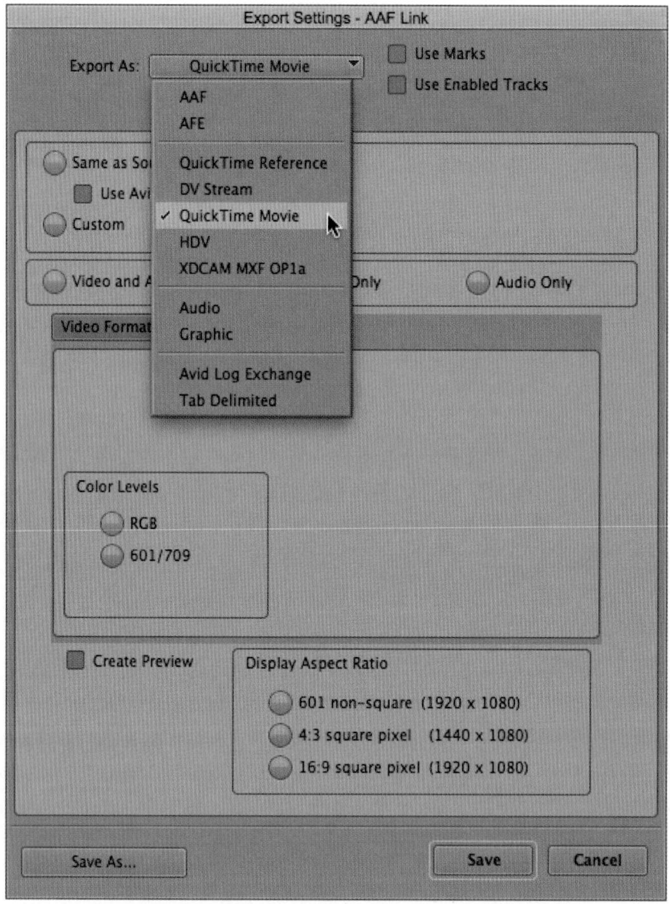

Fig. 5.62 The pulldown menu for Export As gives a number of choices for types of files and media that can be exported.

AFE: AFE is Avid File Exchange. This is essentially like AAF, but only between Avid applications. With the death of DS, this is a limited-use file at this point.

QuickTime Reference: This is a useful format for many people because the file size is very small and the export is quick. However, if you export a QuickTime Reference file, it merely contains pointer information back to the MXF files in the Avid MediaFiles folder, so the resulting file can be used only with apps that can see that original media. If you move this to another computer, the links are broken and the file is worthless. The main use of this is to immediately run the QuickTime Reference file to a compression program on the same system. For FCP editors, this is similar to unchecking the Make Movie Self Contained box in Final Cut Pro.

QuickTime: This is my standard form of export for most sequences. I prefer to export using the Same as Source option, which uses the same compression as the source, so that the export is very quick; usually faster than real time. I also prefer this export because I can use it for archival purposes and can move it off of the original system and not worry about the file becoming useless if the original media goes off-line.

HDV: I rarely use this option, but it allows you to create an HDV stream of media.

XDCAM MXF OP1a: This is valuable for delivering a professional broadcast export of the sequence that can be stored and transported as XDCAM media, which can be read by other external programs. Some broadcasters accept this as a master format for delivery, especially of TV spots.

Audio: I often export just the audio for a sequence so that clients can listen to interviews. I usually export as WAVE or AIF files and then compress them using Squeeze or Compressor to MP3 files.

Graphic: Graphic exports are handy for delivering stills of portions of the sequence.

Avid Log Exchange: Avid Log Exchange is primarily used for exporting bin or clip information.

Tab Delimited: Tab delimited is another method of exporting bin information, but it saves it as a simple tab delimited text file.

When exporting a sequence to a QuickTime file, I like to check the Use Marks and Use Enabled Tracks at the top of the Export window. This gives me specific control over exactly what gets exported. If I want to export only a portion of the sequence, or only certain tracks, I can mark them and enable them in the sequence and then export based on those enabled tracks and selected portion. I almost always choose Same as Source because the export is very quick and because there are better applications for creating high-quality compressed versions, like Squeeze. Avid is not a compressor. Never hammer a nail with a screwdriver. Another issue to be aware of is that Avid Import and Export use the QuickTime engine. This is still 32-bit and is *not* multithreaded, which makes it *much* slower than *any* of the traditional media encoders from Apple, Adobe, or Sorenson. When you choose Same as Source, you don't have to worry about the Video and Audio formatting. Which color level you use depends on the use of the final video. If it is for compression in Squeeze for use on the Internet, then I use RGB. If it is for export for broadcast, then I would use 601/709. Finally, for the aspect ratio, choose 16:9 square pixel. Whenever you create a new export setting, be sure to choose Save As and give it a descriptive name so that you can call it up for use at a later time.

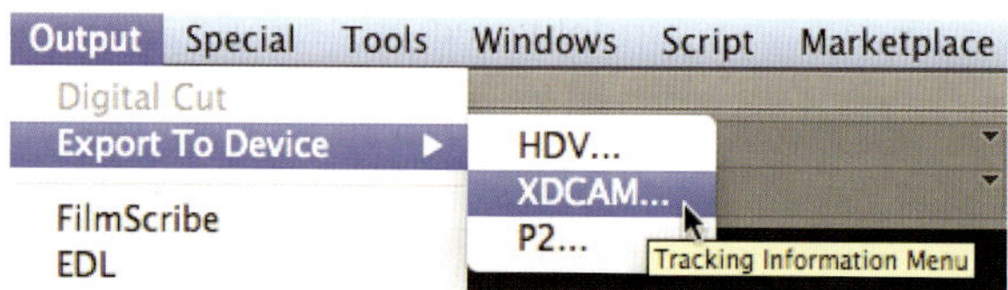

Fig. 5.63 The Output menu provides the option to export to a device, for example, a P2 or XDCAM card.

Export to Device

Export to Device (see Figure 5.63) is useful if you have an XDCAM, P2 drive, or HDV camera. Export to Device is not in the File menu with the other Export options, but is in the Output menu. This allows you to export to that external media. As with the Digital Cut option, you must have the drive, camera, or card connected to the system first.

Finally, we'll address AMA File Export as a special subheading. Despite the name of the menu option, this is really an AS-11 broadcast deliverable export.

Working with New File-Based Delivery—AS11 Support in MC7

AS-11 is a relatively new standard that broadcasters have developed to allow for file-based delivery to broadcasters. Originally developed by broadcasters in the UK (www.digitalproductionpartnership.co.uk), it is designed to be very metadata rich, which should improve the speed and efficiency of delivering the file to the broadcaster and for the broadcaster to be able to ingest and use the file with as little human intervention as possible.

AS-11 is a variation on the MXF OP1a file format that is "vendor neutral" (meaning that implementation of this specification is not limited to Avid) and is designed for the delivery of finished programming to broadcasters. This is a standardized file and specification created by the Advanced Media Workflow Association, or AMWA.

AS-11 files are designed to be completely ready for playout at the broadcaster without the need for transcoding or repackaging, though some servers may need to rewrap to their native format. The AS-11 file contains all of the necessary playback elements, including video, multiple tracks of audio, timecode, Ancillary and Descriptive metadata tracks, and closed captioning in a single file. They also include full MXF Index Tables.

AS-11 is capable of transporting either standard def files (as a D-10 stream at 50Mbit/s) or as high definition (AVC-Intra Class 100, High Intra 4:2:2).

AS-11 HD footage can be delivered as desired and specified by the broadcaster in one of four formats:

1. 1080i50 at 25fps or 50 fields per second, interlaced, with a byte size of 568832
2. 1080i59.94 at 30 or 60 frames per 1.001 second, interlaced, with a byte size of 472576
3. 720p50 at 50fps progressive, with a byte size of 284672
4. 720p59.94 at 60 frames per 1.001 second, progressive with a byte size of 236544

The audio in a standard def AS-11 file can contain an eight-channel AES3 stream. For HD video, AS-11 can contain PCM pairs, AC-3, or Dolby E audio. PCM audio needs to be 48 kHz and up to sixty-four channels.

The AS-11 spec has no limitation on file naming or directory structure of the AS-11 file.

One of the components of an AS-11 file is that the commercial breaks are defined inside the file itself. This is called Program Segmentation, and it's the real reason that Avid added Spanned Marker functionality to the release of MC7 because it allows for the definition of program content segments that can be defined by the editor in Media Composer and then passed on to the Segmentation Track of the AS-11 file. In order to create a compliant AS-11 file, you need to define the program segments of your sequence using Spanned Markers. This procedure is detailed in chapter 3 (page 217), but basically, you mark the beginning and end of an edited program segment with in and out points and hold down Option (Mac) or Alt (PC) while clicking on a Marker. This type of segmentation is known in the spec as "hard parted." The other types are "soft parted" and "single parted." A single-parted program has no breaks. It is a solid block of programming. A soft-parted show has *opportunities* for breaks without specifically regulating that they be used.

The export to AS-11 is very simple and works like almost any other type of export. From the menu bar, select File > AMA File Export (see Figure 5.64), or you can right-click on the sequence and select AMA File Export from the contextual menu. As with other exports, you can select Use Marks and/or Use Enabled Tracks. The Directory Set button allows you to navigate to a specific location to write the exported file (see Figure 5.65).

The Shim Set button basically is the jargon from the AS-11 specifications. A "shim" is just the local specifications of the file from Media Composer. There are a few preset shims that let you properly export using the proper settings. You'll have to navigate to the location where these

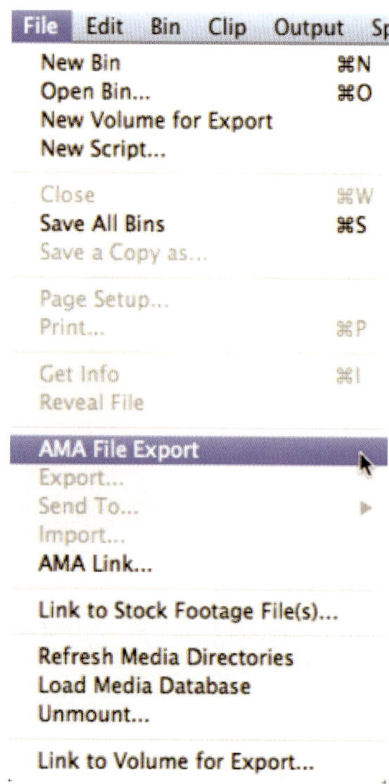

Fig. 5.64 To export an AS-11 file, use the File menu > AMA File Export option.

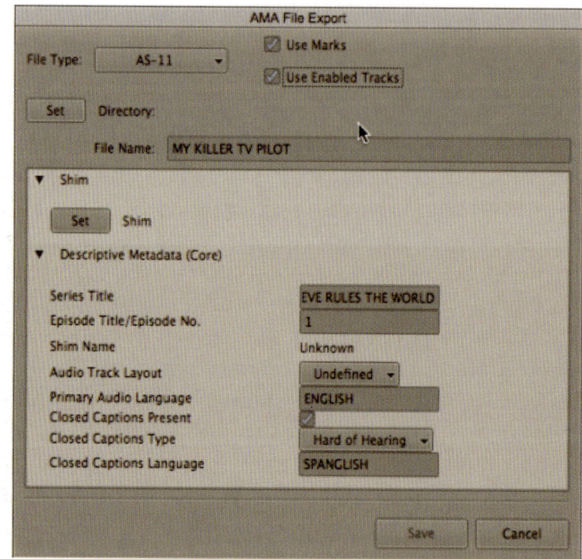

Fig. 5.65 This is the AMA File Export window, with the AS-11 File Type selected.

are stored: for Macs, HD/Applications/Media Composer/Supporting Files; for PCs, Program Files/Avid/Media Composer/Supporting Files. Inside that location are folders for AS-02 Shim Templates and AS-11 Shim Templates. Assuming you want HD, use the AS-11 templates. Then, inside that go to the Avid folder (unless you're prepping files for the United Kingdom), and pick a shim for the type of export you want to do, such as AS-11_Shim_1080i59 .xml and hit Return or the Open button. This Shim Template does most of the heavy lifting in preparing the file for proper broadcast AS-11 specs.

Type in the series and episode titles and information about the audio tracks and closed captioning. Remember, in the United States, closed captioning for broadcast is a requirement. Click Save to export the AS-11, and ship the resulting file off to your broadcast client. When completing the data for a shim, the Save button is not enabled until all required fields are filled in.

GLOSSARY

ESSENCE: This is a relatively new term indicating the actual part of the file containing the media (video and audio) as opposed to the metadata.

Click on Commit and it creates a folder structure and transcodes the essence as defined by the shim. The shim becomes part of the bundle and dictates what the file looks like.

While the metadata in an AS-11 file is embedded, it is possible to extract this information. The UK DPP metadata tool is available for download free for Mac or PC. (http://www.digitalproductionpartnership.co.uk/downloads/metadata-application/).

There is even a plug-in for Avid that allows AS-11 files to be read and viewed using AMA. When AS-11 files are AMA-ed into Media Composer, the spanned markers should be represented and much of the metadata will be available in the bin columns.

Workflows for Archiving and Restoring

AS-11 has actually been touted as an archiving solution, though that's not the purpose or original intent of creating AS-11.

What you archive depends largely on the storage options you have to work with. A single big RAID will fill up quickly. Large numbers of FireWire and USB drives can fail, taking everything with them. LTOs are a good choice, though obviously they are expensive and require some expertise to work with. BluRay disks are a possibility, but DVD use is declining. HD videotape is another option, though the decks are expensive.

Obviously, RAID solutions are the best bet for speed, ease of use and protection, but the costs are high.

No matter what your choice is, it will be best to get your media reduced to the smallest footprint possible. Some things to consider are whether any of the raw media beyond the final sequence is needed for any purpose. If it is, you may want to create a selects reel of additional footage.

Archiving a sequence depends on whether it was edited using an off-line to on-line methodology or whether it was edited together from longer clips at full-rez. If it was done using one of the methods earlier in this chapter (batch capturing from a decomposed sequence, up-rezzed using consolidated media, round-tripped using media rendered from another app, like Resolve), then the sequence is really ready to be saved to disk as it is. Essentially, the process of decomposing or round-tripping through Resolve already acts to consolidate the media. However, the processes we have described in this chapter have largely dealt only with video, so the audio might not be consolidated at all. If that's the case, you may want to duplicate the sequence and delete the video tracks on the duplicated sequence. Then consolidate only the audio as described below.

For projects edited from longer master clips, you will want to consolidate the media before archiving the project. Consolidating breaks the longer master clips into smaller clips that are just the length of the segments in the sequence, plus handles. This allows you to delete the longer master clips because the new consolidated sequence is not referencing the smaller consolidated media clips. Archiving the consolidated media allows you to make some small changes in an edited sequence—trimming slightly, adding transitions, changing graphics, eliminating or revising titles—but limits the amount of media that needs to be stored.

Consolidating is a relatively simple task. Select your sequence and choose Clip > Consolidate/Transcode. Choose the Consolidate option in the top left of the resulting window. Select a drive to which the media will be saved. Choose the number of frames of heads and tails. Choosing to create a new sequence is usually a good idea, unless you have already duplicated the sequence to begin with. I would not choose to automatically delete the original media files. You should do that manually after confirming the success of the consolidation. Then choose the Consolidate button. However, there are some media types that cannot be consolidated and must be transcoded instead.

The process of consolidation is basically selective copying. It makes copies of only the parts of the media that are in the sequence. If you want to save this media to an external hard drive, to a BluRay disk, or LTO drive, you can simply look for the media that was created on the drive you selected with a creation date and starting time of when you started the consolidation process.

The only other real way to identify this media is to use a third-party application called Media Mover from Random Video (http://random video.com/). Media Mover (see Figure 5.66) allows you to search the Avid MediaFiles folders for media that it identifies as coming from specific projects. You can then choose to move or delete the media from specific projects. Avid should really have this ability built in, but Media Mover is worth every penny.

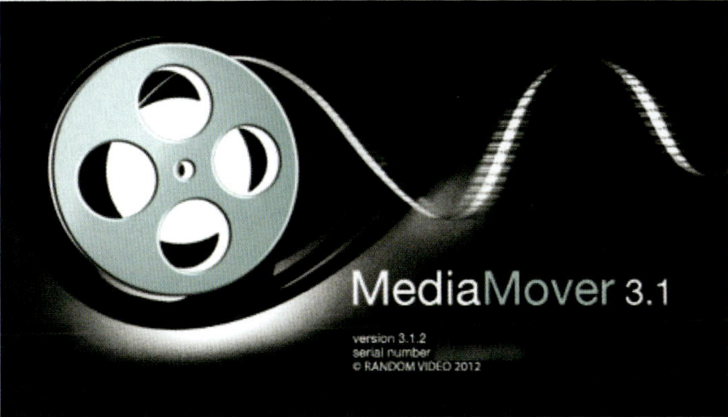

Fig. 5.66 A great third-party application is Random Video's Media Mover, which allows you to identify and move (archive) footage from specific Avid projects in Avid MediaFiles folders on multiple drives.

Make sure to save the Avid project information and any files used to the same hard drive. It's smart to save corresponding projects, like Photoshop files and After Effects projects (not just the rendered media from After Effects, but the projects and corresponding source media), to the drive so that changes and edits to those files can be easily made. Creating "readme" files (simple text documents with relevant information) is also a good choice.

Collaboration with ProTools/Third-Party Audio Mixing

Collaborating with an outside audio house should be relatively simple. Most audio houses will be happy to provide detailed directions for collaborating with them, but for the basics, I contacted my favorite Chicago-area audio house: BAM!

The most obvious things to consider before sending your sequence to an audio sweetening or editing house are to make sure the sample rate, bit rate, and frame rate are all correct and that you've added enough audio handles.

Brian Reed from BAM! walked me through some tips, starting with the request for AAF exports of the sequence information instead of the old OMF exports (see Figure 5.67).

When delivering consolidated, embedded audio, it's best to deliver as much handles as possible, at least three seconds. They often are forced to use these handles for presence. Another tip is to possibly add clips of presence (ambience) for scenes at the end of the sequence.

Don't render any audio or include rendered audio.

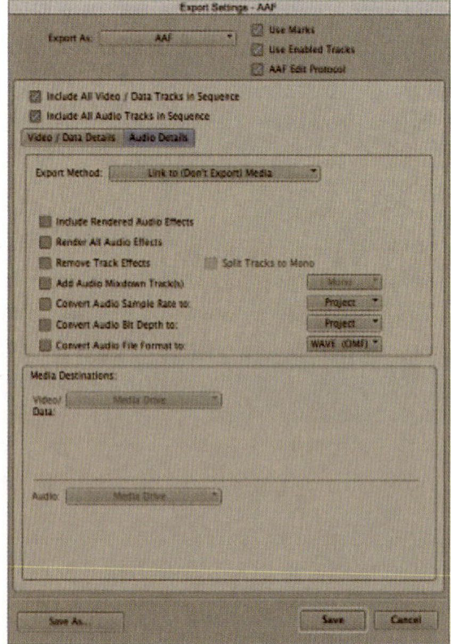

Fig. 5.67 The AAF Export settings are used when sending files off to an audio house for sweetening.

It's rare to embed video when sending a file to an audio house, but, of course, it is useful (more like "critical") for the mixers to be able to see video as they mix. With the latest version of ProTools, Motion-JPEG HD 1280 × 720 is a good choice. ProTools 11 also supports files exported as "same as source" from Avid. The ability of ProTools to easily track along certain video formats changes depending on releases or even point releases, so make sure to check with your audio house about its favorite *current* video format.

It makes it much easier for the audio house if you keep your tracks "pure." In other words, each audio track only has a certain type of audio in it, for example, only music, only sound effects, or only dialog.

If you are using dual mono tracks, a single audio source should be on a single mono track, not two mono tracks. Sometimes editors attempt to gain more volume by doubling a mono track. Don't do that.

If multiple mics were recorded, the audio house would prefer that you send both tracks so that it can choose the one that produces the best sound. If one of the mics is "thin," it's harder to EQ, but if it's full frequency, the audio house can make fixes to that. If the microphone is a bit "boom-y," that's an easier fix.

As with most things, sending a test sequence out and keeping in close communication with the audio house are advisable.

CONCLUSION

I hope that this book has been helpful in learning to fully exploit the power and capabilities of Avid and that the latest workflows are more clear. It's been a privilege to speak with the many great editors from around the world who shared their expertise. Getting to hear first-hand the many ways that these experts are using various tools inside Avid has definitely given me an appreciation for the deep toolset that Avid has developed to best serve the post-production industry. Hopefully, you've been slowly implementing the practices and toolsets that are detailed in this book. Assimilating these skills is definitely a hands-on learning experience that needs to be practiced, just like a musician or athlete.

Thank you for purchasing and reading this book. Please let me know if there are areas of the book that were difficult to understand or gave you new insight, and be sure to share your newfound knowledge by mentioning cool new ideas from this book in social media, like Twitter, Facebook and professional forums.

Index

Note: Boldface page numbers refer to figures.